Narrative of an Expedition to Explore the River Zaire in South Africa in 1816

JAMES HINGSTON TUCKEY
CHRISTEN SMITH

CAMBRIDGE
UNIVERSITY PRESS

CAMBRIDGE UNIVERSITY PRESS

Cambridge, New York, Melbourne, Madrid, Cape Town,
Singapore, São Paolo, Delhi, Mexico City

Published in the United States of America by Cambridge University Press, New York

www.cambridge.org
Information on this title: www.cambridge.org/9781108050517

This edition first published 1818
This digitally printed version 2012

ISBN 978-1-108-05051-7 Paperback

CAMBRIDGE LIBRARY COLLECTION

Books of enduring scholarly value

African Studies

This series focuses on Africa during the period of European colonial expansion. It includes anthropological studies, travel accounts from missionaries and explorers (including those searching for the sources of the Nile and the Congo), and works that shed light on colonial concerns such as gold mining, big game hunting, trade, education and political rivalries.

Narrative of an Expedition to Explore the River Zaire in South Africa, in 1816

In 1816, an expedition to Africa, commanded by Captain James Tuckey (1776–1816), set out on H.M.S. *Congo*, accompanied by the storeship *Dorothy*. The aim was to discover more about African geography – of which relatively little was then known – and in particular the connection between the River Congo, also known as the Zaire, and the Niger Basin. The mission failed when eighteen crew members, including Tuckey, died from virulent fevers and attacks by hostile natives. However, the Lords Commissioners of the Admiralty gave permission for publication of Tuckey's notes, and those of his Norwegian botanist Christen Smith (1785–1816), who also died during the voyage. First published in 1818, the work comprises their narratives of the doomed expedition. At the time it aroused Western interest in Africa, encouraging further research, and it remains of interest to geographers, botanists and scholars of African studies today.

Cambridge University Press has long been a pioneer in the reissuing of out-of-print titles from its own backlist, producing digital reprints of books that are still sought after by scholars and students but could not be reprinted economically using traditional technology. The Cambridge Library Collection extends this activity to a wider range of books which are still of importance to researchers and professionals, either for the source material they contain, or as landmarks in the history of their academic discipline.

Drawing from the world-renowned collections in the Cambridge University Library and other partner libraries, and guided by the advice of experts in each subject area, Cambridge University Press is using state-of-the-art scanning machines in its own Printing House to capture the content of each book selected for inclusion. The files are processed to give a consistently clear, crisp image, and the books finished to the high quality standard for which the Press is recognised around the world. The latest print-on-demand technology ensures that the books will remain available indefinitely, and that orders for single or multiple copies can quickly be supplied.

The Cambridge Library Collection brings back to life books of enduring scholarly value (including out-of-copyright works originally issued by other publishers) across a wide range of disciplines in the humanities and social sciences and in science and technology.

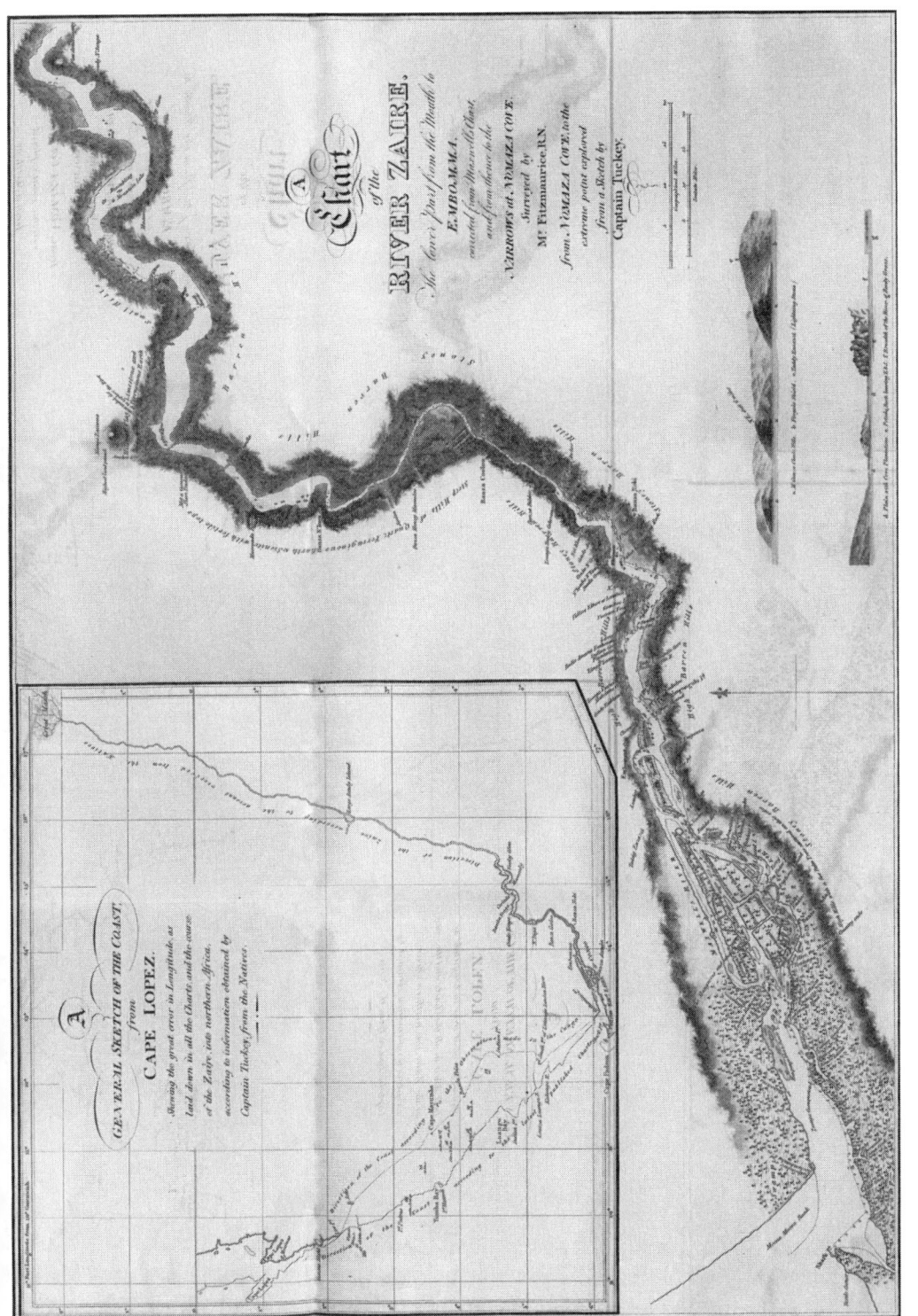

The material originally positioned here is too large for reproduction in this reissue. A PDF can be downloaded from the web address given on page iv of this book, by clicking on 'Resources Available'.

NARRATIVE OF AN EXPEDITION

TO EXPLORE

THE RIVER ZAIRE,

USUALLY CALLED THE CONGO,

IN SOUTH AFRICA,

IN 1816,

UNDER THE DIRECTION OF

CAPTAIN J. K. TUCKEY, R. N.

TO WHICH IS ADDED,

THE JOURNAL OF PROFESSOR SMITH;

SOME GENERAL OBSERVATIONS ON THE COUNTRY AND ITS INHABITANTS;

AND AN APPENDIX:

CONTAINING

THE NATURAL HISTORY OF THAT PART OF THE KINGDOM OF CONGO THROUGH WHICH THE ZAIRE FLOWS.

PUBLISHED BY PERMISSION OF

THE LORDS COMMISSIONERS OF THE ADMIRALTY.

LONDON:

JOHN MURRAY, ALBEMARLE-STREET.

1818.

London : Printed by W. Bulmer and Co.
Cleveland Row, St. James's.

CONTENTS.

CONTENTS.

SECTION III.

SECTION IV.

APPENDIX.

DIRECTIONS FOR PLACING THE PLATES.

ERRATA.

Page 31, for *Hirunda*, read *hirundo*.
32, for *Ettereus*, read *Ættereus*.
33, for *ommon*, read *common*.
40, for *La Marc*, read *La Marck*
for *serrata*, read *serratus*.
65, for *La Marc*, read *La Marck*.
for *sulcata*, read *patula*.
121, for *sephus*, read *cephus*,
289, for *Simio*, read *Simia*.
302, for *decompose*, read *decomposed*.

INTRODUCTION.

The reign of George III. will be referred to by future historians as a period not less distinguished by the brilliant exploits of our countrymen in arms, than by the steady and progressive march of the sciences and the arts. Occupied, as a very considerable portion of that period has been, by a war, longer in its duration, more ferocious in its character, and more extensive in its ravages, than had ever before afflicted Europe, at least in modern times, the advancement of physical and geographical knowledge, though necessarily retarded in its progress, suffered but little interruption, if we except one memorable instance where a French General brutally seized the person and papers of a British naval officer, on his return from a voyage of discovery, and with unparalleled meanness, injustice, and inhumanity, detained the former nearly seven years in captivity, and purloined a part of the latter. With this exception, no war was waged against science; the impulse which had been given to geographical discovery still maintained its direction, and was never lost sight of, even by hostile fleets; witness, among other instances, the interesting and admirable survey of the coast of Asia Minor by Captain Beaufort, while commanding a frigate attached to the Mediterranean squadron, the account of which has recently been laid before the public. " Indeed," as Dr. Douglas has justly observed, " it would argue a most

b

culpable want of rational curiosity, if we did not use our best endeavours to arrive at a full acquaintance with the contents of our own planet." And if those endeavours, which, during war, were so successfully pursued, should be relaxed on the return of peace, we might then indeed have cause to think meanly of the times in which we live, or, to use the words of this eminent writer, " if we could suppose it possible that full justice will not be done to the noble plan of discovery, so steadily and so successfully carried on since the accession of His Majesty, which cannot fail to be considered, in every succeeding age, as a splendid period in the history of our country, and to add to our national glory, by distinguishing Great Britain as taking the lead in the most arduous undertakings for the common benefits of the human race."—Introd. to Cook's Third Voyage.

By following up, therefore, the same system, and being actuated by the same motives, of promoting the extension of human knowledge, the Prince Regent's government has evinced a correspondent feeling; it has moreover proved, by appropriating to the purposes of discovery and maritime geography as great a share as possible of that part of the British navy which constitutes the peace establishment, its laudable inclination to cultivate the useful arts of peace, not from any selfish views, but for the general benefit of mankind. To what purpose indeed could a portion of our naval force be, at any time, but more especially in a time of profound peace, more honourably or more usefully employed, than in completing those *minutiæ* and details of geographical and hydrographical science, of which the grand outlines have been boldly and broadly sketched by Cook, Vancouver, Flinders, and others of our own countrymen; by La Perouse, Dentrecasteaux, Baudin, and other foreign navigators, French, Spanish, and Russian: in ascertaining with greater precision the position of particular points in various parts of the globe—on the shores of Asia Minor—of northern Africa, and of the numerous islands in the Mediterranean—the coasts, harbours, and rivers of Newfoundland, Labradore, Hudson's bay, and that reproach to the present state of European navigation, the existence or non-existence of Baffin's bay, and the north-west passage from the Atlantic to the Eastern ocean—in exploring those parts of the north-west coast of New Holland, which have not hitherto been visited since the time of Dampier—and in obtaining more distinct and accurate information of those great Archipelagos of islands, and those innumerable reefs and islets, which are scattered over the northern and southern Pacific oceans, and the Indian and Chinese seas, many hundreds of which were but the other day discovered, in one spot, by the Alceste, on her late voyage up the Yellow Sea, where not a single island had been even suspected to exist—and, to come nearer home, in filling up and correcting those imperfect and erroneous surveys of our own coasts, and of the seas that surround them—and lastly, in ascertaining with more precision, the extent, direction, and velocity, in different parts and at different seasons of the year, of that extraordinary current known by the name of the Gulf Stream, by which all the currents of the northern Atlantic are more or less influenced. These are objects of general concern in which all Europe and America are equally interested.

By the present improved state of nautical science, by means of His Majesty's ships of war employed on surveys, of the surveying marine of the East India Company, and of the accidental discoveries of commercial vessels, the hydrographical knowledge of every part of the globe is daily extending itself. The line of the coasts which form the boundaries of the continents and larger islands, are traced with more or less accuracy ; the positions of most of the islands or groups of islands are generally ascertained ; and the prevailing winds and currents of the ocean are so much better understood than formerly, that the usual time of an eight or nine months passage to or from China, is now reduced to four months, and rarely exceeds five. It may be said indeed, generally speaking, that, as far as regards maritime discovery, the edge of curiosity has been taken off. Enough however still remains to be done. The deficiency in the detail, and the want of that accuracy so essentially necessary for the advantage and security of navigation, still furnish ample scope for further investigation and research.

But the object of the voyage, of which the narrative is contained in the present volume, though fitted out in the naval department, is nearly, if not altogether, unconnected either with maritime discovery or nautical surveying. It was planned and undertaken with the view and in the hope of solving, or of being instrumental in solving, a great geographical problem, in which all Europe had, for some time past, manifested no common degree of interest ; and, at any rate, in the almost certain means it would afford of adding something to our present very confined knowledge of the great continent of Africa—that ill-fated country, whose unhappy natives, without laws to restrain or governments to protect them, have too long been the prey of a senseless domestic superstition, and the victims of a foreign infamous and rapacious commerce. That great division of the globe of which, while we know that one part of it affords the most ancient and more stupendous monuments of civilized society that exist on the face of the earth, another, and by far the greater portion, exhibits at this day, to the reproach of the state of geographical science in the nineteenth century, almost a blank on our charts ; or what is still worse, large spaces filled up with·random sketches of rivers, lakes, and mountains, which have no other existence than that which the fancy of the map-maker has given to them on his paper. So little indeed has our knowledge of this great continent kept pace with the increased knowledge of other parts of the world, that it may rather be said to have retrograded. If we have acquired a more detailed and precise acquaintance with the outline of its coast, (and in this we are very deficient, as the present expedition has proved,) and with the position of its headlands and harbours, than the Egyptians, the Greeks, or the Romans in their time possessed, it may be doubted whether the extent and accuracy of their information respecting the interior did not surpass ours ; for it cannot be denied that, amidst the fabulous accounts, which fear or fancy is supposed to have created, of regions within and beyond the boundaries of the great desert, many important facts are enveloped, which modern discoveries have brought to light and proved to be correct.

For the greater part of what is still known of the

interior regions of northern Africa, we are indebted to the Arabian writers of the middle ages, and to the information of Arabian travellers of our own times. After them the Portuguese were the first Europeans to penetrate beyond the coast into the interior, where they no doubt collected much information; but, unfortunately for the world, it was their plan to conceal what they discovered, till it has been lost even to themselves. That this nation sent frequent embassies to Tombuctoo, we have the authority of De Barros, which can seldom be called in question, and never, we believe, when he states mere matters of fact, which is the case in the present instance; but though he mentions the names of the persons sent on these missions, he omits all the circumstances and occurrences of the journey, and fails even to describe this renowned city. There are however some circumstances which make it possible that the Tombuctoo of De Barros was no other than the Tambacunda of Park and others, as in all the maps of the sixteenth century, taken from Portuguese authority, Tombuctoo is placed not more than from three to four hundred miles from the coast, which is about one-third part only of its real distance. The Portuguese, however, followed the Arabian geographers in describing the stream of the Niger to flow from east to west, which Herodotus had learned, nearly twenty centuries before, to flow in a contrary direction; an opinion which Ptolemy afterwards seems to have adopted, perhaps on information gained from the same source; though it must be confessed, that Ptolemy is unusually obscure in his geographical delineation of the rise, direction, and termination of this celebrated river.

In the midst of these conflicting opinions respecting the

course of a great river, which was still left undecided in
our times, the authority of an English traveller, from personal inspection, set this question for ever at rest, by determining the direction of the stream to be from west to east.
That part, therefore, of the problem which relates to the
origin and the direction of the early course of this cele-
brated river, has been completely solved; but another and
no less interesting part still continues to be wrapt up in
mystery—where is its termination? As ancient authorities
had pointed out the true direction of the stream, it was but
fair to allow them credit for a knowledge of its termination.
In the examination of this part of the question, by the
first geographer of the age, either in this or any other
country, the authorities of the Arabian writers are weighed
and compared with the geography of Ptolemy; and after
a close and accurate investigation of the various statements of ancient and modern authorities, and a train of
reasoning clear and argumentative, the result of the enquiry appears to be, that the Niger loses itself in the extensive lakes or swamps of Wangara; an hypothesis which
was supposed to have the merit of falling in pretty nearly
with the termination of that river, as assigned to it by
Ptolemy in what he called the *Libya palus*, which lake,
however, Ptolemy only says, is *formed* by the Niger.
In addition to this coincidence, there were also negative
proofs of the disappearance of the Niger in the interior
regions of Africa. It could not, for instance, be a branch
of the Egyptian Nile, as the Arabs generally contend, for
the two reasons adduced by Major Rennell: first, because
of the difference of level; the Nile, according to Bruce's
measurement by the barometer, passing over a country

whose surface is very considerably higher than the *sink* of North Africa, through which the Niger is stated to flow. Secondly, because the Nile of Egypt, in this case, must necessarily be kept up at the highest pitch of its inundation for a long time after that of the Niger, which is well known to be contrary to the fact. Neither was it probable, that its waters were discharged into the sea on any part of the eastern coast, there being no river of magnitude on the whole extent of that coast from Cape Guadafui to Cape Corientes. The hypothesis therefore, of the dispersion and evaporation of the waters of the Niger, in lakes of an ex-extended surface, was the most plausible, and perhaps the more readily adopted, as it fell in with ancient opinion.

The stream of this mysterious river being now traced with certainty from west to east as far as Tombuctoo, so little suspicion seems to have been entertained of the probality of its making a circuitous course to the sea on the western coast, near to which it has its source, that the examination of this side of Africa seems entirely to have been left out of the question. But when Park was preparing for his second expedition to explore the further course of this river, it was suggested, that the Congo or the Zaire, which flows into the southern Atlantic about the sixth degree of south latitude, might be the outlet of the Niger; and as this suggestion came from a person who, in the capacity of an African trader, had not only become well acquainted with the lower part of this river, but had actually made a survey of it, the idea was warmly espoused by Park, who, in a memoir addressed to Lord Camden, previous to his departure from England, assigns his reasons for becoming a convert to this hypothesis; and he adds, that if this should turn out to be the fact, " considered in a commercial point of view, it is second only to the discovery of the Cape of Good Hope ; and in a geographical point of view, it is certainly the greatest discovery that remains to be made in this world."

Park's opinion, it may be said, is entitled to no greater weight, on this point, than that of any other person who had given his attention to the subject ; and so, it appears, Major Rennell thought, who gave him no encouragement to hope for the confirmation of this new hypothesis. But the impression which the facts stated by Mr. Maxwell, and his reasoning on those facts, had made on Park's mind previous to his leaving England, so far from being weakened, appear to have gathered strength on his second progress down the river ; and it can hardly be doubted, that the unknown termination of the stream, and of his own journey, was the unceasing object of his anxious inquiries ; the result of which was, as we are told by his able and accurate biographer, that " he adopted Mr. Maxwell's sentiments relative to the termination of the Niger in their utmost extent, and persevered in that opinion to the end of his life;"—perhaps he ought rather to have said, " to the day of his departure from Sansanding." That no alteration of opinion in this respect had taken place, is quite clear from several expressions in his letters from the Niger, addressed to Lord Camden, to Sir Joseph Banks, and to his wife, in all of which he talks confidently of his reaching England by the way of the West Indies ; not by a painful journey back by land to the Senegal or the Gambia, but by arriving at some other and more distant part of the western coast. This is rendered still more evident from the information he collected at Sansanding, which confirmed the hypothesis of the southern direction of the Niger. " I have hired a guide (he says) to go with me to Kashna; he is one of the greatest travellers in this part of Africa: he says that the Niger, after it passes Kashna, runs directly to the right hand, or to the south; he had never heard of any person who had seen its termination; he was sure it did not end near Kashna or Bornou, having resided for some time in both these kingdoms ;" and to Lord Camden he says, " he was more and more inclined to think that it can end no where but in the sea."

No wonder then that Park, having thus ascertained from " one of the greatest travellers in that part of Africa," the southerly course of the Niger, should be sanguine of proving the validity of Mr. Maxwell's hypothesis, and of reaching the West Indies from the mouth of the Congo. It was not, however, his fate to establish the truth or falsity of this proposition; the problem still remains undetermined; and the termination of the Niger and the source of the Congo, are alike unknown. The probability of their identity, however, appeared to gain ground, not merely because one great river took a southerly direction, and had no known termination, and another came from the northward, nobody knew from whence; but the more the magnitude and character of the latter river was investigated, and its circumstances weighed and compared with those effects which might be expected to happen from natural causes, the greater colour was given to the supposition of their identity. It is only surprising that a river of that magnitude and description which belong to the Congo, should not, long before now, have claimed a more particular attention.

It is true, the first notice of this river is but vaguely given. Diego Cam, in proceeding down the coast, observed a strong current setting from the land, the waters of which were discoloured, and when tasted, found to be fresh. These circumstances led him to conclude, that he was not far from the mouth of some mighty river, which conclusion was soon confirmed by a nearer approach. He named it the Congo, as that was the name of the country through which it flowed, but he afterwards found that the natives called it the Zaire ; two names which, since that time, have been used indiscriminately by Europeans. It now appears that Zaire is the general appellative for any great river, like the Nile in North Africa, and the Ganges in Hindostan, and that the native name of the individual river in question is *Moienzi enzaddi*, or the river which absorbs all other rivers.

All subsequent accounts agree in the magnitude and velocity of this river. In the " Chronica da Companhia de Jezus em Portugal," after noticing the Egyptian Nile, and the common but erroneous notion of its proceeding from the same lake with the Zaire, (namely Zembré, " the mother of waters,") the latter is described as " so violent and so powerful from the quantity of its water, and the rapidity of its current, that it enters the sea on the western side of Africa, forcing a broad and free passage (in spite of the ocean) with so much violence, that for the space of twenty leagues it preserves its fresh water unbroken by the briny billows which encompass it on every side ; as if this noble river had determined to try its strength in pitched battle with the ocean itself, and alone deny it the tribute which all the other rivers in the world pay without resistance." From the following description it is pretty evident that Purchas must have been in possession of this rare book from which the above account is taken; though he has not profited by the information it contains respecting the different sources of the Nile and the Zaire, " for the Portuguese," says this chronicle, " and the fathers of our society who traversed the whole empire of Upper Ethiopia, (which we call Prestê Joaõ) have clearly proved that the Nile does not take its rise in this lake Zembré, and that those authors are mistaken who give it that source." Purchas, however, seems to have no scruples about the truth of what amounts nearly to a physical impossibility,—the flowing of two rivers in opposite directions out of the same lake. " The Zaire," says this quaint writer, " is of such force that no ship can get in against the current but neer to the shore; yea, it prevails against the ocean's saltnesse threescore, and as some say, fourscore miles within the sea before his proud waves yeeld their full homage, and receive that salt temper in token of subjection. Such is the haughty spirit of that stream, overrunning the low countries as it passeth, and swollen with conceit of daily conquests and daily supplies, which, in armies of showers are, by the clouds, sent to his succour, runnes now in a furious rage, thinking even to swallow the ocean, which before he never saw, with his mouth wide gaping eight and twentee miles, as Lopez affirmeth, in the opening; but meeting with a more giant-like enemie which lies lurking under the cliffes to receive his assault, is presently swallowed in that wider wombe, yet so, as always being conquered, he never gives over, but in an eternall quarrel, with deepe and indented frownes in his angry face, foaming with disdaine, and filling the aire with noise, (with fresh helpe) supplies those forces which the salt-sea hath consumed."

The strong current at the mouth of the river, and as far up as ships have been able to proceed—the floating islands carried down by the stream into the ocean—the perceptible effects of the current to a very considerable distance from the shore—have been corroborated by so many concurring testimonies as not to admit of the smallest doubt. Two English frigates, but two years previous to the present expedition, fully experienced these effects. The Honourable Captain Irby, who commanded the Amelia, with difficulty succeeded in getting his ship 48 or 50 miles up the river, the current running down in the middle of the stream at the rate of six and seven knots an hour ; before entering the river the ship was anchored at twelve miles from the southern point of its mouth in 15 fathoms, where the current was running at the rate of four miles an hour, the water being much agitated, of the colour of rain-water, and perfectly fresh. In this situation they observed in the ocean large floating islands, covered with trees and bushes, which had been torn from the banks by the violence of the current. In the journal of the Thais, commanded by Captain Scobell, it is observed, " In crossing this stream I met several floating islands, or broken masses from the banks of that noble river, which, with the trees still erect and the whole wafting to the motion of the sea, rushed far into the ocean, and formed a novel prospect even to persons accustomed to the phenomena of the waters." In Maxwell's chart the current is laid down near the mouth as running at the rate of six miles and seven miles an hour, and the mid channel 100 fathoms deep; at twenty-four or twenty-five miles up the river, where the funnel or estuary is contracted to the natural bed of the river, which is about two and a half to three miles in width, the depth is still 100 fathoms. At fifty miles, the stream is broken into a number of branches, by islands and sand banks. Beyond ninety miles they are again united into one channel, about a mile and a half in width, and the depth, in some places fifty, in others thirty, fathoms, continuing about the same width and depth to the end of the survey, or about 130 miles from the mouth of the river; and it is stated, from information of the native slave dealers at Embomma, that it is navigable beyond the termination of this survey from fifty to sixty miles, where the navigation is interrupted by a great cataract, which they call Gamba Enzaddi. He says, however, in his letter to Mr. Keir, which was communicated to Park, that, according to the accounts he had received from travelling slave merchants, the river is as large at 600 miles up the country as at Embomma, and that it is there called Enzaddi.

All these accounts prove the Zaire to be a river of very considerable magnitude; and though not to be compared with the Amazons, the Oronooko, the Missisippi, St. Laurence, and other magnificent waters of the New World, it was unquestionably the largest river on the continent of Africa. If the calculation be true, that the Zaire at its lowest state discharges into the sea two million cubic feet of water in a second of time, the Nile, the Indus, and Ganges, are but rivulets compared with it, as the Ganges, which is the largest of the three, discharges only about one fifth of that quantity at its highest flood. In point of magnitude, therefore, no objection could be urged to its identity with the Niger.

Many other objections, however, were started against this hypothesis, and in particular the three following, namely,

1. The obstruction of the Kong mountains, which, uniting with the Gibbel Komri, are supposed to extend in one unbroken chain across the continent. 2. The great length of its course, which would exceed 4000 miles; whereas the course of the Amazons, the greatest river in the world, is only about 3500 miles. 3. The absence of all traces of the Mahommedan doctrines or institutions, and of the Arabic language, on the coast where the Zaire empties itself into the sea.

The first objection is wholly gratuitous, as the existence of this chain of mountains has not been ascertained, nor is it easy to conjecture on what grounds it has been imagined. Park saw to the southward of his route, at no great distance from the sea coast, the peak of the cluster of mountains called the Kong, out of which the Niger, the Senegal, and the Gambia, take their rise. The Mountains of the Moon have been placed towards the central parts of Africa; but if Bruce visited that branch of the Nile, which is said to rise out of these mountains, (which is more than doubtful,) they are actually not further removed from the eastern, than those of the Kong are from the western coast. But by what authority they are united, and stretched completely across the continent, like a string of beads, it would be difficult we believe for our modern geographers to point out. There is evidence however to the contrary. All the Haoussa traders who have been questioned on the subject, and who come frequently to Lagos, and other places on the Coast of Guinea, with slaves, deny that they meet on their journey with any mountains, and that the only difficulties and obstructions arise from the frequent rivers, lakes, and swamps which they have to cross. But admitting that such a chain did exist, and that it was one solid, unbroken range of primitive granite, it would be, even in that singular case, the only instance, perhaps, of such an extended barrier resisting the passage of an accumulated mass of waters. Even the Himmalaya, the largest and probably the loftiest range in either the New or Old World, has not been able to oppose an effectual barrier to the southern streams of Tartary. The main branch of the Ganges, it is true, does not, as was once supposed, pervade it, but the Buramputra, the Sutlej and the Indus, have forced their way through this immense granite chain. The *rocky mountains* of America have opened a *gate* for the passage of the Missouri; and the Delaware, the Susquehanna and the Potomac have forced their way through the Alleghenny range. This objection, then, may fairly be said to fall to the ground.

The objection to the length of its course is somewhat more serious, but not so formidable as at first it may appear. The great difficulty, perhaps the only one that suggests itself, arises from the vast height which the source of a river must necessarily be above the level of the sea, in order to admit of its waters being carried over a space of 4000 miles; and from the certainty that Park, (who, it must be observed, however, measured nothing) passed no mountains of extraordinray height to get at the Niger. A critic, in a popular journal, whose arguments

ments in favour of the identity of the Niger and Zaire were probably instrumental in bringing about the present expedition, in answer to this objection, has assumed the moderate height of 3000 feet for the source of the Niger above the surface of the ocean. This height, he observes, would give to the declivity or slope of the bed of the river, an average descent of nine inches in each mile throughout the course of 4000 miles. Condamine," he adds, " has calculated the descent of the Amazons at six inches and three quarters per mile, in a straight line, which, allowing for its windings, would be reduced, according to Major Rennel's estimate, to about four inches a mile for the slope of its bed." And this descent is not very different from that of the bed of the Ganges; it having been ascertained from a section, taken by order of Mr. Hastings, of sixty miles in length, parallel to a branch of the Ganges, to have nine inches of descent in each mile in a straight line, which, by the windings of the river was reduced to four inches a mile, the same as the bed of the Amazons: and this small descent gave a rate of motion to that stream somewhat less than three miles an hour in the dry, and from five to six an hour in the wet season, but seven or eight under particular situations and under certain circumstances. If then, the Ganges and the Amazons flow at the rate of three miles in their lowest, and six miles in their highest state, with an average descent of no more than four inches a mile, while the Niger, according to the hypothesis, would have an average descent of nine inches a mile, the objection to the great length of its course in supposing its identity with the Zaire, would seem to vanish. It has been sufficiently proved, however, that the velocity of rivers depends not on the declivity of their beds alone, but chiefly on the mass and velocity of the water thrown into their channels at the spring head, and the supplies they receive from tributary branches as they proceed in their course. In the Amazons, the Ganges, the Senegal, the Gambia, and in every river whose course, in its approach to the ocean, lies through a low country, it will be found, that the rise of a few feet in the tide is sufficient to force back, up an inclined plane, by its mass and velocity, the whole current of the river to the distance of several hundred miles, and the farther in proportion to the narrowness and depth of the channel beyond its funnel shaped mouth. In estimating the probability, therefore, of the identity of the Zaire and the Niger, as far as the length of their course may be supposed to offer an objection, we should inquire rather into the supply of water than the declivity of the country through which it would have to pass. In this respect, the Niger would be placed under very peculiar circumstances ; its course, lying on both sides of the Equator, and through a considerable portion of both tropical regions, would necessarily be placed, in one part or other, under the parallels of perpetual rains, and consequently receive a perpetual supply of water. Now all the representations that have been given of the lower part of the Zaire, describe it as being nearly in a perpetual state of flood, the height in the dry season being within nine feet of the height in the season of heavy rains; whilst the difference in the height of the Nile and the Ganges, at the two periods, exceeds thirty feet. The flooding of the Zaire is therefore periodical, its highest state being in March, and lowest about the end of August;

a proof that it is influenced by the tropical rains, and that one branch of it, at least, must pass through some portion of the northern hemisphere.

Another objection has been made to the identity of the Niger and the Zaire, grounded on the circumstance of no traces being discovered of the Mahomedan doctrines or institutions on the coast where the latter terminates. It would be a sufficient answer to observe, that as far as our present knowledge extends, the Niger, in Northern Africa, formed the boundary of Mahomedan invasion. What the difficulties may have been, whether moral or physical, or both, " to impede the spirit of enterprise and proselytism which belongs to the Mahomedan character," it would be idle to conjecture; but that they have been impeded, and in a great measure limited to the parallel of the Niger while on its eastern course, is pretty certain ; yet there appears to be neither difficulty nor want of means in crossing this river, though there may be both in descending it. Independently of the lakes and swamps, the sand-banks and rapids, that may occur, the Africans have not at any time, or in any part of the country, been famous for river navigation. But it is far from improbable, that Arab priests or traders may have penetrated into southern Africa ; on the eastern coasts they held, at one time, powerful settlements, and Arabic words occur in all the languages of the negroes even on the western coast.

Some vague objections have been stated to the identity of the Niger and Zaire, from their difference of temperature, the precise meaning of which it is not easy to com-

prehend. In what way, it may be asked, can the temperature of a stream in 16° N. lat. affect the temperature of
the same stream in 6° S. lat. ? There is no assignable ratio
in which it ought to encrease or decrease in its long course;
it may change daily, and many times in the course of the
day according to the temperature of the surrounding atmosphere. Of the temperature of the Niger nothing is
known, for Park does not appear to have noticed it ; but
that of the Zaire was repeatedly ascertained, in the present
expedition, in different parts of its course, and was seldom found to differ more than 2° of Fahrenheit either way,
from the temperature of the atmosphere ; remaining most
commonly about 76, and 77°, which was pretty nearly the
mean day temperature of the atmosphere.

The hypothesis of Mr. Reichard, a German geographer
of some eminence, which makes the Niger to pour its
waters into the gulf of Benin, is entitled to very little attention. The data on which it is grounded are all of them
wholly gratuitous. He proceeds on a calculation of the
quantity of water, evaporated from the surface of the lakes
of Wangara, and the quantity thrown into them by the
Niger, without knowing whether the Niger flows into
them or not, or even where Wangara is situated, much
less the extent and magnitude of those lakes. The Rio
del Rey, the Formosa, and the numerous intermediate
branches that open into the gulf of Benin, are supposed to
join in one great stream beyond the flat alluvial land which
they seem to have formed ; the supposition, however, has
never been verified by observation ; but as far as it is
known, the Rio del Rey proceeds from the northward, and

the other branches have a tendency to the north-west.
Whether, therefore, they unite, or not, the probability certainly is in favour of all the streams, from Guinea to Biafra,
having their sources in the southern face of the Kong
mountains. It can scarcely be supposed that the same
mountains, whose northern sides give rise to the three
large rivers, the Niger, the Senegal, and the Gambia,
should have their southern faces destitute of streams. If
however, we refer these numerous branches to some great
stream crossing the continent, from the north-east, the
Houssa merchants, in their journey to Lagos, must necessarily psas it; but by their own account, though numerous streams, and lakes, and marshes occur, they neither
cross any high mountain, or very large river

In this unsatisfactory state of doubt and conjecture, in
which a most important geographical problem was involved,
two expeditions were set on foot under the auspices of
Government; the one to follow up the discovery of Park
by descending the stream of the Niger, the other to explore the Zaire upwards towards its source. Independently of any relation which the latter might be supposed
to have to the former, the river itself, from all the descriptions which had been given of it, from its first discovery
by Diego Cam down to the present time, was of sufficient
magnitude to entitle it to be better known. To accomplish this object more of difficulty was apprehended in the
navigation, than of danger from the hostility of the
natives, or the unhealthiness of the climate, neither of
which had opposed any obstacle to the progress of the
Portuguese. It was well known both to them and the
slave dealers of Liverpool, who used to frequent this river,

that its navigation was impeded by a cataract at no great
distance from its mouth ; but that was not considered as a
reason why it should not again become navigable beyond it.
Maxwell's information from the slave dealers stated it to be
so for six hundred miles above this cataract.

In exploring the course of an unknown river upwards,
there would obviously be less risk to the parties employed
than in following the stream downwards. In the first case
a retreat could always be secured when the navigation
became no longer practicable, or the state of the country rendered it unsafe to proceed ; in the second, every
moment might be pregnant with unforeseen dangers from
which there could be no retreat. The river might, for instance, suddenly and imperceptibly become bristled with
rocks, and its rapid stream roll with such velocity as to
sweep the unfortunate navigator to certain destruction down
a cataract; or it might spread out its waters into a wide
lake without an outlet, which, becoming in the dry season
a boundless swamp, would equally doom him to inevitable
destruction. No one can tell what the fate of Park may
have been, but no one will believe that this enterprising
traveller finished his career in the manner related by
Isaaco, on the pretended authority of Amadou Fatima.
Some persons indeed are still sanguine enough to suppose
he may be living. It is just possible, and barely so, after
such a lapse of time, that this unfortunate traveller may
have been hurried down the stream of the Niger into the
heart of Africa, and placed in a situation from whence
he had neither the means of returning or of proceeding ;
but what these obstacles may have been, whether moral or
physical, or both, in the total absence of all information

it would be idle to conjecture; it may reasonably be concluded however, that if this intelligent traveller had still been living, he would long ere now have hit upon some expedient to make his situation known.

It was hoped then, even should the immediate object of the expedition up the Zaire not prove succesful, that some more correct as well as more extensive information, respecting the regions through which it flows, would be the result of it. Some doubt was entertained, in making preparations for exploring the river, as to the kind of vessel which might be found most suitable and convenient in all respects to be employed on the occasion. Among other qualifications, two were indispensibly necessary; first, that she should draw but litle water; and secondly, that she should afford sufficient accommodations for the officers and crew, for the Naturalists and their collections, besides an ample supply of provisions and presents, without which there is no getting on among the Africans. It was suggested by Sir Joseph Banks, who, from the lively interest he invariably takes, and the willing assistance he is ever ready to afford, where the advancement of human knowledge is concerned, was the first to be consulted on the present occasion, that a steam engine might be found useful to impel the vessel against the rapid current of the river. Many reasons were urged for and against the employment of a steam vessel. If individuals both in Europe and America find it of advantage to avail themselves of the aid of this powerful agent, it could not be less advantageous when employed on the public service; it would spare the men the labour of rowing when the wind should be foul, or failed. Some difficulties were started with regard to fuel, but these were over-ruled by the well known fact, that on the banks of the lower parts of the river are whole forests of the mangrove, the wood of which possesses the peculiar quality of burning in its green state better than when dry ; but it was not known to what extent these forests might reach, though it was fair to presume, that, in an equinoctial climate, where water was to be found, wood would not be wanting. At the same time it could not escape notice that the labour of felling and preparing fuel for the boiler of a steam engine to the amount of about three tons a day, in such a climate, might be fully as fatiguing, and in all probability more fatal to the crew, than the occasional operation of rowing. If, however, it could have been certain that this vast and rapid river was navigable beyond the cataracts, and its banks well wooded, a steam engine might prove a good auxiliary ; and accordingly it was determined that a vessel, capable of being navigated by steam should forthwith be constructed.

But another difficulty presented itself. The vessel was not only to be constructed so as to be adapted to the flats and shallows that might occur in the river, but so as to ensure her a safe passage across the Atlantic ; this was thought by many persons to be no easy task. The burden of the vessel was not to exceed one hundred tons, her draft of water four feet : of this tonnage it was calculated that the engine of 24 horses power would alone occupy one-third part, and, of her measurement, the whole breadth of the vessel, and twenty feet in length. Such a vessel, however, with so heavy a burden and so small a displacement of water, Mr. Seppings, the surveyor of the navy, undertook to construct, and at the same time to give her such strength and stability under sail as would enable her to be navigated in safety to the southern Atlantic. It had been proposed to send her out in frame, and to set her up in the river, but against this plan there were insuperable objections.

Messrs. Watt and Bolton were put in communication with Mr. Seppings, in order that a proper steam engine might be fitted for the vessel. Unfortunately however, by some misconception, the engine with its boiler was heavier, or the vessel drew more water, than had been anticipated; the consequence of which was, that at the highest pitch of the engine it would not propel her through the water at a greater speed than that of four knots an hour; and when lightened to the draught of four feet three inches, her rate of going never exceeded five knots and a half an hour. It was therefore so obvious that this rate of going never could compensate for the very great incumbrance of a machine that occupied one-third part of the vessel, that it was at once determined to get rid of it altogether; a measure which was earnestly urged by Captain Tuckey, as he thereby would procure a most important addition for the stowage of provisions, and the accommodation of his crew. As the trial of this steam vessel, which had been constructed with the view of combining the opposite qualities of navigating the ocean by the power of the wind, and stemming the current of a river by that of steam, had attracted a very considerable degree of curiosity, the failure brought forward a shoal of projectors, every one ready with his infallible remedy; and Messrs. Bolton and Watt were no less anxious to try the result of another engine with some difference in the application of its power ; but the season for the departure of the expedition had arrived, the month of January had already expired, the officers and men of science were all engaged, the articles for presents purchased, the instruments prepared, and it was most desirable that the expedition should reach the coast of Southern Africa not later than May, or June at farthest, in order that the voyage up the river might commence with the early part of the dry season.

Mr. Seppings and Captain Tuckey were both of opinion that, with a trifling alteration, the Congo (for so she had been named,) might be converted into an excellent sea vessel, and equally proper for ascending the river by the ordinary means of navigation, namely, by sail and oar : and as far as the river navigation was concerned, they were confirmed in this opinion by information collected from the master of a merchant vessel, who had been several voyages up the Zaire to the distance of 140 miles from its mouth, and who stated that he had never found the least difficulty in ascending with a schooner to that distance by the assistance of the sea-breeze, which sets in regularly every day ; the current of the river seldom, by his account, exceeding five miles an hour in the dry season, and in many parts not running more than three.

The engine was accordingly removed without delay from the Congo, and sent, where it was much wanted, to the new works at Chatham dock-yard ; the proposed alteration was immediately made, and on trial of her qualities, Captain Tuckey reported, that " he had no hesitation in saying, that she was, in every respect, fit for the business." The contrary opinion had been held by many sea officers, who still persisted, that she would never cross the Atlantic; but her first essay from the Nore to the Downs fully justified Captain Tuckey's expectations. " I am much gratified," he says, " in being able to inform you, that the Congo justifies, as far as she has been tried, my *obstinacy*, in wishing to keep her in preference to any other. In running, yesterday, from the Nore to the North Foreland, with a fair wind, she kept way with the transport (Dorothy); and what is more extraordinary, in working from the Foreland into the Downs, the wind at west, blowing so fresh that the transport could scarcely carry double reefed topsails, the Congo beat her completely, and indeed every other vessel working down at the same time; in short, she has completely falsified the sinister predictions of her numerous traducers, as far as *sailing* is concerned; moreover, she scarcely feels her sails, and is as dry as possible." It may here be added, that during the voyage, she answered every good purpose, was a dry commodious vessel, perfectly safe at sea, and is now employed on the surveying service in the German Ocean, or North Sea, for which she is admirably adapted. It is the more necessary that this should be stated, as her form, which pretty nearly resembles that of a horse trough, militates strongly against the generally received opinions of naval men, as to the most eligible forms of bodies calculated for moving best through fluids by meeting with the least resistance. It is also worthy of notice, that the principle on which the Congo was built is very similar to that for which the late Lord Stanhope so strongly contended, as being the most proper for ships of war, by uniting in one body, strength, stability, stowage, accommodations for the people, and a light draught of water ; but Lord Stanhope's ideas were rejected by a committee of naval officers as crude and visionary, with the exception, we believe, of one individual. The Congo was schooner-rigged, and three parallel keels assisted in enabling her to hold a good wind.

In the event of meeting with shallows, rapids, or cataracts, of the existence of which no doubt could be entertained, though the accounts given of them were vague and uncertain, it was necessary that some lighter kind of vessel should be provided capable of being transported by land ; Captain Tuckey proposed a double-boat built of light materials, drawing very little water, and which, when screwed together by means of a kind of connecting platform, should be able to carry from twenty to thirty men, with three months provisions ; each boat was 35 feet long, and six feet broad, and when put together a canopy was fitted to keep off the sun and rain. A second double-boat was afterwards provided, and several smaller ones ; and as the size of the Congo was wholly inadequate to the stowage of these boats, with the provisions, water, presents, &c. the Dorothy transport, of about 350 tons, was appointed to accompany the expedition into the river Zaire, when, after transhipping into the Congo all that could be deemed necessary for the prosecution of the great object of exploring the river, she was to return to England.

The armament of the Congo, the quantities and the different kinds of provisions and refreshments, were left to the discretion of Captain Tuckey. Presents of the usual kind, such as iron tools, knives, glass ware, beads, bafts, umbrellas, &c. were put on board in such quantities of each as the Congo could conveniently stow ; mathematical and philosophical instruments for surveying the river, for astronomical, meteorological, and other scientific purposes were also provided, in order that every kind of information might be brought back, in as complete a manner as the present state of the sciences and other circumstances would allow.

The officers and men composing the crew of the Congo consisted of the following persons, the greater part of whom were left to the choice of the commander, and were all of them volunteers :

Captain J. K. Tuckey, *Commander.*
Lieutenant John Hawkey.
Mr. Lewis Fitzmaurice, *Master and Surveyor.*
Mr. Robert Hodder, *Master's Mate.*
Mr. Robert Beecraft, *Master's Mate.*
Mr. John Eyre, *Purser.*
Mr. James Mc Kerrow, *Assistant Surgeon.*
8 Petty officers.
4 Carpenters.
2 Blacksmiths.
14 Able Seamen.
1 Sergeant,
1 Corporal, } of Marines.
12 Private Marines.

making in the whole 49 persons. To whom were added on a supernumerary list,

Mr. Professor Smith, *Botanist.*
Mr. Cranch, *Collector of Objects of Natural History.*
Mr. Tudor, *Comparative Anatomist.*
Mr. Galwey, *a Volunteer.*

Mr. Lockhart, from His Majesty's Garden at Kew.

Benjamin Benjamin, ⎫
Somme Simmons, ⎬ Natives of Congo.

of these 35 took their passage out on board the transport, and 21 were on board the Congo to navigate her to the Zaire.

As so very little was known of the course of the Zaire, and nothing at all beyond the first cataracts, it was at first intended to leave Captain Tuckey entirely to his own discretion, to act as circumstances might appear to require, and to furnish him only with general directions to use his best endeavours for the prosecution of the principal object, and the promotion of general science; but Captain Tuckey pressed with such urgency for specific instructions, that, as he observed, he might be satisfied in his own mind when he had done all that was expected from him, that his wishes in this respect were complied with; and the intructions, of which the following are copies, were given for his guidance, a draft of them having first been sent for his perusal, which met with his entire approbation.

By the Commissioners for executing the Office of Lord High Admiral of the United Kingdom of Great Britain and Ireland, &c.

Whereas, we have thought fit to appoint you to the command of his Majesty's ship Congo, destined for an expedition of discovery up the River Zaire, into the interior of Africa, by command of His Royal Highness the Prince Regent, signified to us by Earl Bathurst: you are hereby required and directed to receive on board the said ship, the persons named on the other side hereof,* with

* *Mr. Professor* Smith, *Botanist.*
Mr. Tudor, *Comparative Anatomist.*
Mr. Cranch, *Collector of Objects of Natural History.*
—- Lockart, *a Gardener.*

their baggage, instruments, &c. (which persons you are to bear on a supernumerary list for victuals only); and having taken under your orders the transport Dorothy, laden with certain stores and provisions, for the use of the expedition, you are to put to sea without delay, and make the best of your way into the river Zaire, commonly called the Congo, in southern Africa, and having proceeded up that river to some convenient place for transhipping the stores, and provisions abovementioned, you are to direct the master of the transport to return to Spithead, sending by him an account of your proceedings, for our information.

On the departure of the transport, you will proceed up the Zaire, and use your utmost endeavours to carry into execution the instructions contained in the memorandum, which accompanies this order; and on your return to the mouth of the river, you are either to proceed to England, to the Isle of St. Thomas, or to St. Helena, as you may judge most expedient for the safety of yourself, and people entrusted to your charge, after a due consideration of the state of the vessel and of your provisions; reporting to our Secretary, for our information, your arrival, and transmitting an account of your proceedings.

Given under our hands, the 7th of February, 1816.

(*Signed*) MELVILLE.
GEO. I. HOPE.
H. PAULET.

To JAMES K. TUCKEY, *Esq. Commander of*
his Majesty's Sloop Congo, at Deptford,

By Command of their Lordships,
(*Signed*) JOHN BARROW.

Memorandum of an Instruction to CAPTAIN TUCKEY.

Although the expedition, about to be undertaken for exploring the course of the river Zaire, which flows through the kingdom of Congo, in southern Africa, was originally grounded on a suggestion of its being identical with the Niger, it is not to be understood, that the attempt to ascertain this point is by any means the exclusive object of the expedition. That a river of such magnitude as the Zaire, and offering so many peculiarities, should not be known with any degree of certainty, beyond, if so far as, 200 miles from its mouth, is incompatible with the present advanced state of geographical science, and little creditable to those Europeans, who, for three centuries nearly, have occupied various parts of the coast, near to which it empties itself into the sea, and have held communication with the interior of the country through which it descends, by means of missionaries, and slave agents ; so confined indeed is our knowledge of the course of this remarkable river, that the only chart of it, which can have any pretension to accuracy, does not extend above 130 miles, and the correctness of this survey, as it is called, is more than questionable.

There can be little doubt however, that a river, which runs more rapidly, and discharges more water, than either the Ganges or the Nile, and which has this peculiar quality of being, almost at all seasons of the year, in a *flooded* state, must not only traverse a vast extent of country, but must also be supplied by large branches flowing from different, and probably opposite directions ; so that some one or more of them must, at all times of the year, pass through a tract of country where the rains prevail. To ascertain the sources of these great branches then, will be one of the principal objects of the present expedition : but in the absence of more correct information, the instructions regarding the conduct to be observed, can be grounded only on probable conjecture.

The unusual phenomenon of the constant flooded state of the Zaire, as mentioned by the old writers, and in part confirmed by more recent observations, would seem to warrant the supposition, that one great branch, perhaps the main trunk, descends from the tropical region to the northward of the Line ; and if in your progress it should be found, that the general trending of its course is from the north-east, it will strengthen the conjecture of that branch and the Niger being one and the same river.

It will be advisable therefore, as long as the main stream of the Zaire shall be found to flow from the north-east, or between that point and the north, to give the preference to that stream ; and, to endeavour to follow it to its source : at the same time, not to be drawn off by every large branch of the river, that may fall into the main stream from the northward, but to adhere to the main trunk, as long as it shall continue to flow from any point of the compass, between the north and east.

It is also probable that a very considerable branch of the Zaire will be found to proceed from the east, or south-east ; as it has been ascertained, that all the rivers of southern Africa, as far as this division of the continent has been traversed from the Cape of Good Hope, northwards, flow from the elevated lands on the eastern coast, across the continent, in a direction from west to north-west : and it may perhaps be considered as a corroboration of the existence of some easy water conveyance between the eastern and western coasts of south Africa, that the language of Mosambique very nearly resembles the language spoken on the banks of the Zaire.

If this conjecture should turn out to be well founded, such an eastern stream, being next in point of interest, will claim the second place in point of attention ; and if it should appear to be navigable through the heart of southern Africa, to the high lands on the eastern coast, it may probably hereafter be considered as the first in point of importance ; by opening a convenient communication through a fine country, from the southern Atlantic to the proximity of the Indian or eastern ocean ; and with the once opulent kingdoms of Melinda, Zanzibar, Quiloa, &c.

With regard to a large branch proceeding from the southward, out of a lake called Aquelunda, so many details, though loose and vague, have been given by the early Spanish and Italian missionaries, than one can scarcely be permitted to doubt of its existence. As this branch is likely to be, from the barren nature of the country to the southward, and along the western coast, the least important and least interesting, it will be adviseable to leave it unexplored until the return of the expedition from examining the others ; unless indeed, what would be contrary to all expectation, and irreconcilable with the peculiar phenomena of the river, this southern branch should turn out to be the main trunk. But though less interesting than the others, this branch will require a more accurate examination than has hitherto been given to it, which however may be left, until the more important branches, whose existence we have supposed, may have been explored.

If, after all, it should be found that unforeseen and invincible obstacles oppose themselves to your penetrating, by any of the branches of the Zaire, to a considerable distance into the interior (obstacles which it is hoped may not occur,) you are, in that case, after collecting all the information in your power, during your descent of the river, to proceed without loss of time to the Bight of Benin, where you will endeavour to ascertain whether the great Delta, supposed to be formed by a river, one branch of which usually known by the name of Rio del Rey, flows into the Atlantic on the eastern, and the other the Rio Formosa, on the western side of the said Delta, be actually so formed : or whether these branches be two separate and distinct rivers. The determining this question is the more interesting, as, on the supposition of the union of these two great streams, the continential geographers have raised an hypothesis that the Niger, after reaching Wangara, takes, first, a direction towards the south, and then bending to the south-west, discharges itself into the Gulf of Guinea. In the eventual prosecution of this discovery, the same instructions will apply as those for your guidance up the Zaire.

Keeping therefore the general principles above mentioned in view, the mode of carrying the examination of either of the rivers into effect must be left, in a great degree, to your judgment and discretion, after a due consideration of all the local circumstances that may present themselves, and the information you may be able to collect on the spot.

It may be observed, however, that the occurrence of rapids, or of a cataract, impeding the navigation of the river, is not to be considered as a sufficient obstacle to the further prosecution of the attempt to discover its source. In the event of meeting with an obstacle of this kind, it will be necessary, in the first place, to look out for a safe and proper situation for mooring the Congo, and then to use all possible means, by trucks and other apparatus with which you will be furnished, to get the double boats, with one or both of the small ones, if necessary, to the upper part of the rapids: and having accomplished this, to divide the officers and crew between the Congo and the double boats, in such manner and in such proportions as you shall deem to be most expedient for the protection of the former, and the management of the latter in their further progress up the river; taking care that, whenever it may be necessary to detach the boats, the gentlemen to whom the scientific departments of the expedition have been assigned, shall always accompany them; in order that no opportunity may be omitted of examining and collecting specimens of the natural products of the country, through which the Zaire, or any of its larger branches, may flow.

On your arrival in the river, you will endeavour to find out a suitable spot for transhipping such stores and provisions from the transport to the Congo as the latter may be able to take on board. The anchorage opposite the *Tall Trees* is understood to be a safe and healthy spot; but there are many reasons that would make it adviseable not to stop at a place so near the slave-trade stations of the Portuguese; and you cannot be too much on your guard against the agents of this nation, concerned in carrying on that trade, whether they are white men or blacks; and as it is understood, that many vessels have been as high up as Embomma, a place of considerable trade, you may, should you find it practicable, carry up the transport to that place, or even to Benda, which is still higher, before you dismiss her down the river; in doing which, you will take care to provide her with such arms and ammunition as may be necessary, for her defence against any attack that may be made upon her by the canoes of the natives; cautioning the master to have no communication with them, but to make the best of his way down the river, and from thence to Spithead; and you will take this opportunity of sending home an account of your proceedings to the date of her departure.

It is understood that, at Embomma, you will be likely to meet with natives who speak the English and Portuguese languages: and it is probable, that you may find it useful to employ one or more of the respectable inhabitants to proceed with you as interpreters, and to explain to their countrymen, the real motives and views of the expedition. You will, of course, avail yourself of the assistance of such persons, and collect from them every information they may be able to give of those parts of the continent through which the river descends; and on your entering the country of any new tribe of people, your first care should be to obtain interpreters, and to make it clearly understood to the chiefs of every tribe, that you mean to make them suitable presents, or pay them such transit duty as may be customary ; and you are to take especial care that no cause of jealousy or quarrels with the natives be occasioned by any of the officers or men under your orders.

It would be unnecessary to go into a minute detail of the various duties you will have to perform on this voyage of discovery, the conduct of which has been intrusted to your charge ; or of the probable objects that will present themselves for your research ; but the mention of some of the more prominent points that should claim your attention, may enable you to prepare yourself on your passage thither, by making arrangements for your future proceedings after the departure of the transport for England.

Among the more important points, then, for observation, may be mentioned, the depth of the river ; the strength of the current in general, and its velocity in particular places ; the quantity of its rise and fall, from land floods and droughts ; the quality of the water as to clearness or muddiness ; the direction of the several reaches ; and the latitude and longitude of every spot remarkable for any particular produce, towns, hamlets, neighbouring mountains, &c. and of the points of junction of branches falling into the main stream ; all of which should be particularly attended to. The variation of the compass should be taken and stated down, as frequently as opportunities may offer for ascertaining it ; and a set of observations of the dip of the magnetic needle is very desirable, to obtain which, their Lordships have directed a very excellent dipping needle, by Blunt, to be supplied for your use.

You will also, with the assistance of the Surveyor, be careful to keep an account of each day's run, to enable you to lay down with tolerable correctness a chart of the river and the adjacent banks, and on such a scale, as will admit of the main features of the country, and all remarkable objects, being marked down upon it: among other things, the ranges of wood along the banks ; the places where those ranges are interrupted, and to what extent; the nature of the prevailing trees, and their quality as fuel in a green state, in order that a competent judgment may be formed of the supply of fuel, should it, at any time hereafter, be thought expedient to navigate the Zaire by steam boats

It is almost unnecessary to observe to you, how important it will be to keep a journal of your proceedings. In this journal all your observations and occurrences of every kind, with all their circumstances, however minute, and however familiar they may have been rendered by custom, should be carefully noted down; and although the gentlemen employed in the several departments of science, will each be instructed to keep their respective journals, it will not, on that account, be the less desirable that you should be as circumstantial as possible in describing, in your own, the general appearance of the country, its surface, soil, animals, vegetables, and minerals; every thing that relates to the population; the peculiar manners, customs, language, government, and domestic economy of the various tribes of people through which you will probably have to pass.

The following, however, will be among the most important subjects on which it will be more immediately your province, assisted by your officers, to endeavour to obtain information.

The general nature of the climate as to heat, cold, moisture, winds, rains, and periodical seasons. The temperature regularly registered from Fahrenheit's thermometer, as observed at two or three periods of the day.

The direction of the mountains, their names, general appearance as to shape, whether detached or continuous in ranges.

The main branches of rivers, their names, direction, velocity, breadth, and depth.

The animals, whether birds, beasts, or fishes, insects, reptiles, &c. distinguishing those animals that are wild, from those that are domesticated.

The vegetables, and particularly those that are applicable to any useful purposes, whether in medicine, dyeing, carpentry, &c. scented or ornamented woods adapted for cabinet work, and household furniture; and more particularly such woods as may appear to be useful in ship building; hard woods fit for treenails, block sheaves, &c. of all which it would be desirable to procure small specimens labelled and numbered, so that an easy reference may be made to their mention in the journal; to ascertain the quantities in which they are found, the facility, or otherwise, of floating them down to a convenient place for shipment, &c.

Minerals, any of the precious metals or stones; how used, and how valued, by the natives.

The description and characteristic difference of the several tribes of people

The occupations and means of subsistence, whether chiefly, or to what extent, by fishing, hunting, feeding sheep or other animals; by agriculture or by commerce.

The principal objects of their several pursuits as mentioned in the preceding paragraph.

A circumstantiul account of such articles, if any, as might be advantageously imported into Great Britain, or her colonies, and those which would be required by the natives in exchange for them.

The state of the arts or manufactures, and their comparative perfection in different tribes.

A vocabulary of the language spoken by every tribe, through which you may pass, using in the compilation of each, the same English words; for this purpose you will receive herewith some copies of printed vocabularies, to fill up, accompanied with the copy of a letter on this subject from Mr. Marsden, which is well deserving your attention,

The condition of the people, as far as can be ascertained; what protection the chief, or the laws afford them; what is the state of slavery among them: whether wars are carried on for the purpose of making slaves: how their prisoners are treated; how disposed of; and every possible information that can be collected, as to the manner and extent to which the slave trade is conducted with Europeans: who those Europeans are; where residing: how their agents are employed; what the articles of barter are; in what manner the slaves are brought down to the coast, &c. The detailed questions furnished by the African Institution, and which accompany this memorandum, will materially assist your enquiries into this interesting subject, and other matters connected with the state of society and the condition of the people.

The genius and disposition of the people, as to talent, mental and bodily energy, habits of industry or idleness, love, hatred, hospitality, &c. The nature of their amusements, their diseases, and remedies, &c.

Their religion, and objects of worship, their religious ceremonies; and the ininfluence of religion on their moral character and conduct.

A description of the manners, appearance and condition of any Mahomedans, that may be found in any of the tribes in southern Africa.

What written or traditionary records, may exist among the latter; any facsimiles of their written character, or copies of any drawings or paintings, they may have attempted, would be desirable.

The several objects hitherto mentioned, which it would be desirable to accomplish are such as relate more immediately to your province as commander of the expedition, and to the officers under your orders, in obtaining which, however, their Lordships cannot doubt you will meet with the willing assistance and co-operation of those gentlemen who have been engaged to accompany you. for the purpose of scientific research. These are,

Mr PROFESSOR SMITH, *Botanist and Geologist;*

Mr. TUDOR, *Comparative Anatomist;*

Mr. CRANCH, *Collector of Objects of Natural History;*

each of whom has specific instructions for his guidance, of which it may be necessary to furnish you with a general outline, in order that you may the better be able to afford such opportunities, and such facilities to all and each of them, as may tend to promote the several objects, which their Lordships have in view.

Professor Smith, *Botanist* and *Geologist,* is directed by his instructions to collect together as many specimens of plants, growing on the banks of the river, as time and opportunities will enable him to do; and these specimens, it can hardly be doubted, will be very numerous, as the country has never yet been explored by botanists, nor have the parallels of latitude, through which the river passes, been yet investigated in any part of the world, with the exception of south America, and these only partially explored. And in order to enable him to execute his laborious duties, a Gardener, from His Majesty's Botanical Garden at Kew, has been assigned to assist the Professor in drying and preserving, as well as in collecting specimens, to afford him more time to note down the prominent characters of the most remarkable plants, which may fall within his reach.

He is also directed to collect the seeds of all new plants, which may offer themselves for the use of the Royal Gardens at Kew; and the supply of these has been limited to two packages of each kind, sewed up, with a view to keep the stock intire, without breaking into it on any consideration; so that the whole collection may be delivered, so sewed up, to the Director of the Royal Botanical Gardens, as soon as possible after the return of the expedition to England.

The Professor is further directed to preserve the fructifications of the more delicate plants in spirits, sewed up in small bags of muslin. It has been thought right to apprise you of the nature of these collections, in order that, in the arrangement and distribution of the stores, provisions, &c., on board the Congo, a sufficient space may be alloted for their stowage.

He will also have occasion to examine into the geology and geognosy of the country, through which the river shall be found to pass, and to collect specimens of such stones and minerals as may occur; and as he is directed to keep a journal, in which will be recorded such remarks as he may think worthy, on the localities, the particularities, and the distinguishing characters and uses to which plants or minerals are applied, you will afford him such assistance with regard to the latitude and longitude of particular places, as he may require, in order that no information may be wanting, on points generally interesting to science.

And as the parts of fructification of trees, in warm climates, are seldom accessible to botanists, on account of the labour of felling them, and consequently the arborescent plants of these climates are least of all known, you will allow one or two of the carpenters to accompany the Professor and Gardener, to enable them, by felling trees of a moderate size, to get at the flowers and fruits of those species of which it may be deemed advisable to collect and preserve specimens.

Though Mr. Smith modestly declines to take upon himself the office of Professional Geologist, yet, having examined the Canary Islands, in company with his friend the Baron Von Buch, who eminently excels in that interesting science, there can be no doubt of his sufficient proficiency in that branch of physical knowledge, to enable him to collect such specimens as may be useful to elucidate the geology of those parts of southern Africa, through which you may have to pass.

To enable you however, and any of the gentlemen who accompany you, to form a competent knowledge of what may be desirable to bring home from the mineral kingdom, a few general directions may be of use.

The objects of the most value, are the metallic ores, fossil bones, teeth, shells, impressions of plants and fishes; those of least esteem, spar, crystal, pyrites, pieces of loose stone or gravel, unless where the last is supposed to contain metallic matter.

The most common substances, (such as flint, chalk, sandstone, coal, clay, limestone, basalt, slate, granite) will be interesting, if labelled on the spot, and kept clean, in separate papers, and not suffered to rub against each other; the label to express the name and situation of the rock or mountain, from which the specimen was detached. The size of a common watch is sufficiently large for each specimen; shape is of little consequence; that of a cube split in two is perhaps the most convenient.

Specimens of rock are always desirable, with the native names for each, where they can be obtained; and the uses to which they are applied.

Fossil bones and shells, of whatever size, should be brought away entire; if an entire skeleton be found it should be brought away; and it is essential to note down whether such bones were found in loose soil, in solid rock, or in caverns.

It will be interesting to ascertain what mines, quarries, or caverns exist, where the different metals, coal, salt, slate, limestone, &c. are found, and if worked, in what manner.

It will be desirable to note down the distance of the mountains or hills on both sides of the Congo, their height, from conjecture, when no means offer for their measurement, their form, which of the sides are steepest, how their strata are disposed, how much they dip, and in what direction; whether they disappear by dipping under the soil, or by the intersection of valleys.

It will also be desirable to note down all the places where two rocks of a different nature may be seen in contact, and to what extent each may be traced.

And for your further assistance in this interesting subject, a printed copy of geological inquiries, published by the Geological Society, accompanies these instructions.

Mr. Tudor, the Comparative Anatomist, is directed by his instructions, to examine the structure and habits of all new and uncommon animals, and it will therefore be desirable that he should always accompany the collector of objects of natural history, when detached, either on the river or on shore.

One portion of his collection, consisting of the internal parts of animals, and of the smaller animals in an entire state, will be required to be preserved in spirits, but of these he is not to preserve more than triplicates of each specimen.

The external parts of animals, as their skulls, skins, feet, &c. he is directed to preserve in a dry state, and the specimens of each, as before, not to exceed three in number.

Any preparations he may have made at the time of the departure of the transport from the Zaire, you are to send home, along with a copy of his journal, in that transport.

Mr. Cranch, Collector of objects of natural history, by his instructions, is directed to commence his operations on the voyage outwards; to fish up out of the sea, by a dipping net from the chains, and by such other means as may be most likely to succeed, whatever sea-weeds or animals may float alongside, particularly of the class of mollusca, which he is directed to preserve in spirit, and to send home by the transport on her return from the Zaire.

On the progress of the expedition up the Zaire, he is instructed to collect all unknown fishes, shells, and crustacea, insects and reptiles, birds, beasts, amphibia, and in short, whatever may occur in the animal kingdom, which he is to preserve with all possible care, by putting the tender animals into spirit, and preserving the larger ones dry, such as the skins of animals, insects, shells, fishes, and other bulky articles, which, when well dried, he is to pack in casks or boxes, so as to secure them against the attack of ants or other insects, with the greatest care. In framing such chests or boxes, as well for his collection, as for those of the comparative anatomist, the carpenters of the Congo will be required to give all possible assistance.

Of the number of specimens of each kind to be preserved, it has been thought necessary to limit him generally to three, lest the stock should occupy more room than could with propriety be allowed in the vessel. If however, it should be found, on approaching the utmost limit of the voyage, or on the return, that sufficient room still remains, he, as well as the others, may all be permitted to extend the number of their specimens beyond that prescribed by their instructions.

The proper times of his going on shore for the purpose of collecting, will of course be regulated by your orders, and be such as not to interfere with the general convenience of the expedition.

But whenever you shall have occasion to stop for the purpose of cutting fuel, purchasing provisions, or holding communications with the natives, or from any other cause, you will afford the several gentlemen in the scientific department the means of going on shore, where it can be done with safety, to give them the pportunity of enriching their collections; and in general, whenever any thing occurs to make a landing on any particular spot very derirable, you will, of course, pay attention to any representation that may be made to you for that purpose, by any of the above-mentioned gentlemen.

And as they are required by their instructions to keep, each of them, a regular journal, in which every thing remarkable relating to the natural history of the objects on which they are respectively engaged is to be entered, it is particularly desirable that they should be able to mark down with precision that spot on which any thing remarkable may be found; you will therefore supply them, on their application to you, with the latitude and longitude of the place where any such rare object may have been discovered.

And as all of them have been given to understand that their journals are, in the first place, to be transmitted to the Admiralty you are to call upon them, whenever an opportunity may occur, to send, along with your own, a copy or an abstract of these journals, according as you may deem the occasion that offers for a conveyance to be a safe or a doubtful one: and to prevent as far as possible, your and their labours from being lost to the world, it is strongly recommended, that triplicates at least be kept of all the journals, and that each person carry about with him a brief abstract of his observations, in order that, in the event of any accident befalling his journals, he may still preserve the abstract to refresh his memory.

Finally, in all your proceedings, you are to be particularly mindful of the health of the officers and men placed under your orders; to this end it will be adviseable to avoid, as much as possible, passing the night in the neighbourhood of the mangrove swamps, which are said to abound on the banks of the lower part of the river; you will prevent, as far as may be practicable, their exposure to the sun in the middle part of the day, as well as to heavy rains, and never permit them to sleep at night in the open air, but under an awning to protect them from the dews, which are always destructive of health in a tropical climate.

You are to spare the people as much as possible, from long and severe exertion, and make use of the sails instead of sweeps or oars, whenever it can be done; considering a moderate rate of progress up the river with the former, as preferable to a rapid one, at the expense of the health of the men by the latter; bearing in mind, that a moderate progress will afford better opportunities of acquiring knowledge in all departments of physical science, than a rapid one. But in the proper execution of all these matters, and the other important duties committed to your superintendance, my Lords Commissioners of the Admiralty rely on your judgment, discretion, and zeal; not doubting that you will use your best endeavours to execute all and every part of this interesting and important mission, in such a manner as will afford full satisfaction to the public, as well as to their Lordships.

By command of their Lordships.

(*Signed*)   John Barrow.

It may not, perhaps, be too much to say, that there never was, in this or in any other country, an expedition of discovery sent out with better prospects or more flattering hopes of success, than the one in question; whether it be considered as to the talents and zeal of the persons selected to carry the objects of the voyage into execution, or the preparations that were made for rendering the means of executing it efficient, and for the health and comfort of those who had embarked in it. Yet, by a fatality that is almost inexplicable, never were the results of an an expedition more melancholy and disastrous. Captain Tuckey, Lieutenant Hawkey, Mr. Eyre, and ten of the Congo's crew, Professor Smith, Mr. Cranch, Mr. Tudor, and Mr. Galwey, in all eighteen persons, died in the short space of less than three months which they remained in the river, or within a few days after leaving the river. Fourteen of the abovementioned were of the party of thirty who set out on the land journey beyond the cataracts ; the other four were attacked on board the Congo ; two died in the passage out, and the serjeant of marines at the hospital at Bahia, making the total number of deaths amount to twenty-one.

This great mortality is the more extraordinary, as it appears from Captain Tuckey's journal that nothing could be finer than the climate, the thermometer never descending lower than 60° of Fahrenheit during the night, and seldom exceeding 76° in the day time ; the atmosphere remarkably dry ; scarcely a shower falling during the whole of the journey ; and the sun sometimes for three or four days not shewing himself sufficiently clear to enable them to get an observation.

It appears indeed from the report of Mr. Mc Kerrow, the assistant surgeon of the Congo, that though the greater number were carried off by a most violent fever of the remittant type, some of them appeared to have no other ailment than that which had been caused by extreme fatigue, and actually to have died from exhaustion. The greater number however of the whole crew caught the fever, and some of them died of it who had been left on board the Congo below the cataracts ; " but these," as he observes, " were permitted to go on shore on liberty, where the day was passed in running about the country from one village to another, and during the night lying in huts or the open air; and though the dews were scarcely sensible at this season, the fall of the thermometer was very considerable, say 15° or 20° below that of the day. Spirituous liquors were not to be obtained, but excesses of another kind were freely indulged in, to which they were at all times prompted by the native blacks, who were always ready to give up their sisters, daughters, or even their wives, for the hope only of getting in return a small quantity of spirits." Perhaps too, the river water may have had its baneful effects, mixed as it is with foreign matter arising from the perpetual decomposition of animal and vegetable substances, by the dead carcases of alligators, hippopotami, lizards, &c. and by the decayed mangroves which for fifty miles occupy the alluvial banks of the river, and which, after their disappearance, are covered with the Cyperus papyrus of the height of twelve feet. Beyond these the Congo was moored, where the river was closed in by lofty hills, and over these woody shores the sea breezes had to pass. Mr. Mc Kerrow seems to think, however, that fatigue and exposure to the sun, together with considerable atmospherical vicissitudes, were the principal exciting causes of the disease which attacked the marching party, and probably those also left in the lower part of the river. Yet Captain Tuckey, so far from complaining of the heat of the sun, observes, as before mentioned, that they scarcely ever got a sight of it; and in a private letter dated from Yellala, the 20th August, after an excursion of several days, he writes, " the climate is so good and the nights so pleasant, that we feel no inconvenience from our bivouac in the open air." The fever appeared moreover in some degree contagious, as all the attendants on the sick were attacked, so that before they got out of the river it had pervaded nearly the whole crew, and extended to that of the transport; Mr. Mᶜ Kerrow was himself last of all attacked after leaving the coast, but he considered mental anxiety and disturbed rest as the sole causes. From the general symptoms that shewed themselves in most of the cases, the fever would seem to be closely allied to the yellow fever of the West Indies, as indicated by " the violent affection of the head, the suffused eye, oppression at the præcordia, great prostration and anxious timidity at the commencement, the yellow suffusion and grumous vomiting, with the indifference and apparent resignation, at the latter stage of the disease." The most prominent features of the disease are thus described by Mr. Mᶜ Kerrow.

" The fever, as I observed it in those who were attacked on board, was generally ushered in by cold rigors, succeeded by severe headache, chiefly confined to the temples and across the forehead, in some cases, pain of the back and lower extremities, great oppression at the præcordia, and bilious vomiting, which in many cases proved extremely distressing ; but in general, where the headache was very severe, the gastric symptoms were milder, and *vice versa*, though in some, both existed in a violent degree. Great anxiety and prostration of strength, the eyes in general watery, though in some the *tunica conjunctiva* was of a pearly lustre ; the tongue at first white and smooth, having a tremulous motion when put out, and shortly becoming yellowish or brown, and in the last stage covered with a black crust; in some cases the face was flushed, though frequently pale, and the features rather shrunk. The skin in some cases dry and pungent, with a hard and frequent pulse; in others the pulse below the natural standard, with a clammy perspiration on the surface. In several a yellow suffusion took place from the third to the sixth or seventh day, in one case livid blotches appeared on the wrists and ankles. The delirium was most commonly of the low kind, with great aversion from medicine. Singultus, a common and distressing symptom. The fatal termination in some, happened as early as the third or fourth, but in others, was protracted even to the twentieth day. With regard to the treatment, I shall here only observe, that bleeding was particularly unsuccessful. Cathartics were of the greatest utility ; and calomel, so administered as speedily to induce copious salivation, generally procured a remission of all the violent symptoms ; when I found it immediately necessary to give bark and wine."

From the accounts of the missions to Congo by Carli, Merolla, and others, it would appear, that bleeding copiously is the common remedy practised by the negroes, in the fevers of the country which are brought on by fatigue, and exposure to the weather. Carli mentions his having been bled no less than ninety seven-times, besides frequent and copious discharges of blood from the nose ; and from the loss of such enormous quantities, he suffers himself to be persuaded, that all the water he drank was turned into blood. Of the fourteen missionaries, who proceeded to the court of Zingha, Queen of Matamba, every one was seized with the fever, in consequence of the fatigue of travelling, and the toil of baptizing the people, and the whole were recovered in the course of four months, by having almost all their blood drawn out of their bodies, and frequent purgatives administered to them, similar, in their violent effects, to those which in Europe are given to horses ; however it is possible that the fever contracted by these pious men may have been of a very different type from that which attacked the expedition up the Zaire.

As a close to this melancholy recital, the editor hopes he may stand excused in putting on record a few brief sketches, which he has been able to collect, of the professional and literary history of those valuable men, who may be said to have fallen the victims of a too ardent zeal in the pursuit of science, which, how much soever we may lament, leaves nothing for us to censure.

JAMES KINGSTON TUCKEY was the youngest son of Thomas Tuckey, Esq. of Greenhill, near Mallow, in the county of Cork, by Elizabeth, daughter of the Rev. James Kingston, rector of Donoughmore, and sister of the present vicar-general of the diocese of Cloyne. He was born in August 1776; and his parents dying during his infancy, he was left under the care of his maternal grandmother, who placed him in the first grammar school in Cork ; here he soon distinguished himself by an ardent and inquisitive mind, and was making considerable progress in his studies, when his inclination took a turn for the sea service, from which it could not be diverted. His thirst after knowledge was ardent, but his mind was romantic in the extreme. With an eagerness natural to youth, he panted after a life of adventure : and the course of his voluntary

reading being directed to the perusal of voyages of discovery, and nautical research, he quickly imbibed a predilection for the naval profession; a predilection whose
growth, fortunately for the British navy, when once it has
taken root, is not easily checked. The period when Mr.
Tuckey fixed his choice of a profession being that of profound peace, and no opportunity being afforded for entering the navy, he was allowed by his friends to undertake
a voyage, on trial, to the West-Indies in 1791; after which
he ventured upon a second to the bay of Honduras, in
which he caught a fever, that had nearly deprived him
of life.

On the breaking out of the revolutionary war, soon after
his return, he was received on board the Suffolk, commanded by Captain Rainier, at the recommendation of
Captain, afterwards Sir Francis Hartwell, a relation by the
father's side. In that ship he proceeded to India, and
was soon rated master's mate; he was present at the capture of Trincomallée from the Dutch, and received a
slight wound in his left arm, from the splinter of a shell,
while serving in the batteries; he assisted at the surrender
of Amboyna, " famous," as he observes in a letter to his
friends," for Dutch cruelty, and English forbearance." On
this occasion, a fate more general, though less horrible in
its complexion, was about to be inflicted on the Dutch,
by the native chiefs, had not the English undertaken their
defence and protection. To assist in this humane purpose, Mr. Tuckey was stationed in a brig to cruise off the
island; and on firing a gun at a party in arms assembled
on the beach, it burst, and a piece striking him on the
wrist, broke his right arm. Having no surgeon on board,

(he writes) I was obliged to officiate for myself, and set it in a truly sailor like fashion, so that in a week after it was again obliged to be broken, by the advice of the surgeon." Mr. Tuckey never completely recovered the use of this arm.

From the intense heat and the suffocating smell of an active volcano, to which they were exposed in Amboyna Roads for ten months, where they experienced the evils of famine and sickness in addition to that of rebellion, they were glad to escape to Macao, where, in the month of January, they found the weather so intolerably cold as several times to have snow. From hence they proceeded to Ceylon; and when at Colombo, on the 15th January, 1798, a serious mutiny broke out on board the Suffolk, then bearing the flag of Rear Admiral Rainier, in the quelling of which Mr. Tuckey exerted himself with so much success, that though wanting eighteen months for the completion of his servitude to qualify him for a lieutenant's commission, the Rear Admiral appointed him, the following day, acting lieutenant of that ship: from her he was removed to the Fox frigate; and when belonging to that frigate, but being at Madras in a prize, intelligence was there received that La Forte, a French frigate, was cruising in the bay of Bengal. His Majesty's ship La Sybille immediately prepared for sea, and Mr. Tuckey, with a small party of seamen belonging to the Fox, volunteered their services in her. In the night of the 28th February they fell in with their opponent, and after a most brilliant action of two hours, frequently within pistol shot of each other, La Forte having lost all her masts and bowsprit, struck to the Sybille. In this action Lieutenant Tuckey commanded on the forecastle. Captain Cooke was mortally wounded, and Lieut. Hardyman, who succeeded to the command, observes, " the scene which presented itself on La Forte's deck was shocking; the number she had killed cannot be accurately ascertained, as many had been thrown overboard during the action, but from every calculation I have been able to make, the number killed must be from 150 to 160 men, and 70 wounded; the first and second captain, the first lieutenant, with several other officers, are included among the number killed. The Sybille had only 3 men killed and 19 wounded, two of whom afterwards died."

La Forte was the largest frigate in the French navy ; she mounted 52 guns, 24 and 12 pounders, and had 420 men. The Sybille mounted 44 guns, 18 and 12 pounders, and had 370 men. In an action with a ship of such superior force, in which so dreadful a slaughter was sustained on the part of the enemy, the vast disparity in the number of killed and wounded affords a striking instance of the great advantage which English coolness possesses over the momentary ardour of French impetuosity, and, at the same time, shews what may be effected by good seamanship and good gunnery. After this action Mr. Tuckey returned to the Suffolk, and received from the Admiral a new acting commission for his meritorious conduct.

In August 1799, he was sent by the Admiral, in the Braave, with dispatches for Admiral Blankett, then commanding a squadron in the Red Sea. At the Seychelles islands they captured a ship proceeding to Europe with an embassy from Tippoo Sultaun to the French Directory. The ambassadors concealed themselves several days in the woods, where they were discovered by Mr. Tuckey, for which he received a French general's sword as the only share for this capture, he being only a passenger in the Braave. On his arrival in the Red Sea, Admiral Blankett had quitted it for India ; and he rejoined his old ship, the Fox, which was left to guard the straits of Babelmandeb. On the return of the Admiral in 1800, he intended to visit Sir Sidney Smith at Cairo, on the supposition of the French having evacuated Egypt, under sanction of a convention with that officer ; and in that idea, sent Mr. Tuckey in the Fox to Suez, to proceed over land from thence with letters for Sir Sidney ; but on his arrival at Suez, he found it in possession of the French, in consequence of Lord Keith's refusal to permit their embarkation. He therefore returned to Bombay. The excessive heat of the Red Sea seems to have laid the foundation of a complaint which never left him. He writes from Bombay, " it may surprise you to hear me complain of heat, after six years broiling between the tropics ; but the hottest day I ever felt, either in the East or the West Indies, was winter to the coolest one we had in the Red Sea. The whole coast of ' Araby the Blest,' from Babelmandeb to Suez, for forty miles inland, is an arid sand, producing not a single blade of grass, nor affording one drop of fresh water ; that which we drank for nine months, on being analyzed, was found to contain a very considerable portion of sea salt. In the Red Sea the thermometer at midnight was never lower than 94°, at sunrise 104°, and at noon 112°. In India the medium is 82°, the highest 94°."

Towards the latter end of the same year he again proceeded with the expedition to the Red Sea, contrary to the advice of the faculty, and arrived at Juddah in

January 1801 ; but in the course of a month his complaint of the liver returned, and his health suffered so many severe shocks that he was reduced to a skeleton, and obliged to make his way back to India, where the physician of the fleet advised him to return home, as the only means of his accomplishing his recovery ; and the Admiral entrusted him with his dispatches.

His native climate had the desired effect ; and immediately on the re-establishment of his health he applied to the Admiralty for active employment ; accordingly in 1802 he was appointed First Lieutenant of His Majesty's ship Calcutta, in which situation he served during the whole of her long and arduous voyage, the object of which was to form a new establishment in New South Wales. Here Lieut. Tuckey had an opportunity of rendering very essential service, which was strongly acknowledged by the Lieutenant Governor, Colonel Collins, who transmitted to the First Lord of the Admiralty a most flattering testimony of his merits; and in particular for a complete survey he had made of the harbour of Port Philip, and for his examination of the adjacent coast and surrounding country. He was also furnished by the Lieutenant Governor with letters of recommendation to Sir Joseph Banks. He reached England in 1804, and published an account of the voyage.

But the favourable testimonies he had received were rendered abortive by the capture of the Calcutta in 1805, on her homeward voyage from St. Helena (whither she had been sent to bring home some ships under her convoy) and by an imprisonment of nearly nine years in France. For the preservation of a valuable convoy entrusted to his charge, Captain Woodriff, with a conduct which, as truly stated by the Members of the Court martial, was, " that of an experienced, brave, and meritorious officer," determined to sacrifice the Calcutta to the safety of his convoy, by first manœuvering so as to draw the attention of the enemy to one point ; and, with this view, he offered engagement to the whole squadron of the enemy from Rochefort, one of which was a three-decker, and four others of the line.   After a sort of running fight with l'Armide, the Magnanime came up, and this ship of the line he engaged for fifty minutes, frequently within pistol shot.   By this time the Calcutta was unrigged and unmanageable, and had six of her crew killed and six wounded ; and the Thetis frigate coming up close under her stern, Captain Woodriff was under the painful necessity of striking his colours ; but the whole of his valuable convoy effected their escape. Captain Woodriff, after an imprisonment of eighteen months, was exchanged for a French officer of equal rank, but Lieutenant Tuckey was kept till the termination of the war.   The Court martial held for the loss of the ship " most honourably acquitted Captain Woodriff, his officers and ship's company ;" and on this occasion the Captain delivered a paper to the court, which was as follows : " I cannot, Mr. President and Members of this Honourable Court, omit to express to you, how much I regret that the captivity of Lieutenant Tuckey, late first of His Majesty's ship Calcutta, should be a bar to the promotion he so highly merits ; his courage, cool intrepidity, and superior abilities as a seaman and an officer, entitle him to my warmest gratitude, and render him most worthy of the attention of the Right Honourable the Lords Commissioners of the Admiralty."

Lieutenant Tuckey was one of about forty lieutenants of the navy, who had cause to execrate the brutal inhumanity of the man, who for so many years tyrannized over France, and the greater part of the continent of Europe; those who had the misfortune of falling into his clutches, felt themselves at once cut off from every hope of advancement in their profession, and many fell the victims of despair. Not so, however, with Lieutenant Tuckey. He still kept up his spirits, and encouraged hope, being, as he expressed himself, on another occasion, " by no means addicted to contemplate the dark side of events; but as cheerful and happy as the possession of health, ease, and a satisfied disposition can make me." He married in 1806, a fellow prisoner, Miss Margaret Stuart, daughter of the commander of a ship in the East India Company's service, at Bengal. She also had been taken by the Rochefort squadron, on her passage in a packet to join her father in India.

Various applications were made at different times, for the exchange of Mr. Tuckey; but they proved fruitless, and he was doomed to remain a prisoner during the war: sad consequence of that implacable spirit of hatred which actuated the ruler of France, and made him careless alike of the lives of his own, and of his enemy's prisoners! How many fair prospects were blighted and destroyed by the unfeeling obstinacy of this disturber of Europe!

In 1810, after considerable difficulties, and repeated refusals, Mr. Tuckey obtained permission for his wife to visit England, for the purpose of looking after his private affairs. Her object being accomplished, she obtained passports from the French government to return to her husband, and was landed at Morlaix ; but counter-orders had been received at this port, and she was detained ; and after many unsuccessful memorials, praying to be allowed to rejoin Mr. Tuckey at Verdun, and after a detention of six weeks, she was sent back to England. We have here another instance, in addition to the many on record, of the capricious cruelty of Bonaparte, which was equally exercised on either sex : and let it not be said by his advocates—strange, that such a man should find advocates, especially among Englishmen—that he knew nothing of such counter-orders. So it was said, with equal truth, in regard to the detention of Captain Flinders ; for it is well known that, in all matters relating to the British prisoners, his ministers stirred not a step without his special directions.

On the advance of the allied armies into France, in 1814, the British prisoners were ordered at a moment's warning into the interior ; and Mr. Tuckey, with his two little boys, was obliged to travel, in the most inclement weather he ever experienced, to Blois. His youngest son was taken ill on the journey, and fell a victim to fatigue and sickness. " I had, indeed," says the father, " a hard trial with my little boy, for after attending him day and night for three weeks, (he had no mother, no servant, no friend, but me to watch over him,) I received his last breath, and then had not only to direct his interment, but also to follow him to the grave, and recommend his innocent soul to his God ; this was indeed a severe trial, but it was a *duty*, and I did not shrink from it." Another severe trial was reserved for him, on his return to his family in England, on the final discomfiture of Bonaparte; he had the misfortune to lose a fine child, a girl, of seven years of age, in consequence of her clothes taking fire, after lingering several days in excruciating agony.

During his long imprisoment in France, Mr. Tuckey suffered considerably from tedious and harassing illness, aggravated by the cruel reflection, that the prime of his life was rapidly passing away, without the possibility of any exertion of his talents being employed for the benefit of himself, or his growing family. In the intervals of sickness, besides the education of his children, which was to him a source of pleasure and constant employment, his chief amusements were reading and composition. Severe as his fate was, he possessed a mind of too vigorous and active a turn to allow his spirits to sink under his unmerited misfortunes; the painful moments of his long imprisonment found some relief, in the laborious compilation and composition of a professional work, " undertaken to pass away the tedious hours of a hopeless captivity, alike destructive of present happiness, and future prospects." This work was published in England, shortly after his return, in four octavo volumes, under the title of " Maritime Geography, and Statistics." It takes a comprehensive view of the various phenomena of the ocean, the description of coasts and islands, and of the seas that wash them; the remarkable headlands, harbours, and port towns; the several rivers that reach the sea, and the nature and extent of their inland navigations that communicate with the coasts. The information thus collected is drawn from the latest and best authorities; to which is added his own " local and professional knowledge, acquired in the navigation of the seas that wash the four quarters of the globe." A brief view is also taken of the history and state of the foreign and coasting trade of the colonies ; the state of the home and foreign fisheries ; of the national, and mercantile marine ; and generally of all maritime establishments and regulations. It is a work of useful reference, and one that may safely be recommended for general information.

In August 1814, Mr. Tuckey was promoted by Lord Melville to the rank of commander ; and in the following year, on hearing of the intention of Government to send an expedition to explore the river Zaire, he made an application, with several other officers, to be appointed to that service ; his claims and his abilities were unquestionable ; he had stored his mind with so much various knowledge and, for the last nine years, had given so much attention to the subject of nautical discovery and river navigation, that he was considered as most eligible for the undertaking ; but his health appeared delicate : he was, however, so confident that his constitution would improve by the voyage, and in a warm climate, and urged his wishes so strongly, that the Lords of the Admiralty conferred on him the appointment. How far his zeal and qualifications were suited to the undertaking, his Journal will furnish the best proof. That document is now given to the public, just as it came from the hands of its author. Not a sentence has been added or suppressed, nor has the least alteration been made therein, beyond the correction perhaps of some trifling error in grammar or orthography. The information it contains must have been procured under very unfavourable circumstances. Had he been permitted to penetrate further into the interior, or to return at leisure, and in health, from the farthest point even to which he ascended, his account of the country would have been so much the more complete; but his zeal to accomplish the object of the expedition had completely exhausted him, and brought on the return of a disorder to which he had long been subject; still he held out to the last; and there is very little doubt, that if the accident which happened to his baggage canoe had not put an end to every possibility of his proceeding much farther up the river, that he would have gone on till he had sunk under sickness and fatigue, and left his remains in the interior of the country.

On the 17th September he reached the Congo sloop, and the following day, for the sake of better accommodation, was sent down to the Dorothy transport, at the Tall Trees. He arrived in a state of extreme exhaustion, brought on by fatigue, exposure to the weather, and privations. He had no fever nor pain in any part of the body; the pulse was small and irritable; the skin at times dry, at others clammy, but never exceeding the temperature of health. On the 28th he thought himself better, and wholly free from pain, but shewed great irritability, which was kept up by his anxiety concerning the affairs of the expedition. On the 30th the debility, irritability, and depression of spirits became extreme, and he now expressed his conviction, that all attempts to restore the energy of his system would prove ineffectual. From this time to the 4th, when he expired, his strength gradually failed him, but during the whole of his illness, he had neither pain nor fever; and he may be said to have died of complete exhaustion, rather than of disease. He had deceived himself, it seems, by the confidence which he felt in the strength of his constitution. The surgeon states that, since leaving England, he never enjoyed good health, the hepatic functions being generally in a deranged state; yet he was always unwilling to acknowledge himself an invalid, and refused to take such medicines as were deemed at the time to be essentially necessary. On his march into the interior, the symptoms became much aggravated, and he was prevailed on by Doctor Smith to take some calomel; afterwards opium was found necessary, and lastly, the bark.

The few survivors of this ill-fated expedition will long cherish the memory of Captain Tuckey, of whom Mr. Fitzmaurice, the master, who succeeded to the command, observes, in reporting his death,—" in him the navy has lost an ornament, and its seamen a father." But his benevolence was not confined to the profession of which he was so distinguished a member. A poor black of South Africa, who, in his youth, had been kidnapped by a slave dealer, was put on board the Congo, while in the Thames, with the view of restoring him to his friends and country, neither of which turned out to be in the neighbourhood of the Zaire, and he was brought back to England. This black was publicly baptized at Deptford church, by the name of Benjamin Peters; having learned to read on the passage out by Captain Tuckey's instructions, of whom he speaks in the strongest terms of gratitude and affection. He was generous to a fault. A near relation has observed, " that a want of sufficient economy, and an incapability of refusal to open his purse to the necessities of others, have been the cause of many of the difficulties which clouded the prospects of his after life;"—that " he knew nothing of the value of money, except as it enabled him to gratify the feelings of a benevolent heart."

In his person Captain Tuckey was tall, and must once

have been handsome ; but his long residence in India had broken down his constitution, and, at the age of thirty, his hair was gray, and his head nearly bald ; his countenance was pleasing, but wore rather a pensive cast; but he was at all times gentle and kind in his manners, cheerful in conversation, and indulgent to every one placed under his command. In him it may fairly be said, the profession has lost an ornament, his country has been deprived of an able, enterprising, and experienced officer, and his widow and children have sustained an irreparable loss.

LIEUTENANT HAWKEY was another of those officers, whose prospect of rising in his profession was blasted by the system of refusal to exchange prisoners of war; a most inhuman system, which doomed young officers to a hopeless captivity, limited only by the duration of the war, or rather, viewing the character of that war, limited by no visible bounds ; with the additional cruelty of an indefinite separation from their country and their friends. They had, moreover, in this hopeless situation constantly before them the melancholy reflection, that, after having spent the first and best years of their lives in the active service of their country, and the middle part of them in a horrid captivity, even when the time of their liberation should arrive, they would have to begin the world again ; and, without a chance of employment in their own profession, as the war would then have ceased, painfully to seek out new means for the support of themselves and their families. Under this unfeeling system, Lieutenant Hawkey suffered an imprisonment of eleven years. A few months after the renewal of the war, in 1803, when serving as a midshipman in La Minerve, under the command of Captain (now Sir Jaheel) Brenton, he was taken prisoner in the gallant defence of that ship, when she was unfortunately, in a fog, run by the pilot on the western point of the stone dyke of Cherbourg. A commission however of Lieutenant had been sent out for him, by mistake, to the West-Indies; which being dated previous to his capture was not cancelled, but forwarded to him in France; and was thus the means, in some degree, of alleviating the evils of captivity. It was in France he became acquainted with his fellow prisoner Captain Tuckey, who, on his appointment to the command of the Congo, requested to have his companion in misfortune to accompany him on a voyage, which held out a fair prospect of gratifying and rewarding their mutual talents.

Lieutenant Hawkey was an excellent draughtsman: he sketched in a bold and artist-like manner; and to a general knowledge of natural history, he united the talent of painting the minuter sea and land animals, with great spirit and accuracy, and in an exquisite style of colouring. A number of specimens of this kind were found in a small pocket book, accompanied with some slight memoranda; but his papers, containing descriptions of those sketches and drawings, and other remarks, in the progress up the river, have unfortunately been lost. He proceeded with the captain to the farthest point of the journey, and though employed in the most active manner, and exposed to the same weather, and the same hardships, as the rest of the party, he had no complaint whatever when he returned to the vessel, on the 17th September; his case was therefore somewhat singular. He continued in good health, and without

any complaint till the 3d October, when the ship was at
sea; he then expressed a sense of lassitude about his loins,
and irritability of stomach; but there was no apparent febrile action; the pulse being about the natural standard,
which with him was only 65°, without the body undergoing any encrease of temperature. The only symptoms
were irritability of stomach, with extreme langour and debility; the next day however, he was seized with vomiting;
on the 6th, became insensible, the pulse scarcely perceptible at the wrist, and the extremities cold; and he continued thus till 11 o'clock in the evening, when he expired
without a struggle.

Mr. Eyre, the Purser, was a young man of a corpulent
and bloated habit. He had no illness, while in the river;
had not been on shore for three weeks, and had taken very
little exercise during the voyage. In the night of the 27th
September, when on the passage to Cabenda, he was attacked with febrile rigors, severe pain in the head, back
and extremities, with general lassitude, prostration and depression of spirits, and on the third day he breathed his
last. Before death a yellow suffusion had taken place,
with vomiting of matter, resembling coffee grounds; this
symptom of extravasated blood into the stomach, which oc-
cured in many of the cases, would seem to confirm the idea
of the disease being the same as that of the Bulam fever.

Mr. Fitzmaurice, the Master and Surveyor, and Mr.
Hodder, Master's Mate and Midshipman, entirely escaped the fever, excepting a slight attack, experienced by
the former, in consequence of a fatiguing march across

the mountains, on a hot sultry day, to view the cataract of Yellala, which after a good night's rest, was entirely removed. Lockhart the gardener was on his legs every day, from morning till the evening, sometimes heavily loaded with the plants he had collected; yet he proceeded to the farthest point, and returned to the ships, without experiencing an hour's illness, and found the climate the whole way remarkably pleasant. Being drenched however with rain in the lower part of the river, he took the fever, and was left in the hospital at Bahia, with the serjeant of marines, both of whom were so much reduced, as to have little hope of the recovery of either. Lockhart however survived, and is now perfectly well in England; but the sergeant died almost immediately after the sailing of the Congo.

MR. CHETIEN SMITH, the son of a respectable land-holder, near the town of Drammen in Norway, was born in October 1785. He was educated at the school of Kongsberg, and finished his studies at the university of Copenhagen; where, under Professor Hornemann, he acquired a taste for botany, and particularly for that branch of the science, of which his native mountains afforded such ample resources, —the mosses and lichens. Though at an early period of life, he had distinguished himself in the study of medicine, and had the care of the sick in the great hospital at Copenhagen, he could not resist the temptation of accompanying his friends, Hornemann and Wormskiold on a botanical tour into the mountains of Norway. In the early part of this tour, the war, which broke out in 1807 between Sweden and Denmark, recalled his companions, and

left Mr. Smith to pursue alone his researches, in the mountains of Tellemarck, where he discovered a great number of
mosses, and other new plants, which gained him celebrity
among all the botanists of the North. In 1812 he made a second excursion across the mountains of Tellemarck and Hal-
lingdal, which were but little known, even to the natives of
Norway ; he ascertained their heights, examined their productions, made a number of curious meteorological observations, and, in short, traversed those solitary regions not only
as a botanist, but as a natural philosopher ; and the narrative which he has given of his proceedings, to use the words
of his friend Von Buch, " will always be considered as one
of the most curious and instructive documents of physical
geography." He has therein exemplified and explained the
immense influence of the proximity of the sea, and the surprising difference, resulting from it, between the temperature of the interior of the continent, and of the coast, and its
effects on the different products of the vegetable world ; the
limits of perpetual snow on the sides of different mountains, and a great variety of interesting facts, connected
with the geography and physiology of plants.

The Patriotic Society of Norway, struck with the zeal
and indefatigable industry of Mr. Smith, engaged him at
its own expense to undertake another scientific expedition
into the clusters of mountains, which, about the 62nd parallel of latitude, separate the valleys of Walders, of Guld-
transdal, and of Romsdal, whose height and extent were
unknown, and many parts of them untrod even by the
hunters of the rein deer. By this excursion the Norwegian
Flora was greatly extended, and from it the geography of
plants acquired fresh facts at once exceedingly curious and

interesting. Nor was this all ; with a true spirit of philan-
thropy Mr. Smith assembled the scattered peasantry of
these high and secluded valleys, explained to them the
characters and the valuable properties of the lichens which
covered their mountains, instructed them how to convert
these mosses into bread that was pleasent to the taste,
nourishing, and wholesome, and prevailed on them to
adopt this bread instead of that miserable resource of bark
bread, which affords but little or no nourishment, and that
little at the expense of health.

The death of his father about this time put him in possession of a little fortune, which he at once resolved to
employ in studying nature in foreign countries. His
nomination to the professorship of botany at the university
of Christiania did not divert him from his plan ; on the contrary, he thought he could not do a greater service to the
cause of science than to consecrate the fruits of his travels
to the new botanical garden of that place. He came to
London, met with a countryman who had been instructed
in the King's Gardens at Kew, and sent him to superin-
tend his favourite garden at Christiania, with abundance
of plants and seeds which he purchased at his own expense. He next proceeded to Edinburgh, from whence
he set out on a tour across the highest mountains of
Scotland to examine their productions. The mountains
of the northern counties of England and of Wales did not
escape his active researches. From Wales he crossed over
to Dublin, scoured all the mountains of Ireland, and returned to London towards the end of the year 1814. It
is needless to add, that so zealous an advocate for the extension of human knowledge engaged at once the friendship

and protection of Sir Joseph Banks.   At the house of this
patron of science he met with the first naturalists of the
age, and among others with that distinguished geologist,
the Baron Von Buch, whose habits and feelings being congenial with his own, they soon projected a voyage of scientific inquiry to the island of Madeira, and to those of the
Canaries.

On the 21st April, 1815, they landed at Funchal, the
capital of Madeira.   " From that moment," says M. Von
Buch, " transported with the sight of so many new objects, Smith knew no repose ; he laid hold of the several
species of Cactus which in the most whimsical forms cover
the rocks, to convince himself that they were real ; he
leaped over the walls to examine those forests of Donax
which the wind agitates above the vines to which they give
support ; he ran from flower to flower, as if in extacy, and
it was with great difficulty he could be prevailed on to enter
the town. Here again his eye was delighted in traversing
the great square, and observing the avenue of large trees of
Justicia, of the Melia Azedarach, and of the gigantic Da-
tura, covered with their large and brilliant blossoms, which
fill the air with their perfumes ; the immense leaves of the
Banana trees waving above the walls, and the superb palm
lifting its lofty head high above the houses; the singular
form of the Dragon tree ; the fragrance of the flowers,
and the tufted foliage of the orange trees threw him into
raptures. The elegant coffee-shrub is found only in the
gardens ; but the pine-apple flourishes in the open fields ;
and the Mimosa, the Eucalyptus, the Melaleuca, the Mam-
mea, Clitorea, Erythrina, Eugenia, of which the dwarfish
fragments only are seen in the conservatories of Europe,

here mount up to large and beautiful trees, and their flowers, glowing with the most vivid colours, are viewed against the most brilliant sky in the world.

His own feelings, on being thus suddenly transported from the moss-grown mountains of his native country to a more genial climate, are thus expressed in a letter to his friend; " how shall I be able to describe to you, how declare to you what I have here felt, what I have here seen! How shall I be able to give you an idea of the variety, of the singularity of those forms, of that beauty and that brilliancy of the colours, of all that magnificence of nature which surrounds me! We ascend the sloping ridges of the mountains which embrace the splendid city of Funchal; we rest ourselves on the margin of a brook, which falls in numberless cascades across thickets of rosemary, of laurels, and of myrtles;—the city at our feet, with its forts, its churches, its gardens, and its roadstead; above us, forests of the stone pine and of chesnuts, interspersed with the flowers of the spartium and lavender. A whole legion of Canary birds makes the air resound with their sweet song; and nothing here, but the snow on the mountain tops, which now and then pierce through the clouds, would recal to my recollection my native country."

M. Von Buch observes, that neither the torrents of rain which fell almost daily, nor the dense clouds which constantly covered the mountains for more than half of their height, nor the snow which enveloped their summits, could restrain them from attempting to ascertain the distribution of vegetation on this island, and the height of its mountains. They found by the barometer, the altitude of Nostra Senhora da Monte to be 1778 English feet above the level of the sea. At the height of 3200 feet they en-countered a wood composed of the beautiful Laurus in-dica, the Laurus nobilis, the Erica arborea or mediterranea, and the Erica scoparia. Through thick fog, and continual rain, they persevered in the ascent; and at the height of 4340 feet, they traversed the valley of Ganada, where Smith viewed with astonishment and delight, a whole forest of myrtles (Vaccinium arctostaphyllos) of trees from 16 to 20 feet high and more. At 5390 feet they fell in with the snow. The summit of Torringas was found to be 5857 feet.

The two naturalists left Madeira on the 2d May, and landed on the 5th at Oratava, in Teneriffe, where they were kindly and hospitably received by one of the most amiable and respectable families in the island, to whom Mr. Smith, by his natural gaiety and the suavity of his manners, made himself particularly acceptable : but the charms of agreeable society did not make him forget the object of his visit to the Canaries. He was always on his feet, and incessantly in pursuit; he never returned from his excursions till late in the evening, and always laden with a rich harvest of plants, the examination and arrange-ment of which, left him but little time for sleep. Some-times, in his rambles, overtaken by night, he passed it in caverns, without food, and harassed by fatigue; but happy in the discoveries he had made, he neither felt nor thought of either. The melting of the snow, on the 19th May, al-lowed him to ascend the summit of the Peak of Teneriffe, from whence he made a tour round the southern part of the island, in which he examined, and determined for the first time, the pine of the Canary islands to be an unde-

scribed species, to which he gave the name of Pinus canariensis; discovered a new species of Ardisia, and collected many other new and interesting plants. Several
botanical travellers have noticed the singular appearance
of the roofs of the houses of Laguna, from their being covered with a sort of house-leek, but none had described
it. Mr. Smith, on examination, found it to be a new species, and named it, from its situation, Sempervivum
urbium.

From Santa Cruz the travellers passed near to Palmas,
the capital of Grand Canaria. It was now the month of
August, and the summer heat had parched the earth and
dried up the plants, excepting some Euphorbias and
others of the succulent tribe. They, however, determined
by the barometer, for the first time, the height of Pico del
Pozo de los Nieves, the most elevated on the island, to
be 6224 English feet above the level of the sea.

On returning to Santa Cruz, Smith and his friend set
out on a journey to the Peak of Teneriffe, along the crest
of the mountain which crosses the island in its greatest
length. Near the summit of the peak the two naturalists
passed several days, traversing the immense stream of
Obsidian, which encircles its western side, examined the
volcano of Chahorra, remaining in the mountains till forced
down from want of food, water, and shoes, the latter of
which were fairly cut in pieces by the glassy lava. Embarking at Oratava, they proceeded to the Isle of Palmas;
examined the immense and almost inaccessible crater,
which occupies the centre of the island; ascended the Pico
de los Muchachos, whose height was found to be 7707
English feet, and which commands a view of the whole

island. They next visited the port of Naos, on the isle of Lancerota, where the vessel was to complete her cargo with barilla, made from the Mesembryanthemum crystallinum. In the mean time, the two naturalists paid a visit to the volcano, which, in 1750, almost entirely destroyed twelve villages, and covered nearly the third part of the island. Instead of one mouth they observed fifteen or sixteen, extending in the same line about six English miles, each having formed a cone from three to four hundred feet high. Across the middle of these mouths were large fissures, exhaling aqueous vapours, which raised the thermometer in a few moments to 180°. On the 27th October, they re-embarked, and arrived at Portsmouth on the 8th December.

It was the intention of Dr. Smith to arrange his Flora for publication, in London, and then return to his native country ; but on Sir Joseph Banks proposing the appointment of botanist, to an expedition then preparing to explore the Zaire, he most readily and unconditionally accepted the same, from a pure love of science, and in the hope of being useful to the world. His Journal, which is now published precisely in the state it was found, is sufficient to evince his great zeal, and qualifications as a botanist. It is but just however to the memory of the deceased to state, that this Journal had evidently undergone no revision, but is a transcript or rather a translation of his original minutes and observations, as they appear to have been entered, from day to day, in a small pocket memorandum book ; written in the Danish language, and in so small and illformed a character as in some places to be perfectly illegible. By the close attention, however, of Doctor Rydberg, to whom the editor is indebted for the translation, the greater part has been pretty well made out. His notes are carried on to the end of the journey upwards, but are not continued on his return down the river. He was taken ill, before they reached the vessels, and came down with the Captain in the last canoe: and was sent with him to the Transport, for the sake of greater convenience: by this time however, he was dangerously ill, and refused to take any thing, either in the shape of medicine, or nutriment. He had tried bark, but his stomach constantly rejected it: and under an idea that his illness proceeded only from debility, he persisted in taking only cold water. On the 21st September he became delirious, and died on the following day.

Mr. Cranch was one of those extraordinary self-taught characters, to whom particular branches of science are sometimes more indebted, than to the labours of those who have had the advantage of a regular education. He was born at Exeter in the year 1785 of humble, but respectable parents; at eight years of age he had the misfortune to lose his father; and as the circumstances in which his mother was left, did not enable her to provide for all her children, John, the subject of the present memoir, was taken charge of by an uncle living at Kingsbridge. The main object in life, and which was nearest to the heart of this relation, was the accumulation of wealth; and his extreme penury denied to his nephew almost the benefit of a common education. The miserable guinea which procured for him a year's instruction in reading, writing, and arithmetic, was wrenched from him with so much grudging, and in a manner so unkind, as to be then severely felt, and never afterwards forgotten.

At the age of fourteen, this provident relation first put him out as an apprentice, to learn " the art and mystery of shoemaking;" a line of life which, from its peculiar monotony of stillness, or in spite of it, seems by no means unfriendly, as experience has shewn, to the progress of intellectual acquirement. The strength of mind for which young Cranch had been distinguished from his childhood was now constantly and obviously struggling with the adverse circumstances of his situation; but every moment, which could be stolen from his daily labour, was devoted to the few books which he had found means to collect. The study of natural history was that in which he mostly delighted; and, even at this early period of his life, he was able to draw up correct and classical descriptions of all the insects he could procure in the neighbourhood of Kingsbridge. Without other assistance than books, he had acquired a sufficient knowledge of Latin and French, to enable him to understand thoroughly those languages, when made use of by zoological writers, and to employ them himself, in describing objects of natural history. He had acquired also a general knowledge of astronomy. But, while thus eagerly endeavouring to grasp at science, every thing tended to depress, and nothing to encourage him. However, he had the fortitude to persevere; and continued, in spite of every obstacle, silently and sedulously, unnoticed and unknown, to nourish his ruling passion, the love of knowledge.

At the expiration of his apprenticeship, he went up to

London, with the professed view of improvement in the art of shoemaking: but in reality with higher objects and better hopes; though he hardly ventured to own them to himself. The manners and morals of his fellow workmen were ill suited to his feelings and pursuits; and served only to encrease his dislike for the profession to which he had been doomed. But it was some consolation to reflect that he was in the great mart of human knowledge; and though unfriended, and a stranger, he found that information flowed in upon him on every side. His mind was filled, but not satisfied. Every museum, auction room, and book stall, every object to which his attention was called, he visited with a rapid and unsatiable curiosity; gleaning information wherever it was to be had, and treasuring it up with systematic care. His account of what he observed in the capital is said to exhibit an obvious and striking proof of an inquisitive, diligent, and discerning mind. A person of this stamp could not long remain in London without meeting with kindred spirits. One of these associates, speaking of Cranch, observes, " our conversations and philosophical rambles near London, have often called forth such observations and disquisitions from him on the various qualities, attributes, combinations, provisions and arrangements of nature, as marked vast comprehension, as well as the most delicate subtilties of discrimination in an intellect, which seemed indeed to be calculated to grasp magnitude and minutiæ with equal address, and which could at once surprise, delight, and instruct."

After a residence of some time in London, he returned to the haunts of his childhood; but it was soon discovered how little chance the " bootmaker from London" had of eclipsing even his humble rivals, who had never lost sight of the smoke of their native hamlet. But he had no alternative; he must eat to live; and work at his trade to be able to eat. His labour however produced him little more than a bare subsistence; and every moment that he could venture to take from it, was dedicated to his favourite pursuit.

Shortly, however, his domestic circumstances were favourably improved by marriage. His workshop was now consigned wholly to his journeymen, while he was sedulously and successfully collecting objects of natural history. No difficulties nor dangers impeded his researches He climbed the most rugged precipices; he was frequently lowered down by the peasants from the summits of the tallest cliffs; he waded through rapid streams; he explored the beds of the muddiest rivers; he sought the deepest recesses. He frequently wandered for whole weeks from home, and often ventured out to sea for several days together, entirely alone, in the smallest skiffs of the fishermen. No inclemency of weather; no vicissitudes of " storms and sunshine," ever prevented his fatiguing pursuits; the discovery of a new insect amply repaid the most painful exertions. Several papers in the " Weekly Entertainer," a little work which accompanies one of the most popular of the western newspapers, were written by him; and by these, and his collection of subjects in natural history, he gradually became better known, and his talents duly appreciated by the most able naturalists. Of this the following extract of a letter to the editor, from Dr. Leach, of the British Museum, bears ample and honourable testimony.

" In 1814, Mr. Montagu and myself, together with Mr. C. Prideaux, visited Mr. Cranch, for the purpose of seeing his museum. We were all astonished at the magnitude of his collection of shells, crustacea, insects, birds, &c. collected entirely by himself, and still more so with the accuracy of their classification, and with the remarks made by this self-educated and zealous individual. He conversed on all subjects connected with natural history, with modesty, but at the same time, with that confidence which is the result of knowledge. Quite delighted with having made his acquaintance, I left him with a resolution to cultivate a correspondence with him on the subject of our favourite pursuits. On the following morning, I received a note from him, offering me any specimens that might be wanting, and that he could supply, to my collection.

" Soon after this meeting I was appointed to the British Museum, when Mr. Cranch applied to me to endeavour to obtain for him some situation in that institution, which would enable him to cultivate the study of natural history on a more extended scale ; but as no vacancy existed, and as I found his demands for employment come within the limits of my pocket, I proposed that he should undertake to investigate the coasts of Devon and Cornwall, for marine productions ; and eventually to make a tour of Great Britain, with the same view ; at the same time, I promised to recommend him to the first situation that might occur, to enable him to attain the object of his ambition.

" On receiving my letter he immediately discharged his journeymen, and converted his manufactory of boots and shoes into apartments for the reception and preservation of such objects of natural history as his daily excursions might procure. He kept up a continual communication with the fishermen of Plymouth, and constantly received from them baskets filled with the rubbish they dredged from the bottom of the sea; and this he examined with diligence and attention, preserving all the new objects that he discovered, and making descriptions of them. He visited, occasionally, the Brixham, Plymouth, and Falmouth fishermen, and made excursions with them. He very often left Kingsbridge in an open boat, and remained absent for a long time together, during which, he dredged when the tide was full, and examined the shores when it was out. At night he slept in his boat, which he drew on shore; and when the weather was too stormy for marine excursions, he would leave his boat and proceed to examine the country and woods for insects, birds, &c. The remarks with which he accompanied the infinity of new objects which he discovered, are invaluable; many of them have been, and the rest shall be hereafter, made public."

In this way was Mr. Cranch employed for the collection of natural history in the British Museum, at the time when the expedition to the Congo was planned: for such an expedition, a person of this description was invaluable, and Doctor Leach recommended him to Sir Joseph Banks, as one in every way fitted for the undertaking. On his part, an appointment so suited to his pursuits, and so flattering to his hopes, was the height of his ambition, and he at once accepted it, though not without some painful struggles to his feelings. It seems he had a sort of presentiment that he should never return, and that the expectation of such an event became weaker and weaker, as his country faded from his view. His conduct, however,

during the voyage out, does not appear to have been influenced by this feeling; nor was his exertions at all relaxed by an occasional lowness of spirits, which was, perhaps, partly constitutional, and owing partly to the gloomy
view taken of christianity by that sect denominated Methodists, of which, it seems, he was a member. He is represented, however, by his friends, as a sincere Christian,
an affectionate parent, and a kind friend.

Mr. Cranch was taken ill on the 23d August, on the
march between the banza or town of Cooloo, and the
banza Inga, and was carried back on the shoulders of the
natives to Cooloo, and from thence in a hammock to the
place of embarkation below the rapids; but it was the
tenth day before he reached the ship in a canoe. The
symptoms, by the surgeon's report, were an extreme languor
and general exhaustion; a restlessness and anxiety,
approaching at times to delirium, but he had no pain,
except an uneasy sensation throughout the abdomen;
the countenance became of a dirty yellow colour, the
pulse was at 108°, and very small. The next day he was
much worse, and on the third day the whole body became
yellow; the countenance assumed a deadly aspect, the
pulse at the wrist imperceptible, and in the evening he
expired, " after uttering," says Mr. Fitzmaurice, " a de-
vout prayer for the welfare of his family, and with the
name of his wife quivering on his lips. He was of that
order of dissenters," he adds, " who are called Methodists,
and if I may judge from external appearances, he was an
affectionate husband and father, a sincere friend, a pious,
honest, and good man." He died in the 31st year of his
age, and was buried at Embomma by permission of the

king, in his own burial ground, where he was laid with military honours by the side of his fellow-traveller Mr. Tudor, who had been interred with the like ceremony, a few days before.

Mr. Tudor was a young surgeon recommended by Mr. Brookes the anatomist, and examined and approved by Sir Everard Home, as a person well qualified to act in the capacity of comparative anatomist. The unfortunate circumstances of the expedition afforded him but few materials to work upon, and but little opportunity to exercise his talent on those few. He was the youngest of the party, and the first who was attacked with fever on shore, being seized on the 15th August, after a march of three days. He was immediately sent back to the vessels, and on the 22d he reached the Congo in one of the double-boats. On his arrival he shewed great debility, anxiety, and impatience. His case was very similar to that of Mr. Cranch, and on the evening of the 29th he died without pain.

Mr. Edward Galwey was second son to the banker of that name in Mallow. He was educated for the university, with a view to qualify for one of the learned professions ; but an eligible appointment offering, in the mean time, to a situation in the East Indies, he was about to proceed thither, when, by the advice of his friends, and a necessity occurring for his assistance in his father's office, he was prevailed on to take his seat at the desk. It was soon however discovered, that the dull routine of such employment was but little congenial with his inclinations, and he escaped from it whenever he could with propriety do so, to indulge his zeal for scientific research, and to cultivate his taste for music, of which he was passionately fond, and in which he excelled. He availed himself of all opportunities to acquire a practical knowledge of botany, and was particularly conversant in all the new discoveries in chemistry, which, with geology, were his favourite studies. He was soon however drawn from his retired and studious habits to seek for health in the south of Europe, having suffered for several months by an oppression and pain in the chest, accompanied with a constant short, dry cough, quick pulse, and all the symptoms of a confirmed consumption ; from all which however he was completely cured before he landed in Lisbon, after a tempestuous and protracted passage in the winter of 1813. Finding himself so well, and conceiving that his uniform of a yeomanry officer would afford him much facility in travelling in the peninsula, he was induced to go into Spain; and the few months he spent in visiting various parts of this country, and the delight experienced by a mind finely stored like his with diversified knowledge, inspired him with so enthusiastic a zeal for foreign travel, that although on his return to Ireland, he re-assumed his station in the bank, it was evident that an opportunity only was wanting to set him out again on his travels. That opportunity soon occurred by the ill-fated expedition to explore the Zaire. On hearing that Captain Tuckey, who was one of his early friends, had got the appointment, he immediately wrote to entreat he might be allowed to accompany him as a volunteer. It was in vain to represent how inconveniently he must be accommodated; and that he could not be allowed even to take a servant; but he pleaded the example of Sir Joseph Banks, as entirely obviating, in his own case, so trifling an objection; his family remonstrated with him on the score of his health being injured from the hardships he would necessarily have to undergo, and from the effects of climate; his argument was, that he had already tried both, and his health had improved by the experiment. In short, remonstrance and persuasion were resorted to in vain: he persisted in his entreaties with the Admiralty and Captain Tuckey; and on the latter expressing a wish to take him, as one likely to be useful, in promoting the objects of the expedition, he was permitted to join the Congo as a volunteer.

Mr. Galwey proceeded with the Captain's party as far up the river as the banza Inga, where he was taken ill, about the 24th August, and sent off from thence to the vessels: but he did not reach the Congo, in his canoe, till the 7th September, being then in a state of great exhaustion; his countenance, by the surgeon's account, ghastly, with extreme debility, and great anxiety; a short cough, with hurried respiration and heaving of the chest, the pulse 108, and very small, the body of a dirty yellow colour. On the following day, all the bad symptoms were encreased, but he was free from pain. On the 9th he became insensible, and expired about the middle of the day. His body was taken to the burial ground of the King of Embomma, and interred with such honours as the dispirited and much reduced party could bestow, by the side of his unfortuate companions Cranch and Tudor.

Mr. Galwey had taken a very active part in collecting specimens, and making remarks on the natural products of the country, and more particularly on its geology; but both his journal and his collections have been lost. They had met in their progress with a party of slave-dealers, having in their possession a negro in fetters, from the Mandingo country. From motives of humanity, and with the view of returning this man to his friends and country, as well as under the hope that he might become useful as they proceeded, and give some account of the regions through which he must have passed, as soon as he should be able to speak a little English, Captain Tuckey purchased this slave, and appointed him to attend Mr. Galwey; but he was utterly incapable, it seems, of feeling either pleasure or gratitude at his release from captivity; and when Mr. Galwey was taken ill, he not only abandoned him, but carried off the little property he had with him, no part of which was ever recovered.

After this gloomy recital of the mortality which befel the officers and naturalists of the expedition, it will be the less necessary to bespeak the indulgence of the public in passing judgment on the present volume. The Journals of Captain Tuckey and Professor Smith, with the collections which have reached England, afford ample testimony how much more might have been expected in less unfortunate circumstances. These Journals will not be deemed the less valuable for being the mere records of facts and impressions, written down without regard to arrangement, the moment they occurred and were made. The few General Observations collected from these Journals, and from detached notes of Lieutenant Hawkey, Mr. Fitzmaurice and Mr. Mc Kerrow, have been thrown together in order to give a connected, though imperfect, view of that particular part of the country, and people visited by the expedition. The papers, No. 2 and 3 in the Appendix, by Doctor Leach, and Sir Everard Home, respecting the parasitic *Vermis* which takes possession of the Argonaut shell, and which clear up a long disputed point in natural history, are reprinted from the Philosophical Transactions, by permission of the venerable President, who is on all occasions most ready to lend his cordial assistance to the dissemination of human knowledge.

The other papers in the Appendix, illustrative of those subjects in the animal, vegetable, and mineral kingdoms which occurred along the line of the Zaire, as far, at least, as the materials sent home would admit, were obligingly drawn up by Doctor Leach and Mr. Koenig of the British Museum, and by Mr. Brown, Secretary to the Linnæan Society, and Librarian to Sir Joseph Banks; to each and all of whom, but more especially to the last mentioned gentleman, for his excellent observations on the plants found in the vicinity of the Congo river, the Editor has to express his grateful acknowledgments.

NARRATIVE OF AN EXPEDITION

TO EXPLORE

THE RIVER ZAIRE.

CHAPTER I.

Passage to, and Notices on, the Island of Saint Jago.

NARRATIVE, &c.

CHAPTER I.

T H E provisions and stores for the expedition having been all shipped on board the Congo, and the Dorothy transport, and the river being free from the ice, which had blocked it up for many days, the two vessels quitted Deptford on the 16th of February, and proceeded to the Nore, where the Congo's crew received six months wages in advance; and on the following morning (25th) we weighed and anchored the same evening in the Downs, where we were detained until the 28th by strong westerly winds. On that day, the wind being at N.N.W., we put to sea, but returning to S.W. when abreast of Plymouth, we were obliged to run into the Sound, where we lay for three days, perfectly sheltered by the Breakwater from the violence of a S. W. gale. Here we completed the Congo's complement, by receiving two marines, and entering two seamen, in lieu of as many who had deserted at Sheerness the same evening on which they received their advance.

On the 5th of March, the wind moderating, though still at S. W., we put to sea in the hope of being able to beat down the Channel; but the return of strong gales forced us into Falmouth on the 6th, from whence we again sailed on the 9th, with a fine breeze at N. N. E.; this, however, failed us on reaching the length of Scilly, and was succeeded by a heavy gale from S. W., with extremely thick and dirty weather, such as, at this season, renders the navigation of the English channel equally disagreeable and dangerous, and which now, by preventing our getting a pilot for Scilly, obliged us to run back to Falmouth.   In standing in for Scilly, we passed the Bishop and Clerk's rocks at the distance of a mile, the sea breaking on them in a frightful manner; we also passed close to the Wolf rock, on which the sea also broke furiously, but without that roaring noise which gave it its name, and which formerly warned seamen of their danger.   On enquiring at Falmouth, I was assured that, some years since, the fishermen of Cornwall employed all their boats a whole summer in conveying stones to fill up the chasm or hole that caused this roaring, which, they alleged, frightened the fish to a great distance.

During a tedious detention of eight days at Falmouth, the winds fluctuated every moment in hard squalls, from W. S. W. to N. W., with heavy showers of rain, snow, and hail; effects ascribable probably to the local situation of this harbour, which, being placed nearly at the narrowest part of the peninsula of Cornwall, and surrounded by high lands, is exposed to the vapours from the Atlantic, and from the English and Irish channels; and these being intercepted and condensed by the hills, produce those frequent squalls and quick succeeding showers.

Falmouth is a neat, clean town, built entirely of stone found on the spot; its market, which is a commodious building of Cornish granite, is exceedingly well supplied with meat, poultry, butter, eggs, and vegetables. The number of meeting-houses indicate the great majority of the inhabitants to be dissenters. The stranger however is most forcibly struck by the strong similarity of features in the Falmouth females, which consists in plump rounded faces, without much expression, but denoting cheerfulness and placidity of disposition, while the bloom of their cheeks sufficiently proves that the humidity of the climate is not unhealthy.

The wind at length coming to the north, we weighed once more, (19th) and at last cleared the Channel, passing Scilly at the distance of 15 leagues. On opening the Lizard, we suddenly, from a very smooth water, got into a heavy swell from N.W., with a cross sea; the swell doubtless proceeding from the late westerly gales, and the cross sea from the meeting of the tides of the two channels at the Land's-end, for as we encreased our distance from the land, the waves subsided, and the sea became more regular.

When abreast of Scilly the wind came to the east, and we now found that the transport rolled so heavily going before it, as to be most uncomfortable even to seamen, while the Naturalists became most grievously sea-sick; in order to remedy this rolling, as far as was now possible, all the lumber stowed in the boats on deck was got out of them; but this had little effect, the cause being either in the ship herself, or in some vice in the stowage of the hold, we were therefore obliged to submit to this discomfort, by which we could neither take our meals, sleep, walk the deck, or even sit down to write with any satisfaction.

The common gull *(Larus canus)* was the only bird that

accompanied us from the Channel, and it did not disappear finally until the 23d, when Cape Finisterre, the nearest land, was 200 miles distant. On the 25th we passed the parallel of this cape, nearly at the same distance.

We now shaped a course for the west end of Madeira, and a pleasant breeze from the N.E. impelling us forward 50 to 60 leagues aday, we made that island at day-light on the 31st, our approach to it on the preceding days being denoted by the wind veering to the west, with frequent squalls and showers of rain, and by the numbers of loggerhead turtles *(Testudo caretta)* seen asleep on the surface of the sea. One of these animals was taken up by the Congo with many clusters of barnacles adhering to the shell; they consisted of two species, the *Lepas anatifera*, and *Lepas membranacea*. The first floating mollusca were also seen on approaching Madeira, and as the scientific gentlemen were now pretty well recovered from their sea sickness, and the weather was warm and fair, Fahrenheit's thermometer at noon being 63°, the tow-net was put overboard, and collected some of these animals, all of the *Vellela* genus.

Passing Madeira to the west at ten leagues distance, we steered for Palma, which, at day-light on the 2d of April, was in sight; running along its west side at the distance of six leagues, we observed the summit of the Caldera mountain

C

patched with snow.   In the afternoon we passed along the
west side of Ferro, also at the distance of six leagues.

As far as the Canaries our route had been very barren of
any event of interest; we saw several vessels, but spoke none.
After losing our English gulls, two birds only were seen on the
day before making Madeira, the one a large bird resembling
a raven, the other an ash coloured gull.   This almost total
absence of sea birds in the vicinity of Madeira and the
Canaries seems the more extraordinary, as it may be supposed that the Dezertas, Salvages, and other rocks would
afford them undisturbed breeding places.

After passing Madeira the winds were generally from
N.N.E. and N.E., blowing moderately with fair weather;
the days rather hazy, but the nights so bright that not a
star was hid in the heavens.   Our route laying to the eastward of the islands of Cape Verde, the trade wind, as we
approached the coast of Africa, lost its steadiness, veering
from N.N.E. to N.W.

On the 5th, in latitude 22°, longitude 19° 9', the sea
being much discoloured, we tried for soundings, but did not
get bottom with 120 fathoms of line.   Cape Cowoeira, the
nearest point of Africa, was at this time 32 leagues
distant.   The atmosphere extremely hazy, and a large
flight of fishing-birds was seen ; both being indications of

the vicinity of soundings. Here I may observe, that should this discolouration of the sea at such a distance from the land be a constant circumstance, it may serve to guide ships which have no means of correcting their reckonings, and thereby prevent the shipwrecks that so often happen on the coast of the Desart in the vicinity of cape Blanco, by their not allowing for the easterly current that seems invariably to set quite from the English Channel to the Canaries, and the effect of which we found to be, in that distance, equal to one degree and a half of longitude.

The towing-net was now become tolerably successful, taking up from time to time various species of mollusca, such as Portuguese men of war, *(Holothuria physalis)*, *Vellela mutica*, (La Marc) *Thalis trilineata*, (ib.) besides some testacea, viz. the *Helix ianthina*, with the living animal; many dead shells of the *Nautilus spiralis*, &c. specimens of all which were preserved by Mr. Cranch.

The holothuria made its first appearance on the 4th instant in latitude 24° 13′, longitude 18° 31′, temperature of the atmosphere at noon being 68°, of the surface of the sea 65°. These animals continued more or less abundant until past the Cape Verde islands, when they entirely disappeared. The greatest abundance of them was however on the

5th and 6th, when nearest to the coast of Africa, where the sea was perfectly covered with them.

The Congo's decks and sides having become extremely leaky, both from the shrinking of the planks with the heat of the weather, and also apparently from imperfect caulking in the severe weather while she was building, I determined to anchor 24 hours in Porto Praya, to caulk her sides, which could not be done at sea; and accordingly steered for Bonavista, with a fresh trade at N. E., and the atmosphere very hazy during the day, as it usually is near the Cape Verde islands. On the 8th we passed along the east side of Bonavista, at the distance of four leagues, and at daylight on the 9th were at the same distance from the west side of Mayo, when we steered for Porto Praya, and anchored in the road in the forenoon.

With exception of the mollusca, &c. taken up by the towing net, our Naturalists had no subjects to employ themselves on since entering the tropic; a single flying fish, (*Exocætus volitans*), the first seen, was found dead on the deck the morning of making Bonavista, but neither dolphin, bonito, albicore, shark, or tropic bird was yet seen.

PORTO PRAYA.

In the afternoon I went on shore, accompanied by several of the gentlemen, to wait on the Captain-general of the islands, who now resided at Porto Praya. On entering the *gateway* of the town, for gate there was none, we were conducted by a negro, to a white-washed house, of tolerably decent external appearance, when contrasted with the miserable hovels that surround it, and on being announced by a ragged centinel, were ushered up a ladder into a large apartment, the rafters, floor, and wainscot of which were as rough as they came from the sawpit ; without paint, or other decoration, save some daubed prints of the Virgin and Saints. Here we found the General at dinner with a large company, among whom were half a dozen greasy monks, wrapped in frize (the thermometer at 84°), whose jolly figures and cheerful countenances denoted any thing but abstinence and penance. The General's lady, a comely European Portuguese, drest à l'Anglaise, was the only female at table, and sat on the right hand of her husband. Not having had any arrivals from Portugal for four months, the General was very inquisitive as to the political appearances in Europe, and as he spoke tolerable French, I was able to satisfy him.

From hence we were conducted to the house of the Governor of the island, who we also found at dinner with his wife, several monks, and officers. The lady was a *half-cast*, and habited *à la negresse*, that is, with nothing but a shift and petticoat, being the only female. The Governor, who can make himself very well understood in English, immediately requested to be employed in procuring the refreshments we required ; at the same time taking great pains to convince us, that he had no interested motives in offering his services; to which, of course, we gave all due credit. Nevertheless, in order to avoid the delay I knew he might cause, and besides the trifling supplies we required being of little moment, I gave him an order for four bullocks and some pumpkins and oranges. This very disinterested officer, who wears the uniform and has the rank of colonel, is however one of the most sturdy beggars I ever met with, and commenced his attack on our liberality, by telling the purser, that his wife desired him to ask if he could sell her some butter; but adding, that he knew English officers never *sold*, but only *made compliment*. This broad hint was followed up by wishes for porter, cheese, and potatoes; and the example of the Governor was followed by his guests ; one of the officers modestly asking me to sell him a pair of old epaulets; another wished he could get a cocked hat; a third, a pair of

English shoes; a fourth, a pair of gloves; at the same time pulling a pair out of his pocket which he assured us were *English* ; but added, with a sigh, that they were not his own, having borrowed them from a brother officer for the day. All these gentlemen expressed themselves in very broken English, and indeed there is scarcely a person in the town who does not speak enough of this language for the purposes of bartering or begging.

Having taken leave of the Governor, we walked over the town, which is situated on a kind of platform or table land, nearly perpendicular on all sides, and quite so towards the bay.   With the exception of half a dozen houses of the chief officers, which are plaistered and white-washed, and of the church, which is without a spire, and externally resembles a barn, this capital of the Cape Verde islands consists of three rows of hovels, constructed of stones and mud, and thatched with branches of the date tree, and chiefly inhabited by negroes.

The fortifications consist of what is here called a fort, but which an engineer would be puzzled to describe; and a line, facing the bay, of sixteen old iron guns, within a half demolished parapet wall.   In a sort of bastion of the fort, the grave of Captain Eveleigh is distinguished by a patch of pavement of round pebbles.   This officer, commanding His

Majesty's ship Acteon, was mortally wounded in a drawn action with a French frigate. On the several high platform points that surround the bay are also mounted some guns, each of these posts being guarded by a single negro family. From the imposing appearance of these *batteries,* it is, doubtless, that the Governor-general expects that all vessels will notify their intention of sailing; nor could I refrain from a smile, when, after informing me that this was a necessary ceremony, even for ships of war, he assured me that on hoisting a flag, he would immediately make a signal to the *batteries* to let us pass!; perfectly satisfied as I was, that the vessels might be almost out of sight of the Island before a gun could be fired.

The bay of Porto Praya, however, possesses the greatest capability of being strongly fortified against shipping, and the town might, by a simple wall in those places where the sides of the platform are not perpendicular, be secured from a *coup de main.* The town must however in this case be supplied with water from the rain collected in cisterns. There are here no regular European troops, a few officers excepted, and the militia ; one of whom may be seen standing centinel every ten yards in the town, perfectly in character with the fortifications, this corps being composed of the most ragged, bare-legged, sans-culotte vagabond-lookingwretches of all shades of colour, from the swarthy European Portuguese to the Negro of Guinea; and, as if it was determined that there should be no incongruity in any part of the military department, not one of their muskets in ten has a lock, and many of the barrels are lashed to the stocks with rope yarns.

From the town we descended by a zig-zag path to the valley on the left, named " Val de Trinidad," over which are scattered some clusters of date trees (Phœnix dactylifera), some mimosas, and other spontaneous vegetation; but the only attempt at cultivation is near the two wells, which supply the town and shipping, where a negro hut is surrounded by a miserable plantation of the cotton shrub (Gossipium herbaceum).   There can, however, be no doubt but that the soil of this valley wants only water to render it fruitful; and it seems equally certain, that water might be had in sufficient quantity by digging wells.   The present possessors of the island must however change their natures, before this or any other improvement is effected.   Though a species of mimosa grows to a large size in the most burnt-up spots, and affords a good shade, they seem never to have thought of planting it in the town, where it would be not only ornamental, but highly useful in moderating the excessive heat, caused by the action of the sun on the ferruginous sand.

The wells we now found surrounded by negro washerwo-

D

men, whose state of *all but* nudity, and pendant flaccidity of bosom, seemed to wake our untravelled companions from the dreams they had indulged in of the sable Venuses which they were to find on the banks of the Congo.

In the afternoon of the 10th we made a more extended excursion; quitting the town, we followed the sides and summits of the hills that bound the valley of Trinidad, for about three miles, when we came to a mean delapidated house, hanging over the precipitous brow of a platform, which we learnt was one of the Governor General's country residences. At the foot of the precipice is what here may be called a garden, containing half a dozen cocoa-nut trees, some manioc, sweet potatoes, cotton shrubs, &c. Near this we measured a Boabab, (*Adansonia digitata*), whose trunk, five feet from the ground, was 21 feet in circumference; it was now without leaves, the branches much resembling those of the chestnut tree.

A mile farther, at the head of a narrow glen, we found the negro hamlet of San Felippe, composed of a dozen huts. The bottom of the glen is covered with huge stones, evidently tumbled from the hills that enclose it; and from the foot of a vast mass of rock issues a fine spring, which serves to nourish a little plantation of fruits and vegetables. A very large tamarind tree, growing out of the crevice of a naked rock, and the profusion of fruit on the cocoa-nut, banana, and papau trees, where there is not a foot of soil, prove that, in this climate, water is the grand principle of vegetation.

The negroes who watched the plantation, and tended a few cows and sheep, received us with much civility, and in return we purchased from them a fine milch goat with her kid, and all the eggs they had to dispose of. The hut of a poor negro slave is not luxuriously furnished ; where there are females, a partition of the branches of the date tree encloses a recess for their use ; the bedsteads are four upright sticks stuck in the clay floor, with transverse sticks for the bottom, over which is spread a mat or blanket ; a solid wooden chest, serving also for table and couch, a wooden mortar to pound their Indian corn, a pot to boil it, some gourds for holding milk and water, and some wooden spoons, form the sum total of furniture and domestic utensils ; the drum made out of a log of wood hollowed, and the rude guitar of three strings, which are seen in every hut, prove however that providence every where " tempers the wind to the shorn lamb," and that if it permits human slavery, it also blunts the feelings of the slave, not only to the degree of endurance, but even to that of enjoying life under its most forbidding form. In witnessing the joyous songs and dances of the negroes, we could scarcely believe that they are subject to be momentarily dragged away to receive the lashes commanded by a brutal owner, were we not painfully convinced by the indelible marks of the whip on their naked bodies.

The strictest precautions are taken against the evasion of slaves on board foreign vessels that touch here, and particularly by not allowing boats of any kind to the inhabitants, the want of which gives to the port the appearance of a deserted settlement.

The industrious pursuits of the islanders appear to be limited by their absolute wants, being confined to producing the stock and vegetables for their consumption, manufacturing a little sugar also for their own use, and weaving the cotton of the island chiefly into shawls for the women.

As there is scarely any thing exported from the island, there is no other ingress of money, than what is paid by ships for refreshments, or that which is sent from Portugal to pay the expenses of the establishment; and this cannot be much if all are paid in the same proportion as the Governor, who told us that his salary was but four dollars a day. Possessed by a more industrious people, and better governed, this island might however be made highly productive of colonial objects ; the sugar cane is equal to that of the West Indies, the indigo plant succeeds perfectly, and the dye it affords for their cottons is excellent ; coffee is also produced for consumption, and with common industry the now burnt-up vallies might be covered with the cotton-shrub. Two or three pitiful shops, containing the most heterogenous assortment of goods, convey the only appearance of domestic commerce ; in them we observed various kinds of English cotton goods and earthen ware ; the other objects, as hats, shoes, &c. being of Portuguese fabric.

Towards the sea shore, where my own observations were confined, St. Jago presents the most forbidding appearance of sterility, the whole surface denoting the effect of some mighty convulsion, which piled matter upon matter in what may be termed a regular confusion. The two prominent forms are those of platforms or table lands generally cut perpendicular as a wall on one side, and level with the neighbouring land on the other ; and series of perfectly conical hillocks diminishing in size by regular gradation. Besides these, vast irregular masses are scattered over the interior of the island, forming shapeless mountains, and long serrated outlines. The whole of the elevated grounds, which I passed over, are covered with loose blocks of stone, basalt, lava, and other volcanic products, and the beds of the numerous torrents, which were now quite dry, shewed a covering of black basaltic sand. With the excep-

tion of the spring at San Felippe, I did not meet a drop
of running water, and all the annual plants were so burnt
up as to be reducible to powder between the fingers. The
only trees seen here are a few melancholy dates, useful only
by their branches, as their fruit does not come to perfection; and some thinly scattered mimosas, serving only to
render the general nakedness more apparent. The lesser
vegetation consists of about a dozen shrubs, on which, as
well as the mimosa, the goats browse, and some herbaceous
plants, particularly a convolvolus, which covers the most
sandy spots, a solanum, a lotus, an aloe, &c.

Professor Smith and Mr. Tudor, who employed the
whole of our short stay here in a botanizing excursion to
the mountains, describe the interior of the island as more
pleasing than the sea shores. The valleys, as they ascended
from the inferior region, being well watered by springs
forming little brooks, and covered with plantations of fruits
and vegetables; the hills well clothed with grass, affording
pasture to numerous herds of cattle and flocks of sheep.
The result of Dr. Smith's botanical researches is thus stated
by him.* " The Cape de Verde islands, though situated

* It may be necessary to observe, that though Dr. Smith understands and
speaks the English language with great correctness, he, as may be expected in a
foreigner, does not write it with equal facility, hence I have been obliged to
put the written observations he has furnished me with into a more correct form
as to *manner*, the matter being entirely his own.

nearly in the middle of the northern equinoctial zone, and separated only by a distance of 120 leagues from the broadest part of Africa, in their climate and vegetation approach nearer to the temperate regions than to the tropical. In the opposite countries of Senegambia, the rains and the hottest season arrive together, and continue during the months of May, June, and July. In the Cape Verde islands, on the contrary, the rains do not set in until the middle of August, (when they are about to cease in Senegambia,) and continue with intermissions until January or February. On our arrival at Porto Praya, the dry season had therefore commenced two months; some of the indigenous trees and shrubs had just lost their leaves, and a few had put forth new ones; all the perrnnial plants were in seed, and all the annual nearly dried up, the vegetation being in the same state as in June and July in the Canary islands.

" Even in the dry season the atmosphere of the Cape Verdes is extremely humid, for the air being heated over the broadest part of Africa, a great capacity for imbibing moisture is thereby acquired, and in passing over the sea it is saturated to the highest point, so that the least diminution of temperature causes it to deposit abundant vapour Not only the highest point of St. Jago, (Pico de San Antonio,)

which has about 4500 feet of elevation, but also the whole
central ridge of hills down to 1400 feet are usually enveloped
in clouds from 10 o'clock in the morning. This humidity
clothes the hills with thick pasture grass, giving to the
country a feature entirely unlooked for in so low a latitude
and of so small an elevation above the sea.

" It is also this moist atmosphere that causes the mean
temperature of the island to be so much less than that of
Senegambia. According to Humboldt's new scale of mean
temperatures, the curve will intersect the latitude of St.
Jago at 27° of the centigrade thermometer, (80° 7' of Faren-
heit,) which is probably the middle between the iso-ther-
mometer of the island and of Senegambia, the latter being
probably not less than 30° centig. (86° of Fahrenheit). On
the 10th of April the temperature of the well in the valley of
Trinidad was 25° centig. (73° Fahrenheit), the well being
two or three fathoms deep, and the afflux of water considerable, as it supplies the whole town. It is probable
that this is about the mean temperature of the well throughout the year, and that we shall not be far wrong in considering it also as the iso-therm. of the lower parts of the
island.*

* On boad the ship in the bay at 2 o'clock in the afternoon, the thermometer
was 70°, while in the town of Porto Praya it was at the same time 84°.

" The nature of the vegetation is here, as well as every where else, the truest criterion of the climate. Few of the undoubtedly indigenous plants can be called tropical ; on the contrary, the flora of the island is poor in some families which occupy a large portion of tropical vegetation in general ; such, for instance, as the *Composita*, while it is much richer in others chiefly found in temperate climates, as the *Labiatæ*. Few of the plants of the hotter regions of Africa are found here ; but a much greater number of species similar, or allied to, those of the temperate extremities of this continent, and of the Canary islands in particular. The number of these encrease in ascending from the low grounds to the hills of second magnitude, where they are succeeded partly by European and partly by other Canarian plants to the height of 3000 feet, which was the greatest elevation I reached."

" St. Jago, though enjoying the genial influence of a tropical sun, seems to be poor in indigenous plants, as is indeed usually the case with islands at any considerable distance from a continent. But though nature is not here spontaneously productive, she has adopted every plant, which has been brought to the island either by accident or design. Thus the most prevalent species are exotic, and chiefly introduced from the other Portuguese colonies, particularly from the Brazils and the Malabar coast. The

E

Jatropha curcas, probably first brought here for its seeds, which afford a good oil, forms thickets in the vallies and on the sides of the hills; the *Anona tripetala* is also common in the same wild state, and in similar situations. The *Justicia malabarica* covers all the fields and bottoms of the vallies. The *Argemone mexicana* is dispersed in every direction; and the *Cassia occidentalis* is scattered amongst the rocks round Porto Praya. Three species of *Sida*, the *canariensis* being most abundant (and even more so than in the Canaries) and one species of *Malva*, also very prevalent, are probably from America.

" The principal indigenous plants in the lower region, are a species of *Mimosa*, which I have named *glandulosa; a Convolvolus*, which is doubtful, and may also be American; a *Zizyphus* resembling the *vulgaris. Spermacoce verticilata*, said to be common in the West Indies and Africa: *Momordica senegalensis*, and *Cardiospermum hirsutum*, both of Senegal, and *Lotus jacobæus.* In the hilly region some indigenous plants cover large tracts, resembling, by this character of aggregation in one place, the vegetation of temperate climates. A new species of *Pennisetum* covers all the hills, having no other resemblance to tropical grasses, than its height and ramified stems. Among the many Canarian plants, I looked long for the family of *Euphorbia*, so prevalent in the Canaries; but at the height of about 1600 feet I at length found the sides of the hills and small vallies covered with large bushes of a *Thymalea*, resembling the *piscatoria*, but the identity difficult to be established. A *Sideroxylon*, I was told, formed thickets on the highest mountains, but I saw only one sterile plant resembling the *Marmulana* of Madeira.

" I have, in the following table,* divided the vegetation of St. Jago into two regions only, and doubt if the plants of the Pico de San Antonio differ sufficiently to form a third. I must however observe, that a two days excursion, in the dry season, and in one of the least fertile parts of *one* island only, is by no means sufficient to establish a physical arrangement of the flora of the islands in general, where such difference of localities exist as in the burning peak of Fogo, and the wooded mountains of the island of San Antonio. Indeed, from the little I had time to observe, I am convinced that a botanist would have his labour well repaid should he give a sufficient time to the examination of the vegetable reign of these islands; nor can I help being surprised that no one has yet turned his attention towards them."

With respect to the cultivated vegetables, Dr. Smith observes, " Cultivation is only seen in the glens or ravines,

* Inserted in Professor's Smith's Journal.

which are watered by rills from the mountains. In the upper and wider part of the valley of Trinidad, we first met with plantations of Indian corn, cassava, sugar cane, *Arum esculentum*, and pine apples. Cotton and indigo had also been formerly planted in some spots of the valley, but being neglected, a few plants run wild are now only to be seen. On the sides of the brooks grow luxuriantly the fig, lemon, orange, papaw, *(Anona triloba,)* custard apple, *(Anona africana,)* the tamarind, guava, plaintain, and banana, *(Cassia fistula,)* and prickly pear *(Cactus opuntia)*. Near one of country houses we saw some *Ailanthus glandulosa Ximenia americana*, and a few grape vines. Besides the date palm, which grows in abundance in the sands near Porto Praya, some tall cocoa palms are scattered here and there, and bear ripe fruit at the elevation of 800 feet above the sea. A single palmyra *(Borassus flabelliformis)* was seen.

On some spots of the elevated grassy hills, roots and vegetables are cultivated with great success ; we saw no traces of other *Cerealia* than Indian corn, but were told that wheat succeeds perfectly when sown in the dry plains in the rainy season, as does rice in the lowest and wettest grounds ; but the islands being supplied with corn from America, in return for their salt and mules, the indolent inhabitants do not think of cultivating either. The inhabitants we conversed with were entirely ignorant of any tree affording dra-gon's blood, though the *Dracæna draco* is said to be found in these islands, as well as Madeira and the Canaries.

The deep valley of St. Domingo, on the east side of the island, which we saw beneath us from the mountains, and that of Ribeira on the south-west side, we were assured are better watered, more fertile, and more extensively cul-tivated than that of Trinidad."

Dr. Smith remarks of the geological features of the island, that " the Cape Verdes, like all the African Atlantic islands, are of sub-marine volcanic origin, and mostly of the ba-saltic formation. Few of them seem to have had super-marine eruptions, and perhaps the cone of Fogo, which rises above 7000 feet, and still smokes, is the only one. The forms of the four high north-western islands, and of Brava, as represented in the charts, lead to the belief that they do not differ essentially in structure from the basaltic mountains of St. Jago, and it is probable that Mayo is similar to the inferior region of the latter island.

" The south-east and south coasts of St. Jago are sur-rounded by steep and often perpendicular rocky cliffs of a few fathoms in height, from which the land rises towards the mountains, in a generally flat surface, with a few hills covered by loose fragments and furrowed with ravines.

" The valley of Trinidad, the largest and deepest ravine in the south side of the island, commences at the sandy beach of Porto Praya, and runs S. S. W. and N.N.E., with its upper extremity bent to the E. N.E. until it is lost in sloping hills. It is generally covered with volcanic fragments.

" The central ridge of hills follows nearly the largest diameter of the island from S.E. to N.W., but nearest to the eastern coast, with sloping sides to the west, and having many steep basaltic rocks, and well watered vallies or ravines to the east. The peak of St. Antonio rises above the other mountains in an oblique, conical, sharp-pointed form, to the height of about 4500 feet.

" The sea rocks round Porto Praya expose five strata to view; 1st, or lowest, a *conglomerat*, passing into pumice tufa; 2d, *pumice*; 3d, a thin layer of *porous basalt*; 4th, *columnar basalt*; and 5th, or uppermost, a *basalt-like substance*, which from its concentrical and globular forms, seems to have been in a semifluid state. Farther inland, the basaltic strata sometimes contain *olivin* and *augite*, and more rarely *amphibole*. About a league up the valley, on its western border, are huge rocks, which cause a bending in its direction, and which are composed of a deep red *quartz*, with crystals of *feltspar*; about two leagues up are found loose masses of lava, the cells sometimes empty, sometimes filled with crystals of *mesotype*. To the west, I observed at some distance a discoloured appearance, not unlike a lava stream, and not far distant from some conical hills, in the direction of the Peak of Fogo; but the stinted time did not admit of examining if these were the vestiges of an eruption. In two or three places I met beds of a compact *felspar*, mostly decomposed into a white earth. I was also told of a bed of shells among some hills, not far distant from the place named *Toara*, but which the same reason prevented me from verifying."

The island appears to be scantily supplied with birds, either as to species or numbers; those seen were three species of *falco;* the first a fishing eagle, common at Porto Praya; the second ash-coloured, of a large size, seen only on shore; and the third, which was shot on shore, nearly resembled the sparrow-hawk. The small birds, of which specimens were shot, were a fine king fisher *(Alcedo)*, very common; the common swift *(Hirunda apus;)* a sparrow differing little from the European house sparrow; a bird resembling the lark; and a very small warbler, the only one that appeared to have any song. Some covies of Guinea fowl were seen, but too shy to be shot at; and the common quail was also seen. The greater tropic bird,

(Phaeton etherus) breeds in the crevices of the elevated rocks near the shores, but was not at this time numerous.

Fish are tolerably abundant in the bay, and the seine may be hauled with good success, either in a sandy cove on the west side of the east point of the bay, or on the beach west of the town; the latter appearing preferable, the former being subject to a sudden rise of surf, when the sea breeze blows fresh. Of nine species of fish which we took, three only were familiar to us, viz. a young white shark, *(Squalus carcharias)* barracoota, or barracuda and grey mullet. The others we were prevented from examining by a mistake of the cabin steward, who (supposing they were selected and put by for the purpose) caused these specimens to be drest for dinner. Although the most rigid catholics, the inhabitants seem to make fish a very small portion of their general food, a single boat alone going out to fish in deep water; and the few fish we observed on shore were taken as we understood by hook and line from the rocks. The Governor, however, on learning that we had hauled the seine with success, let us know that it was customary to pay him the compliment of a dish of fish, which through ignorance we had omitted. Of crustaceous fish, we only took a prawn four inches long, a few small crabs among the rocks, and a species of land crab. The testaceous mollusca collected among the rocks were not numerous, consisting of *patella, buccina, turbo, trochii,* and dead shells of cones. Two species of sea egg *(echinus)* were also found on the rocks.

The insects seen (besides the common fly of a small size, and neither numerous nor troublesome,) were several kinds of grasshoppers *(grylli),* three or four species of *coleopterous* insects, among which was a small beetle *(Scarabæus),* and some moths and butterflies. The only reptile seen was the common stone lizard.

Porto Praya has been so often visited by our navigators, that it may be supposed they have left little room for new nautical observations ; the directions for knowing the bay are indeed so minute and various, as to confuse rather than assist a stranger ; it seems however to have been forgotten, that one marked and prominent feature is a better guide than a number of trivial appearances, which may change with the position of the observer.

It seems to me to be quite sufficient to inform the navigator, that the S.E. point of the island is seen as a very long and very low point in coming from the north or south ; that to the west of this point, three or four miles, is a bay with a brown sandy beach, a building, and a grove of date (not cocoa-nut)* trees ; that this first bay must not be mis-

* This mistake is made in all the directions for Porto Praya that I have seen ; the trees are however sufficiently different in appearance, to render the correction proper.

F

taken for that of Porto Praya, as its east point is surrounded by rocks that do not *always* break ; that after passing this bay you may keep along shore towards Porto Praya within $1\frac{1}{2}$ mile, or in 10 fathoms. This last bay is first distinguished by a battery of earth or brown stones on its west point, off which the sea *always* breaks to some distance. In standing on, round the east point of the bay, (which is safe, and should be rounded in seven or eight fathoms, or within a cable's length,) the brown sandy beach opens, on which is first seen a house or shed, then a grove of date (not cocoa-nut) trees, and shortly after the fort itself.

With respect to anchorage, it may be proper to observe that a large ship should lay well out, and near the east shore, in order to ensure her weathering the west point of the bay, should the wind be light, or far to the east, as is often the case. The best birth I conceive to be with the flag-staff of the fort N.W. by W., the east point of the bay E.S.E. and the S.W. point W.S.W. in seven or eight fathoms. The ground is coarse sand and gravel, that does not hold well; consequently it requires a good scope of cable to bring the ship up in a fresh sea breeze. It is also advisable to drop a kedge anchor to the west to steady the ship, and keep the bower anchor clear, when at times in the forenoon the wind is light from the west.

There is always some surf on the beach, so that it is

proper to have grapnels in the boats going on shore. When the surf is high, there is a good landing place at a rock east of the town, where a path-way is seen. It is also very necessary to be cautious in carrying sail in boats, the puffs of wind from the high lands being very dangerous, as we experienced by the oversetting of the gig, by which Lieutenant Hawkey was nearly drowned. Two other boats were also nearly lost in the surf, by which unlucky accidents my own watch and four others were totally spoiled, causing, in our situation, a very serious and irremediable evil.

Refreshments for a ship's crew are by no means to be procured at Porto Praya on reasonable terms; for lean bullocks of 250lb. weight they at this time expected 40 dollars; for long-haired African sheep, 4 dollars each; milch goats, 2 to 3 dollars; pigs of 50lb. (a long-legged and long-sided breed), 5 dollars; large turkies, $1\frac{1}{2}$ dollar each; small long-legged fowls, 6 for a dollar. A few Muscovy ducks were seen in the country, but no geese. For bullocks or sheep, bills or cash are alone taken; but all other stock, as well as fruit and vegetables, which usually belong to negroes, may be most advantageously procured in exchange for any articles of wearing apparel, or for blankets. Monkeys are offered for sale by every negro, and unless a prohibition is issued, the seamen will always fill a ship with these mischievous animals. The only species here is the green monkey (*Cercopithecus sabæus*).

CHAPTER II.

Passage from Porto Praya to the Mouth of the Zaïre.

CHAPTER II.

Having completed the Congo's caulking in the evening of the 10th, I should have quitted Porto Praya the following morning, but it being Holy Thursday, consequently a great festival with Catholics, all the free inhabitants, drest in their best attire, were occupied the whole day in church ceremonies, which not permitting them to attend to worldly concerns, we could not get our business settled on shore, and were therefore obliged to defer sailing until the next day, in the afternoon of which we again got to sea.

In compliment to the religion of the place, we this morning, it being Good Friday, hoisted the colours half-mast, the fort having done so, and the Portuguese vessels putting themselves in mourning by topping their yards up and down.

At sun set the Peak of Fogo was seen nineteen leagues distant.

A moderate trade-wind between N. E. and E. N. E. continued until the 18th, when in latitude $7\frac{1}{2}°$, longitude 18° W., we lost it, and got into the region of light variable breezes and very sultry weather, the thermometer rising in the afternoon to 82° and 84°; the temperature of the sea being 80° and 81°;

during the nights constant faint lightning without thunder. Many porpoises *(Delphinus phocena)*, flying fish, and tropic birds were now seen, and a swallow rested on the yards when 250 miles distant from the land. From the 15th to the 19th the sea represented a continual succession or riplings, and on trying the current with a boat, it was found to set to the S. E, at the rate of $\frac{3}{4}$ of a mile an hour, nearly agreeing with our chronometers.

The towing net, which was kept constantly overboard, gave us for the first time on the 18th, great numbers of perfectly diaphanous crustacea, resembling insects of glass ; they were of four different species, and considered by Dr. Smith, as belonging to the genus *Scyllarus*. (La Marc,* p.156.) We also took a small squalus, of a species new to us, and which from the form of its teeth may be named *Squalus serrata*.

On the 19th the first deluge of rain was experienced in a heavy squall from N. E., and was the commencement of that succession of squalls, calms, and rains, which would seem to be entailed as an everlasting curse on this region of the Atlantic ; in consequence of which, from this time till we passed the meridian of Cape Palmas, our progress was exceedingly slow, never exceeding 40 miles a day, and sometimes making no progress at all. The winds, when there

---

* Similar crustacea were taken during the rest of the passage in greater or less numbers until we made the continent of Africa.

were any, were between E. N. E. and S. W., but mostly southerly Our only amusement now was the taking of sharks, all of the white species *(carcharias)*, except one of the blue *(glaucus)*, and the only one seen during the passage ; the largest of the former was a male, ten feet long, the latter a female impregnated, seven feet long; she was unattended either by pilot-fish or sucking-fish, while the white sharks had many of both accompanying or attached to them. It was observed of the pilot-fish *(Gastorosteus ductor)*, that they took especial care to keep out of the way of the shark's mouth, generally playing over the hinder part of his head. The shark was also observed to lift the head above water and seize objects floating, without any change of position. One shark was seen to leap out of the water and seize a small albicore while it was itself in pursuit of a flying-fish.

The first bonitos *(Scomber pelamis)* were seen on the 25th, in latitude 5° 53., and many cavally or shipjack sported after showers of rain, while flocks of tropic and other oceanic birds hovered over the riplings they caused, in order to seize the flying fish frightened from their element.

On the 26th, in latitude 6° 16′, longitude 13° 45′, the temperature of the sea at the depth of 220 feet was 64°, that of the surface being 80°, and of the air 81°.

G

Since the commencement of the rains every additional precaution was taken to guard against the effects of the damp sultry weather on the people; they were never exposed to the rain when it could be avoided, and when unavoidably wetted, they were obliged to put on dry clothes as soon as possible, occasionally receiving a small glass of spirits when shifted. The humidity of the air between decks was dried up by frequent fires, and the bedding often aired. The large quantity of water I had shipped in the river enabled me to afford a proportion for washing the people's clothes twice a week, until now, when the rain water saved by the awnings was put by for the purpose.

From the very commencement of the voyage, I had much difficulty in *forcing* the observance of general regulations for cleanliness, and the consequent preservation of health on the transport's crew; for the master and mates, like the generality of merchant seamen, considering all such regulations as useless, took no steps to enforce them, nor could I even get the hammocks brought on deck after our arrival in the warm latitudes, until I had recourse to coercion, and the punishment at the gangway of one of the most refractory of the crew, which effectually broke up the confederacy that seemed to have been formed to resist all my orders on this subject.

The currents, from leaving Porto Praya until in latitude 6°, longitude 15°, set to the south and S. E. ; they then changed to the N. E. and E. N. E., with various degrees of velocity, from 8 to 40 miles a day, and retained this direction until we made Prince's Island.

The winds until the 5th of May, when we crossed the meridian of Cape Palmas, at the distance of 15 leagues from that Cape, were very light and variable, between south and S. W. The greatest heat of the atmosphere was 85° in a clear calm at 2 P. M., and the least 74° after heavy rain ; the rain water as it fell being at 75°. The various trials of the temperature of the sea gave between 81° and 82° at the surface, and 63° to 64° at the depth of 200 fathoms. A large shoal of the bottle-nose porpoise or dolphin of naturalists, (*Delphinus delphis*) was seen ; flocks of tropic birds, and a few men-of-war birds (*Pelicanus aquila*) now also accompanied our course. It was observed that the former bird fishes in the manner of the gull, flying low, and seizing its prey only at the surface, and often sitting on the water ; while the man-of-war bird soars very high, hovers on the wing like the kite, and darts perpendicularly on its prey, diving after, and carrying away, the largest flying-fish into the air.

After passing Cape Palmas, the light southerly air was

succeeded by moderate breezes from S. S. W. and S. W. with which we stood close hauled across the Gulf of Guinea; but the strong N. E. currents prevented our making any southing. The weather, in crossing the Gulf, was always extremely cloudy, with frequent drops of rain, and much less sultry, the thermometer varying between 80° and 78°.

May 6th. Until this time the Naturalists were obliged to content themselves with the small animals the towing net afforded them, but they were now gratified by the capture of albicore and bonito, many of both being taken by the grains and hook. The most apparent distinctive characters of these two species of the *Scomber* are the following. The albicore *(Scomber thynnus)* has 14 rays in the first dorsal fin, 8 small false fins on the back, and the same number on the under side; the dorsal, anal, and false fins are strongly tinged with orange, the under part of the sides of the fish marked with transverse whitish stripes, the palate studded with boney points. The foremost dorsal fin of the bonito *(Sc. pelamis)* has 16 rays, the false fins are eight on the back, and only seven beneath. These fins have no orange tinge; the under sides are marked longitudinally with four black stripes, and the palate is quite smooth.

If the esteemed tunny-fish of the Mediterranean and the albicore of the Atlantic be the same species, there seems to be an enormous difference in their sizes. The tunny-fish arriving at the weight of 8 to 12 cwt. while the largest albicore I have ever seen taken in the Atlantic weighed but 160lbs. and the most common weight was between 30 and 40lbs. and these latter were evidently full grown fish.

On the 11th we had full moon, and the same day and the next, such heavy rains fell, that I feared the wet season had already set in to the north of the line, we being on this day in $2\frac{1}{2}°$ N. and $1\frac{1}{2}°$ E. By a rain guage made on board, we found that, on the morning of the 12th, between 1 and 4 o'clock, the water that fell from the heavens was equal to $3\frac{2}{10}$ inches. On this day died Joseph Burgess, seaman, of the Congo ; on opening him, his death was found to have been occasioned by a disease of the heart caused by the ancient rupture of a blood vessel

Though the rains lasted but two days, seven of the transport's crew were already attacked by fevers, more or less serious, all of which were to be traced to their sleeping on the wet decks, and to the neglect of changing themselves after being exposed to the rain during the day. The almost inevitable bad consequences of carelessness in these respects, may be estimated by the state of the thermometer at night in various parts of the ship. In the space called between deck, where the people slept, it was 88°, in my cabin 79° or 80°. On deck 73° to 77°. The great evaporation from the decks, &c. after rain, being found to lower the thermometer a degree or more below the temperature of the rain in falling. With respect to my own people, I obliged them to wear flannel next their skin, in addition to the other precautionary regulations; and the good effect of these precautions was fully evinced in the continued good health of the crew, one or two only (and these were proved to have neglected them) being slightly attaked with symptoms of fever, which gave way by immediate bleeding, and gentle cathartics.

May 14. The bird named booby *(Pelecanus sula)* now frequently settled on the yards in the dusk of the evening, and two of them were taken; the external characters of these birds seem by no means to authorise their being placed in the genus of Pelican. Of the two individuals now taken, the largest measured 18 inches from the point of the bill to the extremity of the tail, and weighed seven ounces; the plumage a rusty brown, deepest and rather glossy on the upper side of the wing quill feathers, the crown of the head only being of a dove colour, lightest towards the forehead. The upper sides of the wing quill feathers black, the under side a dirty white; the bill conical, slightly curved; the nostrils very open, being two wide longitudinal slits on the sides of, and about the middle of the upper mandible; the eye a dark brown approaching to black, surrounded by a circle of minute white feathers ; three toes full webbed, the fourth toe behind very small, and quite free ; bill and legs black. This specimen on examination proved to be a full grown male.

The second specimen, which was found to be a young female, was somewhat less than the first ; the dove colour on the crown of the head was deeper, nearly mixing with the general brown ; and the circle of minute feathers round the eye was black ; it differed in no other respects from the male. These birds were observed generally in pairs; they fly close to the water with the neck stretched out and the tail spread.

On the 16th, at day-light, Prince's island was in sight, bearing S. E. 12 or 14 leagues ; our approach to it the preceding day having been denoted by great numbers of fishing birds, apparently different species of gulls.

The swarms of albicore round the ship were now such as almost to justify the hyperbole of their obstructing the ship's way ; and twenty a day was the usual success of our fishery with hook and line, the flying-fish found within them serving as bait. The proportion of bonito appeared to be small, not one being taken to 10 albicores. The flying fish, in endeavouring to escape from their cruel enemies, skimmed the surface like flights of birds, and it was observed, that when they rose in the direction of the wind, they could reach a considerable distance, but when against the wind, they dropped again almost immediately; when the rise was in an oblique direction to the wind, they sometimes described a considerable curve, until they got before the wind, and this without any assistance from the wings, the only movement of these members being at the moment of their quitting the water, when they had for a few instants a quick fluttering motion. Four different species of these fish were taken.

After passing Cape Palmas and entering the Gulf of Guinea, the sea appeared of a whitish colour, growing more so until making Prince's island, and its luminousity also encreasing, so that at night the ship seemed to be sailing in a sea of milk. In order to discover the cause of these appearances, a bag of bunting, the mouth extended by a hoop, was kept overboard, and in it were collected vast numbers of animals of various kinds, particularly pellucid *Salpæ*, with innumerable little crustaceous animals of the *Scyllarus* genus attached to them, to which I think the whitish colour of the water may be principally ascribed. Of *Cancers*, we reckoned 13 different species, eight having the shape of crabs, and five that of shrimps, and none more than a quarter of an inch in length; among them the *Cancer fulgens* was conspicuous. In another species (when put into the microscope by candle light), the luminous property was observed to be in the brain, which, when the animal was at rest, resembled a most brilliant amethyst about the size of a large pin's head, and from which, when it moved, darted flashes of a brilliant silvery light. Beroes, beautiful holothurias, and various gelatinous animals were also taken up in great numbers. Indeed the Gulf of Guinea appears to be a most prolific region in these sort of animals ; and I have no doubt but the marine entomologist would here be able to add immensely to this branch of natural history. As it was found impossible to preserve the far greater number of these animals by reason of their delicate organization, the spirit of wine dissolving some, and extracting the colours of others, and as most of them require the aid of a microscope to describe them, a great portion of them were lost on us, from the want of a person either to describe or draw them from that instrument.

Light baffling winds from south to S.W. kept us in sight of Prince's island until the 18th, when a hard squall from the S.E. brought to our view that of St. Thomas, which at day-light on the 19th bore S. by W., distant 19 leagues. We were again plagued with light winds for two days off this island, when another squall from the S.E. ran us clear of

H

it to the west; but the wind soon returning to south, and
blowing fresh, we were unable to weather it, and I thought
it advisable to stand off to the W. S. W. in the hope of making
southing; accordingly we crossed the line in this course
on the 23d, and in the meridian of $4\frac{1}{2}°$ E.

From the time of our making St. Thomas, we experienced
a current setting to the W. N. W., encreasing in velocity as
we went to the westward, until on the line it set 33 miles in
24 hours.   Finding we made little southing, the wind still
hanging obstinately at south, we tacked on the 24th to the
eastward, and on the 27th passed to the south of St. Thomas,
within 5 leagues, our latitude being 0° 17′ S., so that we had
gained but 45 miles southing in 6 days, owing to the strong
northerly currents, although in this track the latest chart of
the Atlantic marks a strong southerly current.

While in sight of the two islands above mentioned, the
weather was so very cloudy that we could see little
more than their outlines.   Towards the south end of
Prince's island are two whitish ravines; but whether this
colour is from the nature of the ground, or from the excrement of birds (of which there are immense numbers round
the island) we could not ascertain.   St. Thomas, which we
approached within 7 or 8 miles, appears to be wooded up
to the summit of what is rather improperly called the Peak

of St. Anna, being little conical, but rather a round topped mountain, of the probable elevation of 7 or 8000 feet, with a gap in the summit. Off the north end, the rock or islet named *Mono Cacada* (significant I suppose of its being co-vered with the dung of birds), leaves a considerable open space between it and the main island.

Our chronometers gave the longitude of the north end of Prince's island 7°; the variation, by the mean of many observations, 21° 22′ W. The same watch makes the N. W. point of St. Thomas in 6° 31′, and Rolle's island, at the south end, in 6° 44′; the variation at this end of the island 22° 7′.

The winds now came more westerly, but were at the same time so light, that our progress was most tiresomely slow; I therefore determined to make the continent, in the hope of finding land and sea breezes in shore; and ac-cordingly we first saw it on the morning of 3d of June, and at noon were three leagues off shore in 16 fathoms, latitude observed 2° 10′ S.; the land very low and entirely covered with wood.

The atmosphere for the two days before making the land, had become so saturated with moisture, that the hygrome-ter at noon marked 5°, and the thermometer stood at 71°. At 7 o'clock in the evening a dew, little less penetrating than rain, began to fall, and continued the whole night, with so sensible a degree of cold, that instead of melting under an equinoctial sun in the lightest cloathing, as our gentlemen expected, they were glad to resume their woolens.

The albicores which had accompanied us in vast shoals to the edge of soundings, and were taken in such numbers, that besides being consumed fresh to satiety, the crews of both vessels pickled and salted several barrels, now entirely disappeared, and with them the sea birds; the white colour of the water changed to the oceanic blue before we struck soundings, the marine animals much decreased, and the sea lost a great portion of its luminosity.

From the 3d to the 8th we were plagued with light airs, veering towards midnight to the west as far as S. W., and having for an hour or two sufficient strength to send the ship two or three miles an hour, then again dying away to light airs, which in the morning veered to south and S.S.E.; these variations being the only signs of the mutual re-action of the land and sea on the atmosphere; and indeed we experienced similar variations morning and evening since making Prince's island.

The nature of this part of the coast is doubtless the cause of the want of more marked alternate breezes from the land

and sea: here the land is very low, and entirely overgrown with wood, which causes the atmosphere over it to preserve nearly an equal temperature day and night; this temperature by reason of the great evaporation from the wood (which, as I before observed, saturates the atmosphere with moisture), seems even for the greater part of the 24 hours somewhat less than that of the sea; and hence the light breezes that blow from the land, or between south and S. E. for 18 hours of the 24, or from six o'clock in the morning until midnight, when the evaporation having ceased for some hours over the land, the temperature becomes a little higher than that of the sea, and produces a short and weak breeze from the latter.

The general range of the thermometer while in with the land was at 6 A. M. 71°.; at 2 P. M. 73°.; at 9 P. M. 70°.; the temperature of the sea at 2 P. M. 72°. The hygrometer varied during the day from 5° to 15°.

The dredge was put over board, and brought up two or three species of *echini*, some small *cancri*, bits of coral, &c. While in soundings no fish were seen, nor any birds except an occasional solitary tropic bird or pair of boobies.

The longitude of the coast in the latitude of 2° 10′ S. our chronometers make 9° 40′, and by ☉ and ☽ 9° 51′. The bank of soundings stretches off about 10 leagues from the land, deepening regularly as follows.

Fathoms.

About 9 miles off shore, 16, oozy sand.
        18   ditto,   30, brown sand.
        24   ditto,   47, ditto and broken shells.
        28   ditto,   67, ditto.
        30   ditto, no bottom at 120.

Although we took every advantage of the variations of the wind, to stand off and in shore, the lightness of the breezes, and a daily current of fifteen miles to the north and N. N. E. permitting us barely to hold our ground, I determined again to stand off out of soundings, in the hope of losing the current and getting fresher breezes. In both respects I was, however, disappointed ; for though the current became more westerly as we went off shore, its velocity encreased at the same time to 30 and 40 miles a day. The winds still remaining very light in the morning from S. S. E., and from S. S. W. in the evening, while the transport being extremely leewardly, and both she and the Congo sailing very badly in light winds, our progress was slower than ever.   In this choice of difficulties I again stood in for the land, hoping, that as we were now past the low land to the north of Loango, we should meet more regular land and sea breezes.   We in consequence made the land on the 18th in 3° 24' ; thus having gained but 75 miles southing in 15 days, by working out of soundings.

In this most tedious fortnight we found little to amuse us ; birds and fishes seemed to have forsaken this region ; a single swallow or martin being the only one of the former seen ; the towing net, however, again afforded us abundance of marine animals, amongst which were many of the paper nautilus *(Argonauta sulcata)*, with the living animals, which, in contradiction to the opinion of the French naturalists, proved to be perfect *Octopi*.* When forty leagues from the land, several floating patches of reeds and trees passed us, proving, if our chronometers had not shewn it, the existence of a strong western current. The day we made the land a dead albatross *(Diomedea exulans)*, was picked up floating in a putrid state ; which seems to shew that these birds wander farther towards the equator than is generally supposed. The same day a whale (apparently a species of the *Physeter*, having large humps behind the back fin), struck our rudder with his tail in rising, and one of these fish rose directly under the Congo ; and, according to the expression of those on board her, lifted her almost out of the water. These animals indeed were now extremely numerous.

This day a vessel was seen for the first time since leaving Porto Praya ; from her warlike appearance and superior

---

* L'animal qui forme cette coquille ne peut être un poulpe La Marc, Animaux sans Vertèbres, p. 99.

sailing, she was at first supposed to be a ship of war; but on approaching us she hoisted English merchant colours, and keeping half-gun shot to windward, we were unable to speak her; nor did she seem to have any desire to communicate with us. This circumstance, together with her apparent force and preparation for defence, having 18 guns run out of her between-deck ports, with the tompions out, left little doubt of her being employed in a forced and illicit slave trade. Her anchor a-cock-bill, and her tacking with the variations of the wind, proved her to be working along shore to the south.

The land and sea breezes, though now more regular as to time, (the former setting in about four in the morning from N. E. to S. E., and the latter from two to four in the evening from S. W.) were so faint and of so short duration, that neither afforded us a run of more than ten miles, while the current setting one mile an hour to the north, we remained in sight of Mayumba bay until the 24th, anchoring whenever we found we lost ground.

The land to the north of this bay presents an undulating line, Cape Mayumba being the highest point, and forming a little hummock. Point Matooly, the south point of the bay, also forming a hummock, descending gradually to the south into a line of low even land. The bight of the bay is also low land, with a saddle hillock in the centre of the back ground. The whole of this land is covered with wood, but is proved to be inhabited by the numerous fires seen on the shore, and which were probably intended as signals for us to land.

We now, while at anchor on a sandy bottom, took a good number of fish of the *Sparus* genus, named by the seamen sea-bream, and light-horsemen, the latter, from a reddish protuberance on the back of the head (fancifully thought to resemble a helmet) ; they were taken with the hook close to the ground, and baited with fresh pork or their own livers; the largest weighed 18lbs., and though rather dry and insipid, were infinitely preferable to the albicore and bonito with which we had been surfeited in the gulf of Guinea. Sea birds had also entirely disappeared, with the exception of an occasional tropic bird, and a few of *Mother Carey's* chickens (storm petterel). Numbers of insects of the genus *Tipula* were taken from the surface of the sea.

The weather, though now much less damp than when we made the land to the north, was still very hazy, and the cold even encreased, the thermometer in the day never rising above 73°, and falling in the night to 67°. As the moon approached the full, the current diminished, and on the 24th a more favourable sea breeze than we had hitherto

I

experienced, carried us along shore until the evening, when we anchored in ten fathoms. The land south of Mayumba to 3° 50′, has an agreeable appearance, rising in a series of three or four gentle elevations from the sea inland, the farthest and highest not deserving the name of hill; the whole covered with wood, except in some spots which were bare of wood, and resembled spots of burnt-up grass. A sandy beach margins the sea, which breaks in a surf that must prevent the access of an European boat, unless some of the many projecting points give shelter to coves where a landing may be effected. The soundings are here very regular, altering about a fathom in a mile, and the depth at six miles off shore ten fathoms. The bottom is extremely various, but sand predominates, brown, black, white, with sometimes quartz pebbles, small lumps of yellow ochre, bits of corals, and fragments of shells of the cockle and venus genera.

Never did lover wait more anxiously for the hour of assignation with his mistress, than we now did for that of the usual setting in of the sea breeze, on which alone we found we must depend to finish this eternal passage, for the land winds were so faint as not to render us the smallest service, and the currents the day after new moon returned with encreased velocity.

On the 28th we had reached the latitude of 4° 30', and found the land we passed from 3° 50', more picturesque than to the north; the variety of elevations being here greater, and the clear spaces more numerous; these we were however now led to think the signs of barrenness rather than of fertility, having, when viewed near, the appearance of tracts of naked reddish clay.

We were now opposite to Loango bay, the red hills on the north side of which (formed by clay of the appearance above mentioned) we anchored off, in 16 fathoms mud, at about 8 miles distance. The next afternoon, when the sea breeze set in, we weighed; but it again dying away, we found ourselves carried towards the land by the current, and again let go the anchor in 12 fathoms; but before the ship brought up we were in 8 fathoms on a reef of rocks, over which the current ran to the N. N. E. two miles an hour. The south point of the bay (Indian Point) bearing S. E. The sea breeze freshening, we cut our cable, and leaving the stream anchor behind us, made sail and deepened gradually over the rocky bottom until in 12 fathoms, when it again became soft and mud.

This reef is in about latitude 4° 30' (an observation at noon possibly erring 2 or 3 miles, the horizon being bad), and it lays seven miles off shore; towards which latter we sounded for three cables lengths, and found $7\frac{1}{2}$ fathoms, nor is it probable that there is much less until near the shore, between which and the ship many whales were seen sporting, and they doubtless would not go into very shoal water.

The position of Loango bay is most erroneously laid down in the latest charts, the latitude of Indian Point being 4° 37′ (we were in 4° 39′ at noon of the 29th, by good observation, when the pitch of the point bore E. $\frac{1}{4}$ S. true bearing). The description of the land however in Laurie and Whittle's chart is sufficiently exact, and particularly so with respect to Indian Point, which strongly resembles the Bill of Portland, but of a greater length. The north side of the bay is formed by reddish land of moderate elevation, with ravines or fissures resembling chalky cliffs discoloured by the weather. These high lands descend gradually to the low land at the bottom of the bay; Indian Point also falls gradually towards the south into low land entirely covered with wood. Here the water was first observed to have a deep red tinge as if mixed with blood, but on being examined in a glass was found perfectly colourless; the bottom however seems to account for this appearance, being a soft mud composed of a reddish clay without the smallest mixture of sand, and so smooth that it might be laid on as paint. The only fish taken since we have been in muddy ground were two toad fish (*Diodon*) and several eels, one of which measured in length 4 feet 10 inches and in circumference 7 inches.

On the 30th June we anchored in the evening off Malemba point, in 15 fathoms, and on the morning of the 1st, were surprised by a visit from the Mafook or king's merchant of Malemba, accompanied by several other negro gentlemen, and a large cortege of attendants in an European built four-oared boat and two canoes, one of which latter preceded the boat to announce the great man, and the officer in her introduced himself by letting us know, that " he was a gentleman, and his name was Tom Liverpool." The first question put by the Mafook on his coming on board was " if we wanted slaves ;" nor could we for a long time convince him in the negative, observing that we were only merchant ships, and particularly from our numerous boats. Having at last made him understand the motives of the expedition, and informed him that no nation but the Portuguese were now permitted to trade in slaves ; he very liberally began to abuse the sovereigns of Europe, telling us that he was over-run with captives, whom he would sell at half their value, adding, that the only vessel that had visited Malemba for five years was a French ship about a year before this time ; and according to him, the Portuguese government had prohibited their subjects from trafficking in slaves to the north of Cabenda, where there were now nine vessels bearing their colours, and one Spaniard. The Mafook however acknowledged that they sometimes sent their boats from Cabenda to Malemba to procure slaves, and indeed we saw an European boat sailing between the two ports. From the description of the vessel hoisting Spanish colours at Cabenda, there could be no doubt of her being the ship we passed on the 18th.

The Mafook finding we did not want slaves, offered to supply us with fresh provisions; and as I knew we should, as usual, be obliged to anchor in the evening not far from our present station, I accepted his offer of sending his boats on shore for that purpose, he himself desiring to remain on board for the night with eight of his officers, doubtless in the expectation of having a glorious dose of brandy, which in fact they swilled until they could no longer stand.

The dresses of these gentry were a singular medley of European and native costume; the Mafook had on a red superfine cloth waistcoat; his secretary, an English general's uniform coat on his otherwise naked body; a third a red cloak edged with gold lace like a parish beadle's, &c. &c. The native portion of the dress consisted of a piece of checked or other cotton cloth folded round the waist, and a little apron of the skin of some animal, which is a mark of gentility, and as such is not permitted to be worn by menial attendants. A striped worsted cap, or else one of their own manufacture and of very curious workmanship, on the head, completed the useful part of their dress. Their ornaments consisted of rings of iron and copper on the ancles and wrists, welded on so as not to be taken off; and many of the copper ones having raised figures tolerably executed. This metal we understood was abundant in their country. Besides necklaces of beads, the general neck ornament was circles or rings of the bristles of the elephant's' tail, called by them morfil, and which seemed to be multiplied in proportion to the puppyism of the wearer, the graver or middle aged men having but one or two, while some of the young ones had so many, that they could with difficulty move the head, and reminded us of our Bond-street bloods with their chins hid in an enormous cravat.

All were loaded with fetiches of the most heterogeneous kinds; bits of shells, horns, stones, wood, rags, &c. &c.; but the most prized seemed to be a monkey's bone, to which they paid the same worship that a good catholic would do to the *os sacrum* of his patron saint. The *master fetiche* of the Mafook was a piece of most indecent sculpture representing two men, surrounded by the tips of goat's horns, shells, and other rubbish, and slung over the shoulder with a belt of the skin of a snake. The features of these sculptured figures, instead of being Negro, as might be expected, were entirely Egyptian; the nose aquiline and the forehead high. The canoes are of a single tree; each had five men, who worked them with long paddles standing up. At night our visitors were satisfied with a sail in the 'tween-decks, where they all huddled together, and from which they started at daylight to light their pipes and resume their devotions to the brandy bottle.

As I had expected, we were obliged to anchor, by the failure of the sea breeze opposite to Cabenda, from whence, in the forenoon, a boat came off with another cargo of *gentlemen;* but, as I had been quite sufficiently plagued by my Malemba guests, I excused myself from not being able to receive them on board; the sea breeze being about to set in, and as there was no appearance of the Malemba boat bringing off the stock, I, much against their inclination, sent off my visitors in this boat.

The information we picked up respecting the coast from Loango Bay to the mouth of the Zaire, proved, as we expected, that it is very erroneously laid down in the most recent charts. The only river between Indian Point and

Cabenda is the *Loango-Louise*, and is that marked in the charts by the name of *Kacongo*, being by our observations when at anchor nearly opposite to it, 5° 17'. Its opening is between two high lands, and appears to be wide and clear. The country is divided into petty sovereignties, tributary to the king of Loango; the northernmost of these states, after passing Laongo bay, is named Boal, to which succeeds Makongo, of which Malemba is the port; then that of N'Goy, whose port is Cabenda, and which extends along the north side of the entrance of the river Congo. The king of Makongo, or Malemba, resides inland at a town named Chingelé (evidently the Kinhelé of the charts,) but which is *not* situated on a river. From our visitors I procured a vocabulary of their language; they all speaking English to be perfectly understood, and several of them French still more correctly.

While at anchor this day, I sent two boats in shore to look for the bank of Belé, said by Grand Pré to be situated south-west of Malemba, and which, according to him, shoals suddenly from seven fathoms. The boats, however, could not find any bank, but on the contrary, the water shoaled very regularly from where the ship was anchored in 15 fathoms to 5 fathoms within about three miles of the shore, all soft muddy bottom. One of the natives on board

K

assured us that he had been on the bank in question, with Mr. Maxwell, and that it lies within a short distance of the shore, and nearer to Malemba than Cabenda.

Having weighed with a tolerable sea breeze, we were enabled to stand along shore until eight in the evening, when being in 8 fathoms, the anchor was let go, and the current was found running N.N.W. $1\frac{3}{4}$ mile an hour; an officer being sent in shore to sound, reported that the water shoaled very gradually to three fathoms within half a mile of the shore, near to which is a lengthened reef, with the sea breaking violently, but which seems to shelter the beach within it, and thereby affords landing to boats.

The coast from 4° 50' is moderately elevated, forming reddish gray cliffs, similar to those near Loango bay; until past Cabenda, when the coast descends to low land covered with wood, (apparently the mangrove) and our view this day terminated on the Red Point of the charts, (Chabaroca point of the natives) which they informed us was the entrance of a little river.

CHAPTER III.

Passage up the River to the place where the Congo was left, and from whence they proceeded in the double-boats.

CHAPTER III.

As we were now approaching the scene of action, I thought it right to issue to the Officers and Naturalists the following memorandum of regulations for our conduct while in the country.

" Although it is impossible to foresee all the circumstances which, in the progress of the expedition, may call for the exertion of the utmost prudence and presence of mind in those who may have intercourse with the natives; nevertheless the following observations are offered, with the certainty that an attention to them will be the means of avoiding the ill effects, which may as certainly be expected from a different line of conduct to that which they recommend.

" Though we are not to expect to find in the natives of Africa, even in the most remote region, that state of savage nature which marks the people of other newly discovered countries, with whom the impulse of the moment is the only principle of action, it is nevertheless highly necessary to be guarded in our intercourse with them; that, by shewing we are prepared to resist aggression, we may leave no hope of success, or no inducement to commit it.

" In doing this, it is, however, by no means necessary to exhibit marked appearance of suspicion, which would probably only serve to induce the hostility it seemed to fear; it is, on the contrary, easy to combine the shew of being guarded, with marks of the greatest confidence.

" In the event of the absolute necessity of repelling hostility for self-preservation, it will certainly be more consonant to humanity, and perhaps more effectual in striking terror, that the first guns fired be only loaded with small shot.

" Although we may expect to find the idea of property fully known to all the people we shall have intercourse with, it is not to be the less expected that they will be addicted to theft, the punishment of which in savages has been one of the most frequent causes of the unhappy ca-- tastrophes that have befallen navigators; it is therefore urgently advised, not to expose any thing unnecessarily to the view of the natives, or to leave any object in their way that may tempt their avidity.

" In the distribution of such presents as may be entrusted to those going on shore, great caution is requisite to ascertain the rank of the persons, to whom they are given, and to proportion the value accordingly, in order, as much as possible, to prevent jealousies.

" A great cause of the disputes of navigators with un-civilized people is in unauthorised freedoms with their females ; and hence every species of curiosity or familiarity with them, which may create jealousy in the men, is to be strictly avoided ; taking it for granted, that, in a state of society where the favours of the women are considered as a saleable or transferable commodity by the men, the latter will be the first to offer them.

" As one of the objects of the expedition is to view, and describe manners, it will be highly improper to interrupt, in any manner, the ceremonies of the natives, however they may shock humanity or create disgust; and it is equally necessary, in the pursuits of the different Naturalists, to avoid offending the superstitions of the natives in any of their venerated objects. Hence, in inhabited or enclosed places appearing to be property, permission should be first sought to cut down trees (particularly fruit-bearing ones,) which, as well as animals, are often held sacred. When no superstitious motive interferes, a few beads will, probably, always purchase the required permission.

" As it is probable that the different pursuits of the

scientific gentlemen may be as well carried on in company as if separated, it is therefore strongly recommended to them to keep together as much as possible for their mutual support and safety. Should they however think proper to separate in their excursions, it is to be understood that the *two* or *three* marines, who will always be appointed to accompany them, are to remain with the gentleman having the direction on leaving the vessel; and, in order to avoid the possibility of any dispute for precedence in this last respect, the succession in which the Lords of the Admiralty have given me the names of the scientific gentlemen, is to be considered as the established rule, viz. Mr. Professor Smith, Mr. Tudor, Mr. Cranch; and when it shall be thought necessary to send a midshipman or other petty officer to command the escort which may accompany the naturalists, he is strictly directed to comply with the wishes of the gentleman having the direction of the excursion, as far as his ideas of safety will authorise.

" The health of the persons accompanying the naturalists in their excursions will of course be a particular object with those gentlemen, by taking care not to expose them unnecessarily to the sun in the hottest hours of the day, or to the rain, if shelter can be had, and by carrying them as little as possible into swampy tracts.

" It is most particularly enjoined to every person who may be on shore to return on board, as soon as possible, on seeing the signal for that purpose."

The scantiness and short duration of the sea breezes and the current kept us nearly stationary, until the 5th, when in the afternoon a fresh sea breeze sprung up at W.S.W. with which we stood to the south, and soon shoaled our water from 22 to 13 fathoms, which depth we carried without alteration until 8 o'clock, when we deepened to 18 fathoms, and the next cast had no ground with 150 fathoms of line; whence it was evident we were in the deep channel of the river Congo or Zaire, and thus had overshot my intention, which was, in consequence of the expected velocity of the stream, to anchor on the edge of the bank, and take the next sea breeze to cross it. We had now however no alternative but to stand on, and the breeze lasting for near an hour, carried us across the fathomless channel, and we struck soundings in 23 fathoms on this side, as suddenly as we had lost them on the other; the wind at the same time failing, we anchored in that depth, and found no current whatever; indeed in the deep channel of the river it must have been insignificant, in comparison with what we had been led to expect, certainly not above two miles an hour.

L

At daylight of the 6th we found Cape Padron bearing S. b. E. $\frac{1}{2}$ E., and Shark Point S. E. $\frac{1}{2}$ S.; the latitude at noon in the same situation being 6° 5′. At noon weighed with a pretty fresh sea breeze, and ran in for the land between the above points, until within half a mile of the shore, when we had 20 fathoms water. We then bore up towards Shark Point, and immediately lost soundings, nor did we again get bottom with the hand lead until it suddenly struck the ground in 5 fathoms; the anchor with the chain cable was immediately let go, but finding the ship did not bring up, and was drifting from the buoy, I concluded the chain had snapped, and directed another anchor to be let go; but before this was done the ship was in 36 fathoms and still drifting; both the chain and cable were now veered away, and she at last brought up; but fearing she would again go adrift, the kedge anchor, backed by a smaller one, was run out. The Congo sloop, which had let go her anchor in 4 fathoms, also drove, and fell along side of us, but without any other ill consequence than the loss of her anchor and cable, which, by some mismanagement, was suffered to run out end for end. When the ship had brought up we found that she tailed on a mud bank with but 7 fathoms, while under the chains was 14, and under the bows 36. Where we first let go the anchor in 4$\frac{1}{2}$ fathoms, there was no current whatever, but at the place to which we had drifted it ran $2\frac{1}{2}$ miles an hour to the N. N. W.; but it was here also considerably affected by a twelve hours tide, being almost still water at 5 o'clock in the evening and 6 the next morning.

On heaving up the chain we found that the anchor had broken at the crown. In the forenoon, while waiting for the sea breeze, the Mafook of Shark Point came on board with half a dozen of his myrmidons, and though the most ragged, dirty looking wretch that can be well conceived, he expected as much respect as a prince; first complaining that the side ropes were not proper for a person of his quality (they were only covered with canvas); then insisting on a chair and cushion on the quarter deck; with the latter of which being unable to comply, he was satisfied with spreading an ensign over the former. Seating himself at the taffarel, he certainly made a very grotesque appearance, having a most tattered pelisse of red velvet, edged with gold lace, on his naked carcase, a green silk umbrella spread over his head, though the sun was completely obscured, and his stick of office headed with silver in the other hand. It being our breakfast hour, he notified his desire to be asked into the cabin, to partake of our meal; but he smelt so offensively, and was moreover so covered with a cutaneous disorder, that my politeness gave way to my stomach, and he was obliged, though with great sulkiness, to content himself on deck. To bring him into good humour, I however saluted him with one swivel, and gave him a plentiful allowance of brandy. He seemed indeed to have no other object in coming on board than to get a few glasses of this liquor, which he relished so well that he staid on board all night and the five following days. From him we learnt that there were three schooners and four pinnaces (all Portuguese) at Embomma, procuring slaves. He also affirmed that the transport could not, at this season, ascend higher than the tall trees, on account of the little water in the channel.

At 2 P. M. of the 8th, a fresh sea breeze coming in, we weighed, but the moment the ship came abreast of Shark Point, she was taken by the current and swept right round ; with difficulty her head was again got the right way, and she rounded the point in $4\frac{1}{2}$ fathoms. We then stood on S. S. E. by compass, carrying a regular depth of 7 and 8 fathoms for about 2 miles, when finding her go astern, blowing a fresh breeze with all the studding sails set, let go the anchor in 8 fathoms, and veered 30 fathoms of chain. Shark Point bearing W. b. S. about 2 miles. At 8 o'clock, found the ship driving, and that through the neglect of the men attending the lead ; she was already in 16 fathoms, and the sea breeze being still fresh, we loosed and set all the sails, and let go another anchor, which brought her up. During the night the wind remained light at S. W., and the ship was steered as if under way, though riding taught with all sail set. The Congo, without difficulty, went over the current, and might have run up to the Tall trees, had her signal not been made to come to, and she accordingly anchored opposite Sherwood's Creek.

At four o'clock this afternoon, a schooner appeared off the point, hoisted Spanish colours, and fired a gun; after laying to for some time, she hoisted the royal colours of Spain, fired a shot, which fell near the transport, and ran in and anchored. A boat was immediately sent from her to ask what we were, and on being informed, they made some excuse for firing the shot, intended, as they said, to assure the colours; their vessel, by their account, was from the Havannah for slaves; but it was perfectly evident, from their answers to my questions, that she was illicitly employed in this trade, and prepared to carry it on by force, being armed with 12 guns, and full of men: this was indeed put out of doubt on the return of her boat on board, by her getting under way and again running out of the river; doubtless from apprehension of the sloop of war, which they were told was gone up the river. This vessel was destined to take off 320 slaves; her burthen being 180 Spanish tons.

The sea breeze setting in fresh at 2 P. M. on the 9th, and finding the ship under all sail go ahead of her anchor, we weighed, but, though the current was running scarcely three miles an hour, she at first barely stemmed it, and soon went astern; deepening the water so rapidly, that finding we should be out of soundings before we could bring up, I had no alternative but to run out again and try to anchor under Shark Point; but so little effect had the sails, even when going with the current, that she drifted bodily on the Moena Moesa bank, on which I was just about to let go the anchor in 7 fathoms, when the sea breeze becoming suddenly and providentially very strong, we got her to stay, and again just fetched Shark Point, where the wind again failed, and we were driven round by the current, and again obliged to stand out. The wind however once more freshening at sunset, after making a short tack off, we stretched in, and were fortunate enough at 7 o'clock to get to an anchor under Shark Point in $4\frac{1}{2}$ fathoms; an attempt, which I should have considered highly imprudent had I not examined the bank, and taken accurate marks; nor indeed had I any other alternative but that of finding myself off Cabenda in the morning had I kept to sea.

The three succeeding days, there being either no sea breeze, or only such as was too weak to attempt any thing with our brute of a transport, we were obliged to remain at anchor, rolling gunnel in, from the ground swell on the bank; consoling ourselves however that, of the three evils which threatened us, we had escaped the two worst, either being obliged to anchor on the Mazea bank, or being driven by the current to the northward, God knows where.

Our Shark Point visitors were now succeeded by the Mafook of Market Point and a *gentleman* from Embomma, who told us he was sent by the Great Mafook of that place to accompany us up the river, in order that no accident should happen to us; and though I would very gladly have foregone the pleasure of their company, I could not refuse receiving them on board. I however endeavoured to make them clearly understand that they were not to expect the same attentions on board a *King's ship* (I was sorry to be obliged to disgrace the name by applying it to the detestable transport,) which they had been used to receive from slave traders; and the uniforms of the officers, and the marines, seemed to give them the proper feeling of our consequence.

Several of the Sonio men who came on board were Christians after the Portuguese fashion, having been converted by missionaries of that nation; and one of them was

even qualified to lead his fellow negroes into the path of sal-
vation, as appeared from a diploma with which he was furnished.   This man and another of the Christians had been
taught to write their own names and that of Saint Antonio,
and could also read the Romish litany in Latin.   All these
converts were loaded with crucifixes, and satchels containing
the pretended relics of saints, certainly of equal efficacy
with the monkey's bone of their pagan brethren; of this we
had a convincing proof in each vociferating invocations to
their respective patrons, to send us a strong wind; neither
the fetiche or Saint Antonio having condescended to hear
their prayers.   The Christian priest was however somewhat
loose in his practical morality, having, as he assured us, one
wife and five concubines; and added, that St. Peter, in con-
fining him to *one* wife, did not prohibit his solacing himself
with as many handmaids as he could manage.   All our visitors, whether Christians or idolators, had figures raised on
their skins, in cicatrices, and had also the two upper front
teeth filed away on the near sides, so as to form a large
opening, into which they stuck their pipes, and which is so
perfectly adapted to the purpose that I thought it expressly
formed for it; until on enquiry I learned, that, as well as
the raised figures on the skin, it was merley ornamental,
and principally done with the idea of rendering themselves

agreeable to the women, who, it seems, estimate a man's beauty by the wideness of this cavity, which in some measured near an inch, the whole of the teeth, and particularly the two front ones, being enormously broad, and very white.

Our Sonio visitors were almost without exception sulky looking vagabonds, dirty, swarming with lice, and scaled over with the itch, all strong symptoms of their having been *civilized* by the Portuguese, and in their appearance and manners forming a striking contrast, not unimportant to the study of national manners, to our Malemba guests, who were chearful, clean, drest even to foppishness, and *choquéd* their glasses with us ; in short, quite gentlemen *à la Française*, the nation with which they have had most intercourse. Some canoes brought on board a few pigs, goats, fowls and eggs, for sale, but the prices they asked were so exorbitant, that for fear of spoiling the market up the river, by their reports of our facility, I confined myself to the purchase of a few fowls and eggs. The value they here set on our different articles for barter was by no means in the proportion of their respective English prices ; for an empty bottle, a looking-glass, or knife, invoiced at 3d, we got a full grown fowl, while for a bunch of beads that cost 2s. 10d. they offered but two ; and for a small goat, they wanted four fathoms of blue baft. We however purchased for a mere

M

trifle, a fresh water turtle weighing 40lb., which, when drest, we found equal the green turtle.

The method of closing a bargain, and giving a receipt, is by the buyer and seller breaking a blade of grass or a leaf between them, and until this ceremony is performed, no bargain is legally concluded, though the parties may have possession of each other's goods; this we only learned by experience, for having bought, and, as we thought, paid for a couple of fowls, they were immediately slaughtered for dinner, but the owner taking advantage of the omission of the ceremony, pretended that he had not concluded the bargain, and insisted on another glass, which we were obliged to give him, but profited by the lesson.

During our forced detention at Shark point, the Naturalists made some excursions on shore, and were gratified in their respective pursuits, particularly Dr. Smith, who procured many interesting plants. Mr. Cranch shot some birds, amongst which were an eagle, an anhinga, several varieties of the king fisher, a toukan, and many small birds. Near the shore, these gentlemen saw, close to a place where had been a fire, human skulls and other human bones. Observing the natives take considerable numbers of fish with nets, we sent two boats to haul the seine at day-light of the 11th. On the outside of the point they were entirely unsuccessful, not taking a single fish, but on the inside, in one haul, thirty large fish were taken, some weighing 60lbs; these were all of one kind, of the *Sparus* genus, and named *Vela* by the natives. They were found to be excellent in taste and firmness, much resembling the cod. The only other species taken were a single large cat fish *(Lophius)* and a few small mullets.

July 12. I now determined to lose no more time in the attempt to get the transport up the river, but to extricate myself from this exquisitely tantalizing situation by the immediate transhipment of the provisions and stores to the Congo; for which purpose the double boats, and all the ship's boats were hoisted out, on the evening of the 12th, and the double boats being put together and rigged, the whole were next morning loaded; when therefore the sea breeze set in at two o'clock in the afternoon, I took my leave of the transport, with the Naturalists, leaving the master and purser to see her discharged, or get her up the river if an opportunity offered. I had now the satisfaction to find the double boats answer my best expectations in their fitness for this service; for though the breeze was very light, and the current running round Shark point three miles an hour, they without difficulty doubled this, to us more redoubtable promontory than that of Good Hope to our early navigators, and in two hours and a half we reached the Congo sloop, laying about ten miles from the point.

The berth she occupied I found was about half a mile from the south shore, nearly opposite Sherwood's creek, (Fuma of the natives). The current here at its maximum ran $3\frac{1}{2}$ miles an hour, but was subject to very great irregularities, apparently from the combined effects of a regular tide, and of eddies formed by the points of land or banks. These effects were frequently so great as to entirely overcome the stream, and create perfect slack water of various duration from half an hour to five minutes. The rise and fall of tide by the shore, as marked on the roots of the mangrove, was $2\frac{1}{2}$ feet. The water thus high is too brackish for use, and though perfectly colourless in a glass, has the same red appearance as we remarked off Cabenda.

Hitherto the river has presented no appearance to inspire the idea of magnitude equal to that of a river of the first class; unless we were indeed to consider the estuary formed between the Sonio and Moena Mazea shore, as the absolute embouchure of the river, than which certainly nothing would be more erroneous; the true mouth of the river being at Fathomless Point, where it is not three miles in breadth; and allowing the mean depth to be 40 fathoms, and the mean velocity of the stream $4\frac{1}{2}$ miles an hour, it will be evident that the calculated volume of water carried to the sea has been greatly exaggerated.

The peninsula of Cape Padron and Shark Point, which forms the south side of the estuary, has been evidently formed by the combined depositions of the sea and river, the external or sea shore being composed of quartzy sand, forming a steep beach ; the internal or river side, a deposit of mud overgrown with the mangrove ; and both sides of the river towards its mouth is of similar formation, intersected by numerous creeks, (apparently forming islands) in which the water is perfectly torpid. This mangrove or alluvial tract appears to extend on both shores about seven or eight miles inland, where the elevated and primitive soil then occurs, and the outline of which is frequently caught from the river, through vistas formed by setting fire to the mangrove, or over the creeks. This mangrove tract is entirely impenetrable, the trees growing in the water, with the exception of a few spots of sandy beach. Small islands have in many places been formed by the current, and doubtless in the rainy season, when the stream is at its maximum, these islands may be entirely separated from the banks, and the entwined roots keeping the trees together, they will float down the river, and merit the name of floating islands. At this season however, they are reduced to occasional patches of a few yards of brush wood, or reeds, which, gliding gently down the stream, convey the idea of repose rather than the rush of a mighty river.

Liutenant Hawkey proceeded up the creek opposite to which the Congo was anchored, and describes it " as dividing into two branches, one having a direction E. by S. and the other W. by N., the former of which he followed, and found it extremely tortuous; after passing twenty reaches in directions almost opposite, he reached the primitive land, composed here of sandy precipitous cliffs; the soil in some spots bare, in others covered with wood, particularly with the Adansonia or boabab. Here we met with the excrement of elephants, tygers, and other animals both herbivorous and carnivorous; the skeleton of the head of a wild hog was picked up, and an antelope was seen; on the sandy beach close to a pond of stagnant fresh water were many birds, where the river turtle had deposited their eggs."

Many canoes visited the Congo, with pigs, goats, fowls, and eggs for sale, but being almost as exorbitant in their demands as at Shark Point, we did little business; some trading canoes with 10 to 20 men in each, going up and down the river, also stopped along side to satisfy their curiosity. Their general cargoes were salt, and palm nuts, from the latter of which they extract oil. The salt is procured from the north shore, in the district of Boolambemba, near Fathomless Point. In one of these canoes were also an elephant's tooth and a boy for sale.

From the natives who were on board we learnt that the King of Sonio resides at Banza Sonio, on a fresh water river, the entrance of which is the creek marked in Maxwell's chart " Raphael's creek," and that a boat would be twelve hours ascending to the town, though the current is trifling. We also now learnt how the human bones came in the place w here they were seen by the Naturalists nearShark Point, and which, without an explanation, might have led to the supposed cannibalism of the natives ; we were however assured that they were the remains of criminals, who had suffered for the crime of poisoning, this spot being the place of execution of a certain district. When a common man is convicted of this crime, his head is first severed, and his body then burnt ; but the punishment of a culprit of superior rank is much more barbarous, the members being amputated one by one, so as to preserve life, and one of each sent to the principal towns of the kingdom, to be there burnt. The trial is always by a kind of ordeal.

This afternoon the transport weighed with the sea

breeze, but being as unmanageable as ever, she ran on Shark Point, where she lay half an hour, when the tide flowing, she went off without the smallest damage, and by the great exertion of the master of the Congo, was brought to an anchor two miles within the point.

We had now visitors arriving hourly, all of whom pretended that they were sent by the Mafook of Embomma, to see the vessels safe up the river, and each of these gentlemen assured us that all the others were impostors, and only came on board to get brandy, so that I had a difficult task to keep clear of offending them, and at the same time avoid imposition. I however succeeded in getting rid of them all, by telling them that they should remain on board until we reached the town, when the Mafook would decide who were and who were not impostors, and doubtless would punish the latter. From them we learnt that an express had been sent from Cabenda to notify our approach, and that on this intelligence all the Portuguese vessels at Embomma, had precipitately left it, and quitted the river, passing us no doubt in the night. I had however expressly declared to the Malemba and Cabenda people who visited us, that I should not in any manner interfere with the slave traders, of whatever nation they might be.

The transhipping the stores and provisions being finished

on the 18th, the double boats were loaded and every thing ready to proceed up the river, but there being only a very faint sea breeze this day, we were obliged to continue at anchor. The Mafook Sina, or chief king's merchant of Embomma, came on board this morning, but as I had been frequently deceived by gentlemen Mafooks, I received him so cavalierly that he quitted the Congo, and went on board the transport, where his quality being acknowledged by several natives then on board, he sent back his interpreter and head man to me, and on finding that he was really the person he pretended to be, I desired the transport to salute him with four guns, which made up for my first bad reception, and he visited me in the afternoon, bringing with him a retinue of twenty rascals, all of whom he expected to be gorged with brandy ; and as I knew he had great influence at Embomma I endeavoured to gratify even his immoderate wish, and lent him the Congo's jolly boat to return to Embomma, and my own boat cloak to keep him warm.

The 19th, there being no sea breeze, we continued at the same anchorage, but the next day were more fortunate, and succeeded in getting the Congo up abreast of Halcyon island (Zoonga Campendi). The banks of the river, along which we passed sometimes within a stone's throw, are entirely covered with mangrove, intersected by creeks, the first

N

of which of any consideration after passing Fuma, is Kanga-
vemba, (Alligator's pond of Maxwell), which seems to be a
large expanse of water, but according to the natives, goes but
a little way inland. The next considerable creek is that
whose entrance contains the three islands called by Maxwell,
Bonnet, Knox, and Halcyon ; the first having its name from
a clump of trees, and is called by the natives Zoonga,
Casaquoisa ; Knox's island the natives describe as a peninsula. The eastern part of the entrance of this creek forms
an excellent little haven, where the Congo was now anchored, entirely out of the stream of the river in five fathoms.
This inlet, the natives say, goes up to the town of Loocansey,
the distance from the mouth being about three hours rowing
of a boat.

July 21. This morning we sent a party to haul the seine on
one of the banks which lie close to Knox's island, and took
great abundance of fish of four species ; one being a *Sparus*
of a large size, a mullet a *(Surmuletus)*, and an old wife
(Ballistes). A brig under Spanish colours, with 12 guns and
50 men, cleared out from the Havannah, arrived this day in
the river for slaves; her nominal mate, but real captain (named
Sherwood) and a number of their crew being English and
Irish, though pretended Americans, left no doubt of her
being either English or American property.

The precariousness of the sea breezes by which alone we could get the Congo up the river, and the necessity of my losing no time in endeavouring to arrange matters at Embomma, made me determine on proceeding thither in the sloop's double-boat; and I accordingly quitted the Congo with the Naturalists (except Mr. Cranch, who preferred the accommodations afforded by the Congo), at 4 o'clock in the evening, keeping within boat's length of the shore; we found no current until reaching the point named Scotsman's Head, where it ran $3\frac{1}{2}$ miles an hour; and the breeze being very weak, we barely stemmed it. In the hope of meeting a counter current on the opposite shore, I now crossed the stream, and it being dark when we reached it, I anchored on one of the banks in six feet, entirely out of the current. This evening's sail along the banks was particularly agreeable, the lofty mangroves overhanging the boat, and a variety of palm trees vibrating in the breeze; immense flocks of parrots alone broke the silence of the woods with their chattering, towards sun-set; and we learnt that those birds make a daily journey across the river, quitting the northern bank in the morning to feed in the Indian corn plantations on the south side, and returning in the evening.

July 22. The shoals and low islands near which we anchored are composed of a border of sand and clay, with a muddy swamp in the middle, the islands being covered with reedy grass. By the natives they are named Monpanga, or look-out. They were covered with fishing eagles, terns, white herons, and other beach birds, of which several were shot, and Dr. Smith collected no fewer than thirty new species of plants. Our bearings at anchor were, west end of Tall Tree island nearly shut in with the north shore, west, and the entrance of Maxwell's river, N.E.

At noon we had a light breeze from W.S.W.; weighed and ran along the edge of the shoals in one and two fathoms; at four, a fresh breeze; and being past the low reedy islands and shoals, we ran along the bank of the mangrove land, nearly touching the trees in 3 or 4 fathoms until 7 o'clock, when the darkness obliged us to anchor in 2 fathoms.

July 23. At daylight, we found that we had anchored within 20 yards of a dry shoal; being also close to the island named Draper's island by Maxwell, and Zoonga Kampenzey or Monkey island by the natives; bearings as entered in yesterday's journal (by mistake). The land, for about 3 miles west of the entrance of Maxwell's river, is thickly covered with palms intermixed with the mangrove, and other trees; and here a great quantity of palm wine is made for the Embomma market. Hordes of Negroes came down to the bank as we passed, and learning that we had one of their countrymen on board returning from slavery, they greeted us with cheers, after their fashion, and clapping of hands. A great quantity of shell fish, of the Mya genus, are taken out of the mud round Kampenzey island by the natives; and the fish, stuck on wooden skewers, as the French do frogs, and half dried, are an object of traffic; their state of half putrefaction being entirely to the taste of the Negroes. In a raw state they are uneatable, having no flavour of the oyster, though confounded with that fish by the English who have visited the river.

At noon, we weighed with a light breeze at S. W., and ran along the main bank until opposite the entrance of Maxwell's river, when, by the advice of a native on board, we attempted to pass between the two easternmost of Draper's islands, but found them joined by a bank with only 6 feet, where deepest; and keeping too close to the eastern island we grounded in $2\frac{1}{2}$ feet, but shoved the boat off without difficulty; and by sending the gig ahead, passed round the shoals through a very winding channel in 2, 3, and 4 fathoms; then ran along the south side of Monkey's island of the chart (Zoonga Chinganga of the natives), and the islands east of it, in 6, 7, 8, and 9 fathoms. Here we lost the mangrove tract, and the soil became a stiff clay, cut into perpendicular low cliffs at the margin of the river, covered with high reedy grass and scattered palm trees. We passed two Negro villages, and at 7 anchored within a few yards of the bank in 8 fathoms.

July 24. At daylight, having a light breeze at S. W., we weighed and ran along the edge of Stocking island, composed (as yesterday) of reedy grass and thinly scattered palms: at 8, the breeze dying away, anchored in a little cove in 9 feet, close to the Negro village of Peter Mesougy, where we purchased a few fowls. Here, in searching for something in the boats' cabin, I put my hand on a snake coiled up on a bag of clothes; on killing it with a cutlass, it proved to be a water snake, and apparently not venomous; though the natives asserted that its bite is mortal.

At noon, we weighed with a light westerly breeze, and crossed the channel named " Mamballa river" (not distinguished by any other name than " Boat's channel" by the natives); the middle of it is filled with dry shoals, the channels between which are very winding, so that we kept the gig constantly ahead, and had from 1 to 5 fathoms water until we gained the shore of Farquhar's island, where there is 7 and 10 fathoms close to the bank. Here we saw the first plantation consisting of Indian corn about 2 feet high, and tobacco. In crossing the channel, a hippopotamus was seen, and, from the shoalness of the water, it must have been walking on the bottom, the head only appearing above the surface. The natives tell us that the irregularity of soundings which we found is caused by these animals assembling in a spot and making holes with their feet. Two women, an old and a young one, came on board from one of the plantations; by their dress and ornaments they appeared to be of a superior class; I therefore gave them some beads and a glass of rum, which they swallowed as greedily as the men; and, in return, the old lady offered, through our interpreter, to leave the young one on board, pour m'amuser; a civility which, under existing circumstances, I thought proper to decline; though the young lady seemed much chagrined at such an insult to her charms. At 3 o'clock the sea breeze set in fresh, and we again crossed the channel; and at 7 anchored on the bank of the east end of Stocking island in 2 fathoms.

July 25th. At day-light we observed the Fetiche rock bearing W. by S., and the Beacon rock N. E. I visited the Fetiche rock, which is a collection of masses of the oldest granite, mixed with quartz and mica, running into the river perpendicularly, and entirely isolated, the land behind it being a plain with reedy grass and some corn plantations. It completely commands the passage of the river, being about

$1\frac{1}{2}$ mile distant from the opposite bank. Some of the natives on board could not be prevailed on to accompany me in the boat, dreading the whirlpools off the rock, as much as the ancients did Charybdis; a few very insignificant eddies, close to the rock, were however now the only signs that some whirlpools may exist in the rainy season, and the current, which here ran stronger than in any other part of the river, did not exceed $2\frac{1}{2}$ miles an hour. The prospect of the river from the summit of the rock is extensive, but in no other respect prepossessing, the hills which bound the view being naked, except a few Adansonia, and apparently of the same formation as the Fetiche rock. Just as we reached the rock, two hippopotami were observed about 100 yards from the shore, with their heads above the water, snorting in the air; a ball fired at them sent them off.

M'Gonza Cheela hills, the middle hill N.E. 6 miles.

The three hills in the fore-ground are those named Tunkloo in the chart of Maxwell, and *a* is his Fingal's shield; they have the same appearance on all bearings, and consequently are nothing like the representation in the chart. *b* Taddy Enzazzi (lightning stone).

From a sketch by Lieut Hawkey.

James Fittler sc.

CAPTAIN TUCKEY'S VOYAGE IN AFRICA.

The Fetiche Rock looking down the river.

Page 96

Published by John Murray, Albemarle Street, London, Nov.r 1st 1817.

Taddy d'ya M'wangoo, or Fetiche rock, W. by S.

Fetiche rock south ½ mile.

a a Plain, and corn plantations.   Fetiche rock E. by S.   *b* Breadth of the river, about 1½ mile.   *c* Reedy island.

On the very summit of one of the M'Gonza cheela hills, named Fingal's shield in the chart of Maxwell, is a very singular pyramidal stone, which has all the appearance of an artificial building, resembling a watch tower or a light-house, but is a natural block of loose granite with another perched upon it.   It is called by the natives *Taddi enzazzi*, or the lightning stone, and is held as an object of great veneration.   It was sketched by Lieutenant Hawkey, and appears to rise out of the circular summit of the hill, as under.

O

At 3, weighed with a fresh sea breeze, and at 6 anchored opposite the village of Lombee, where the Fuka or king's merchant resides, who was to accompany me to the Chenoo, or king of Embomma.

Simmons, a black man whom I had received at Deptford from Sir H. Popham's flag-ship for a passage to his country, here first met with some of his family. His father and brother came on board the sloop. The transport of joy at the meeting was much more strongly expressed by the father than by the son, whose European ideas, though acquired in the school of slavery, did not seem to assimilate with those of Negro society, and he persisted in wearing his European jacket and trowsers ; he however went on shore with his friends, and throughout the night the town resounded with the sound of the drum and the songs of rejoicing. The story of this man, which I had before never thought of enquiring into, and which was partly related by his father, adds one blot more to the character of European slave-traders. His father, who is called Mongova Seki, a prince of the blood, and counsellor to the king of Embomma, entrusted him, when eight or ten years old, to a Liverpool captain of the name of —————, to be educated (or according to his expression to learn to make book) in England ; but his conscientious guardian found it less troublesome to have him taught to make sugar at St. Kitts, where he accordingly sold him; and from whence he contrived to make his escape and get on board an English ship of war, from which he was paid off on the reduction of the fleet. During our passage he performed, without any signs of impatience or disgust, the menial office of cook's-mate.

July 26th. Lombee is a village of about a hundred huts, and here is held the market of the banza or King's town, no trading operation whatever being carried on at the latter; all trading vessels also anchor opposite Lombee. The reason assigned for the market being held here is, that as a great concourse of country people frequent the market, if any dispute were to arise between them and the banza people, the banza would run great risk of being burned, and the person of the Chenoo himself would not be safe. Mr. Simmons this forenoon paid us a visit, in so complete a metamorphosis that we could with difficulty recognize our late cook's mate; his father having dressed him out in a silk coat embroidered with silver, which seemed by its cut to have adorned the person of a stage fop in the days of Sir Roger de Coverley; this piece of finery worn over his own dirty banyan and trowsers; on his head a black glazed hat with an enormous grenadier feather, and a silk sash, which I had given him, suspending a ship's cutlass, finished his costume. He was brought to the boat by two slaves in a hammock, an umbrella held over his head, preceded by his father and other members of his family, and followed by a rabble escort of 20 muskets. His father's present to me consisted of a male goat, a bunch of plantains, and a duck. I had now no small difficulty in keeping the sloop from being constantly crammed with visitors, every Fuka, which appears to be a common title of honour, having his linguister, (linguist) and his two or three gentlemen, all equally voracious for brandy, and without whom it is impossible he could move a step.

The market here we found miserably supplied, being only able to procure a few fowls, a dozen eggs, and some plantains, in exchange for beads, that made them come dearer than in a London market. The staple article of trade here seemed to be salt, in which there were both wholesale and retail dealers, the former having 40 to 50 baskets, which he sold to the latter by the basket, who retailed it to the consumer by the handful, two handfuls for a money mat.

At four o'clock the sea breeze setting in, we ran up and anchored before the creek of the banza, in 8 fathoms, close to the shore, saluting with four swivels. Here I found a hammock sent by the Chenoo, to convey me to his presence; but it being too late, I sent his majesty an excuse by the Fuka Sina, who appeared to be extremely angry, because I could not visit the Chenoo in the dark.

July 27. At ten o'clock I quitted the sloop, with the Naturalists and Mr. Galwey, and with an escort of four marines; the hammock I found to have some resemblance to the native palanquin of India, but in a miserable dirty plight, so that I ordered it to follow; and after the walk of an hour, for the first mile over a plain, covered with reedy grass, except in some spots where Indian corn and a kind of French bean were planted, and which is under water in the rainy season, and then over a fatiguing hill, we reached the banza, at the entrance of which I got into the hammock, and was set down under a great tree, the ground having been swept clean. Here the first objects that called our attention were four human skulls, hung to the tree, which we were told were those of enemy's chiefs taken in battle, whose heads it was the custom to preserve as trophies; these victims, however, seemed to have received the *coup de grace* previous to the separation of the head, all the skulls presenting compound fractures. After waiting half an hour under the tree, we were led to the Chenoo's habitation, where, in a court formed by a fence of reed mats, and which was crowded with the king's gentlemen, I found a seat prepared of three or four old chests, covered with a red velvet pall, an old English carpet with another velvet pall being spread on the ground. Having seated myself, in about five minutes the Chenoo made his appearance from behind a mat screen, his costume conveying the idea of punch in a puppet-

show, being composed of a crimson plush jacket with enormous gilt buttons, a lower garment in the native style of red
velvet, his legs muffled in pink sarsenet in guise of stock-
ings, and a pair of red Morocco half-boots; on his head an
immense high-crowned hat embroidered with gold, and surmounted by a kind of coronet of European artificial
flowers; round his neck hung a long string of ivory beads,
and a very large piece of unmanufactured coral. Having
seated himself on the right, a master of the ceremonies with
a long staff in his hand enquired into the rank of the gentlemen, and seated them accordingly. The doctors (Messrs.
Smith and Tudor) having the first places, and then Mr. Galwey, whom they stiled chief mate; the serjeant of marines
they metamorphosed into a boatswain, taking all the titles
of officers from the trading vessels to which only they had
been accustomed.

All being seated (the crowd of king's gentlemen squatting
on bullocks hides,) I explained to the Chenoo, by Simmons, the motives of my mission; stating that " the king
of England, being equally good as he was powerful, and having, as they already had heard, conquered all his enemies,
and made peace in all Europe, he now sent his ships to all
parts of the world, to do good to the people, and to see
what they wanted, and what they had to exchange; that
for this purpose I was going up the river, and that, on my

return to England, English trading vessels would bring them the objects necessary to them, and also teach them to build houses, and make cloth," &c. &c. These benevolent intentions were however far beyond their comprehension; and as little could they be made to understand that curiosity was also one of the motives of our visit; or that a ship could come such a distance for any other purpose but to trade or to fight; and for two hours they rung the changes on the questions " are you come to trade," and " are you come to make war." At last, however, they appeared to be convinced that I came for neither purpose; and on my assuring them that though I did not trade myself, I should not meddle even with the slave traders of any nation, they expressed their satisfaction by the frequent performance of *sakilla,* one of the chief men first starting up and making gestures with his arms, like a fugle man at exercise, and all the company striking their chests at the termination of every motion. This ceremony they afterwards repeated whenever any thing was said that pleased them; and with redoubled energy when I shook hands with the Chenoo. The keg of spliced rum which I had brought as a part of my present to the Chenoo, was now produced, together with an English white earthen-ware wash-hand bason covered with dirt; into which some of the liquor was poured and distributed to the company; the king saying he drank only wine, and retiring, as he told me, to order dinner. The moment he disappeared the company began to scramble for a sup of the rum, and one fellow, dropping his dirty cap in the bason, as if by accident, contrived to snatch it out again well soaked, and sucked it with great satisfaction.

While dinner was preparing we walked over the banza, accompanied by some of the chief men. It is situated on a small plain on a summit of a hill, and consists of about 30 dwelling places or tenements, each composed of two or three huts, within a square enclosure of reeds matted; the huts are composed of the same materials, and consist of two sides and two end pieces, which they call walls, and two other pieces for roofs; so that a house, ready to put together, may be purchased for the same price as four fowls, and in five minutes may be made ready for occupation; the entrance is by a square door in one of the sides, just large enough to crawl in at, and opposite to it is a window; both of which openings are closed at night with shutters of the same fabric as the walls. The Chenoo's tenement differs in no other respect from the common ones, than in containing one large apartment, a little better lighted and aired, and in being surrounded by a double fence, forming a succession of outer and inner courts; of which the sketch underneath will give a sufficiently just idea.

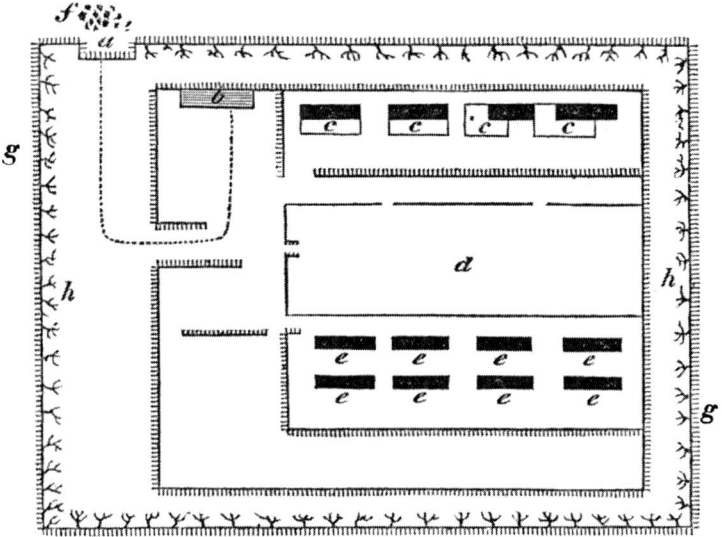

a. The grand entrance, being an opening in the outer fence about three feet high.

b. The audience seat.

ccc. The women's huts.

d. The grand apartments.

eee. Huts.

f. Heap of stones for Fetiche.

gg. Fences or screens.

hh. Young trees inside the outer fence.

Besides a large heap of common stones opposite the grand entrance of the Chenoo's tenement, a fetiche was seen in every hole and corner, consisting of sculptured figures in wood and stone, one exactly resembling the figures we see in England of Bacchus astride on a barrel, with the

P

addition of a long pipe in his mouth and a spear on his shoulder. The two figures of men below, representing two war fetiches, the one armed with a musket, and the other with a broad sword, will serve to shew in what a low state the art of sculpture is among these people, which indeed could hardly be otherwise where writing is utterly unknown. In no one shape whatever do they seem to have profited by the multitude of missionaries that flocked to Congo in the 16th and 17th centuries.

Our repast was laid out in the grand apartment, where some chests covered with carpets served for seats and tables. A few plates and mugs of earthenware, and some Venetian gilt glass were placed on the table, together with a few silver spoons and forks, evidently of French workmanship. The meats consisted of a soup of plantains and goat's flesh, a fowl cut in pieces and broiled, and roasted plantains in lieu of bread ; a large silver tankard filled with sweet palm wine, and a bottle of the rum I had brought, were placed as our beverage.

While we were at table, I learnt that the Chenoo held a palaver in another part of the tenement, where Simmons was most strictly questioned with respect to the motives of our visit, and obliged to swear in the most solemn manner of the country, to the truth of his assertion. This palaver being finished, the Chenoo sent to me, to say that he would again speak to me; and we accordingly were all seated in the audience court as before, with the addition of an old man, who, we were told, was the Chenoo's uncle, and who seemed to be a chief counsellor. After again tiring me with questions as to my motives, the old man, starting up, plucked a leaf from a tree, and holding it to me, said, if you come to trade, swear by your God, and break the leaf; on my refusing to do so, he then said, swear by your God you don't come to make war, and break the leaf; on my doing which, the whole company performed a grand *sa-killa*, and the assembly broke up; the King retiring into an inner hut, where the present I had brought was carried ;* for on my first telling him that I had brought him a present

---

* It consisted of a piece of furniture cotton, some beads, a plated tankard and goblet, and a silk umbrella.

from the King of England, he begged it might not be produced until all his gentlemen were dismissed.

While we were seated in the audience court, the King's
women (of whom he has fifty), were peeping out of one of
the squares, and before retiring, the King very politely
offered me the choice of all his daughters, while his courtiers
as civilly proferred their wives; so that I began to fear
I should find myself in the same dilemma as Frere Jean (in
Compere Matthieu); fortunately, however, the gentlemen
who accompanied me were not so fastidious as the Frere's
companions. I however learnt that the ladies, though apparently nothing loth to change husbands, resisted all solicitations to consummation during day-light, under the apprehension that the fetiche would kill them. The language
of the men in offering their women was most disgusting
and obscene; being composed of the vilest words picked up
from English, French, and Portuguese. The faces of many
of the women were by no means unprepossessing, and their
forms extremely symmetric. Among the men we saw one
marked with the small-pox, another with a short leg, and a
third with a withered arm. Great numbers of the boys had
a large knot at the navel. A cutaneous disorder seemed to
be very general, and, like the itch, chiefly on the wrists;
and the hands of several of the men were perfectly bleached
as if from a scald.

July 28. This morning the Chenoo returned my visit on board the sloop, accompanied by half a dozen of his sons and gentlemen; his modesty was much greater than that of any of our inferior visitors, expressing himself perfectly satisfied with my present, and only hoping that, when I came down the river, I would build him an English house, leave him a boat, and give him a musquet; with the latter request I complied immediately, to his great satisfaction. I learnt from one of his sons, who spoke tolerable English, that a palaver had been held all night, at which the Fuka Sina and all the traders insisted that our intentions could not be good, and that the King should order the sloop to quit the banza, and not let me ascend the river. The King, however, and the court party, would not listen to his suggestions, saying, they were satisfied I came to do them good, and that I might go where I pleased. Indeed the King now repeated the same thing, adding, that if I came to make the country (signifying to form a settlement), as the Portuguese had done at Cabenda, he would grant me all the land I required; in short, we parted the best of friends; and on his landing I saluted him with four swivels, the report of which (though they had been warned) struck all the crowd on shore with such a panic, that they ran off precipitately, tumbling over each other; which shewed us at least that we had little to fear from the warlike disposition of this people.

This morning I dispatched a canoe to the Congo, with instructions for bringing her up the river, and employed the afternoon in sounding the river.

On the following morning (29th), I visited the Chenoo, at his request, unceremoniously, taking only two marines. I found him seated on a mat, in a court of his tenement, distributing palm wine to a family party consisting of about 40 men and boys, of all ages ; a seat being placed for me, he sent for a small box full of papers, which he requested me to read, and which I found to be all Portuguese, generally certificates of the Chenoo's good conduct ; and one letter from the governor of St. Paul de Loando, complaining that the Sonio men had killed some of the missionaries, and cut off a Portuguese trading pinnace ; its date 1813. After a palaver of half an hour, in which I was obliged to repeat my assurances of not coming to prevent the slave trade, or to make war, the Chenoo led the way to a pen in which were six fine cows, a young bull and a calf; and one of the largest and fattest cows was selected as a present for me. This animal, I found, had been introduced by the Portuguese, and was now considerably multiplied, though no care whatever seems to be taken to encrease them, the cows in calf being indiscriminately killed with the bulls; nor do the natives make any use of their milk. The Portuguese, we also understood, had brought several horses, but none now remained. Near the pen was an 18lb. carronade, with which I had been saluted on first landing, and which I now learnt had belonged to an English vessel, burnt at the Tall Trees some years since by the slaves on board her, and that the rest of her guns (which had been fished up by the Sonio men), were now in the possession of the Fuka Sina, who offered to sell them.

July 30. Prince Machow Candy, known to English traders as Fuka Candy, (he being Fuka of Market Point when the English traded here), paid me a visit; and, as he is considered as having great influence, I gave him a piece of chintz. He is a mulatto of French extraction, and said to have made a great fortune by trade, while he filled the office of Fuka of the Point.

July 31. This day and Wednesday, I employed in taking a sketch of the reach of the river.

August 1. This forenoon I was visited by Mr. Sherwood, the ostensible mate of the brig under Spanish colours, and who had been an old slave trader out of Liverpool in this river. He was accompanied by four Portuguese masters of trading vessels now at Cabenda, and part of those that had quitted Embomma, on intelligence of an English King's ship approaching. Their visit was for the purpose of assuring themselves if I meant to interfere with the slave trade, and desired to show me their papers. I declined however looking at them, declaring explicitly that I should not meddle with trade or traders in any manner, which seemed to satisfy them, and they went off, as they said, to bring their vessels back from Cabenda. The Fuka of the Point and all the trading men seemed to be also rejoiced at learning this declaration, as they had still doubted, it seems, my assurances made to themselves. There seems to be no reason to doubt but that the chief slave trade to this river is *bona fide* Portuguese. Two persons of this nation visited me, saying they were from Rio Janeiro; I endeavoured to learn *en passant* the amount of the trade, and by combining their answers with the accounts of the natives, think it may be averaged at 2000 slaves a year.

The price of a slave at this time, as stated by the natives, is as follows

2 Muskets.

2 Casks of gunpowder

2 Guineas (1 fathom each).

12 Long Indians (10 fathoms each).

2 Nicaneas (6 fathoms each).

1 Romaul (8 fathoms).

1 Fathom woollen cloth.

1 Cortee, or sash of cloth.

2 Jars of brandy.

5 Knives.

5 Strings of beads.

1 Razor.

1 Looking-glass.

1 Cap.

1 Iron bar.

1 Pair of scizzars.

1 Padlock.

I have no doubt however but that slaves are now sold for one half this valuation.

The Congo this evening succeeded in getting up to an anchorage under Leyland island.

August 2d. I this morning shook hands with the Chenoo, giving him, as a parting token of friendship, one fathom of scarlet cloth, an amber necklace, two jars of spirits, and some plates and dishes. We found a seat placed in the audience court for us, and the Chenoo seated opposite on a mat with fifty or sixty of his friends ranged on each side. On being told that I came to take leave, he retired with me into the grand apartment, where he endeavoured to persuade me to

Q

defer my departure, until he consulted his great men ; but in fact I suppose he was sorry to lose his daily bottle of spirits, for as he sent me every morning a bottle of palm wine, I returned him one of rum. Finding I was determined, he ceased all solicitation, and gave me three of his sons, and two pilots to accompany me to Binda ; I had also engaged four boys as a boats crew, finding them extremely useful, in saving my own people a great deal of trouble by going backwards and forwards with the Naturalists.

In returning from this visit, we passed a hut in which the corpse of a woman was lying, drest as when alive ; inside the hut, four women were howling, and outside, two men standing close to the hut, with their faces leaning against it, kept them company in a kind of cadence, producing a concert not unlike the Irish funeral yell. These marks of sorrow, we understood, were repeated for an hour for four successive days after the death of the person. This scene induced me to enquire for the burying ground, and the natives at first seemed very unwilling to let us see it ; after a little persuasion, however, two or three of them led us towards it, and we found it not above 200 yards from the banza, amongst a few rugged trees and bushes, and over-run with withered grass. Two graves were now preparing for *gentlemen,* their length being nine feet and their breadth five. At this time they were nine feet deep, but we were told they would be dug to the depth of the tallest palm tree, preserving the same length and breadth as at present ; the soil, we observed, was a superficial layer of black earth 18 inches deep, and all the rest a compact yellow clay ; the graves are dug by the same hoes that are used to till the ground, and the excavation is carried on in the neatest manner. One of the old graves had a large elephant's tooth at each end, and another, which we understood to be a child, had a small tooth laying on it ; all had broken jars, mugs, glass-bottles, and other vessels stuck on them ; some shewed that there had been young trees planted round them, but all were dead except one plant of the Cactus quadrangularis. The graves seemed to be indiscriminately dug to all parts of the compass, and no attention appeared to have been paid to them since their first being filled in.

Simmons requested a piece of cloth to envelope his aunt, who had been dead seven years, and was to be buried in two months, being now arrived at a size to make a genteel funeral. The manner of preserving corpses, for so long a time, is by enveloping them in cloth money of the country, or in European cottons, the smell of putrefaction being only kept in by the quantity of wrappers, which are successively multiplied as they can be procured by the relations of the deceased, or according to the rank of the person ; in the case of a rich and very great man, the bulk acquired being only limited by the power of conveyance to the grave; so that the first hut in which the body is deposited, becoming too small, a second, a third, even to a sixth, encreasing in dimensions, is placed over it.

August 3d. This morning at daylight, I rowed the sloop round Booka Embomma island to the south entrance of the creek, where I anchored to wait for the Congo coming up with the sea breeze.

The reach of the river formed by the main land and the island Booka Embomma on the north, and the islands Hekay (Molyneux) and Booka (Leyland's) on the south, is a bason surrounded by elevated hills composed of primitive granite, or schistus ; in general the first formation is naked of trees, and the second covered with brush wood, and large trees in the crevices of the rock. The hills are all extremely rugged, forming deep hollows, separated by natural causeways, and much resembling, but on a greatly larger scale, the road which passes the Devil's Punch Bowl between Portsmouth and London; the flattened summit on which the banza stands is, as we observed in the groves, an under stratum of compact clay covered with black mould, on which the ridges

From a sketch by Lieut. Hawkey.

C.Heath sc.

CAPTAIN TUCKEY'S VOYAGE IN AFRICA.

The market Village near Cooloo mouy.

Page 213.

Published by John Murray, Albemarle Street, London, Nov.ʳ F.1817.

marked that it had been cultivated; and there can be no doubt but it would be capable of producing the finest wheat. The plain, which we passed over to reach the foot of the hills in going to the banza, is equally proper for the production of rice, and would probably afford two crops a year; one by the natural watering in the rainy season, and the other by a very small degeee of labour, in introducing the river water upon it in the dry season.

The Booka Embomma, which is separated from the main land by the creek named Logan by Maxwell, is entirely of schistus, except an exuberantly fertile level that borders the creek; this latter is about 5 yards broad, and has 3 and 4 fathoms depth up to a ledge of rocks, which crosses it near its south entrance, and through which there is an opening barely capable of admitting a canoe. The only part of this reach, in which the current has any considerable force, is at the east side of Zoonga Booka, where it runs in little whirlpools over the rocky bottom about 3 miles an hour; the stream in the mid-channel (where the depth is 15 fathoms) is from $1\frac{1}{2}$ to 2 miles an hour, and in shore on both sides it is often stagnant; and sometimes a small counter current is experienced. The island Booka Embomma would be the most eligible place thus far for a settlement. The trees we had occasion to observe here are generally of a soft spongy nature, unfit for fuel while green, and useless as timber; one species only affording a wood as hard as lignum vitæ, and proper for the same purpose; the largest size we found it arrive at, was that of a man's body.

The miscellaneous information I was able to collect here, I shall give without attention to arrangement of the matter, which my time at this moment does not permit.

This is the winter of the country, the thermometer in the day seldom rising above 76, and at night, when there are occasionally (not always) heavy dews, falling to 60. The mornings, from sun-rise to 9 or 10 o'clock, are dark, hazy, and sometimes foggy. The winds in the morning are often light from south to S. W. The sea breezes set in very irregularly from noon till 4 o'clock, from west to W. S. W.; they have seldom any considerable force more than once a week, and are stronger after a hazy morning, succeeded by a hot sun; they die away from sunset to 10 o'clock. The natives feel the changes of temperature very severely, shivering with cold when the sea breeze sets in fresh.

Salt is the great object of trade at the Market point, and is made near the river's mouth, and brought up by canoes in baskets of the substance that covers the trunk of the palm trees, of about 7lbs. each, one of which fetches about two fathoms of blue baft. The other objects of petty traffic are palm oil and palm nuts, from which the oil is extracted. Indian corn, pepper (chiefly bird pepper), and mat sails for canoes. The small money in use is little mats of the leaf of the bamboo, about 18 inches square, 20 of which will purchase a fowl. The name of Zaire is entirely unknown to the people of Embomma, who call the river " Moienzi énzaddi," the great river, or literally the river that absorbs all the lesser ones ; this title must however be derived from its receiving tributary streams higher up, as we could not understand that there is a stream of any consideration thus far ; and the only springs we observed were two very insignificant ones issuing from a rock near the banza ; there is also said to be good rock water at the Market point, and at Tall Trees ; and while at anchor at Sherwood's creek, the natives brought us a cask of excellent water from a creek near Kelly's point. The river water is at this season but little muddy, and after being boiled and allowed to deposit its sediment, is not found to affect the people.

There are several varieties of the palm trees here, three of which afford palm wine ; the first, the sweet kind, is given by that named Moba, and the second by the Mosombie ;

the liquor is extracted as in the West Indies. The sweet wine is allowed to ferment, and produces an intoxicating beverage; when quite fresh it is very pleasant and whole-some, taken moderately, keeping the body open. The Masongoi tree also affords a palm wine, considered of supe-rior quality; an inebriating liquor is also produced from Indian corn, and named baamboo.

The cultivation of the ground is entirely the business of slaves and women, the King's daughters and princes' wives being constantly thus employed, or in collecting the fallen branches of trees for fuel. The only preparation the ground undergoes is burning the grass, raking the soil into little ridges with a hoe, and dropping the Indian corn grains into holes. The other objects of cultivation that we saw near the banza, were tobacco and beans of two sorts. Fruits are very scarce at this time, the only ones being long plantains, small bitter oranges, limes and pumpkins. There are no cocoa nut trees, nor, according to the natives, is this tree found in the country. The only root we saw is the sweet cassava which the natives eat raw and roasted. Sugar cane of two kinds was seen.

The only vegetable production of any consequence in commerce is cotton, which grows wild most luxuriantly; but the natives have ceased to gather it, since the English left off trading to the river, the Liverpool ships formerly taking off a small quantity.

The domestic animals are sheep spotted black and white, with pensile ears and no horns, goats, hogs of a small breed, a few dogs resembling the shepherd's dog, and cats. The black cattle brought by the Portuguese cannot be considered as fully established, no care being taken of them, though, from their very fine appearance and their excellent meat, no part of the world seems more proper for their multiplication. Common fowls of a small breed, and Muscovy ducks are the only domestic poultry

The wild animals of whose existence we have any certainty, are elephants in small numbers, this hilly country being unfavourable to them. Buffaloes, which are said to be abundant. Antelopes, of which a few have been seen; wild hogs, the skeleton of the head of one being found. Tigers and tiger cats, the skins being seen with the natives. Monkeys in abundance, (*Simia sephus*). The hippopotamus and alligator appear to be numerous. The only species of fish we have seen to be peculiar to the river is a kind of cat-fish, and some small ones resembling the bleak.

Among the birds are the grey and other parrots, the toucan, the common royston crow, a great variety of king-

R

fishers, and many of the falcon tribe. A species of water-hen is also very numerous.

Insects (with the exception of ants,) are not numerous, there being no common flies, and very few musquitoes; some moths, and beetles.

The natives speak of a large species of snake, and some of the early catholic missionaries make mention of them from twenty to thirty feet in length, but we have seen no other reptile than the water snake which I killed in the boat, and small lizards.

The natives are, with very few exceptions, drest in European cloathing, their only manufacture being a kind of caps of grass, and shawls of the same materials; both are made by the men, as are their houses and canoes, the latter of a high tree, which grows up the river, and appears to be a species of the ficus, resembling that of the *ficus religiosa*. These vary in their size, but they appear to be generally from twenty to twenty-four feet long by eighteen to twenty, and even twenty-four inches wide. Their drinking vessels are pumpkins or gourds, and their only cooking utensil earthen pots of their own making, in which they boil or stew their meats, but more generally boil them. They take no wild animals for food, a few birds excepted, but they are very inexpert in the use of the musquet; and their natural indolence seems to suppress any fondness for the chace. Their musical instruments consist of a large drum and a kind of guitar, or rather perhaps a lyre, of which the following is a representation.

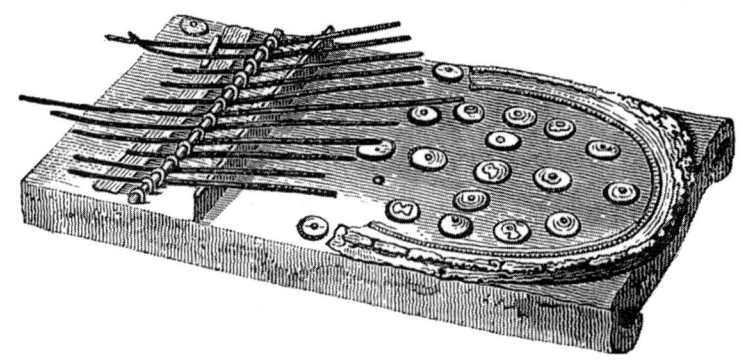

Some pains have been taken, and no small degree of labour bestowed, in collecting the materials for this instrument. The body is of wood much lighter than deal; the bridge and the eleven bars which it supports are of iron; they are confined by a strip of bamboo fixed to the body or frame by strings of leather; and they rest also on a piece of skin. The circular ornament is part of the brass frame of a print or looking-glass; and the circles are French brass buttons with the head of Louis XVI. on them. The tones are soft, and by no means unpleasing.

Both men and women shave the head in ornamental figures, according to fancy, and the brides are always close shaven before they are presented to their husbands; this operation being performed on them by an old woman. The women seem to consider pendent breasts as ornamental, the young girls, as soon as they begin to form, pressing them close to the body and downwards withal with bandages. They also sometimes file the two front teeth away, and raise cicatrices on the skin as well as the men.

The common ceremony of closing a bargain, of giving a receipt or an assurance, is by breaking a leaf, which is considered as then irrevocable; and this ceremony we found necessary to perform with the seller of every fowl.

Excepting one knife, which was stolen by a boy, we met with no instances of theft; and on one of the great men being informed of the loss in this case, the whole of the persons present were called under the great tree, and asked individually if they had taken it; when a boy confessed and produced it.

There being now a general peace in the country all the men go entirely unarmed, except when they go down the river in canoes, when there is usually a musquet in each canoe.

Among the number of their superstitions is that of refraining from different kinds of food at certain times and occasions; thus the men will not eat the flesh of a fowl until a woman has tasted of it, to take off the fetiche, as they express it. Pumpkins and eggs are objects of similar superstitions; and when we killed the cow, the king sent one of his men to take the fetiche piece, as we learnt, for the Ganga or priest; and they seemed to know the best piece, carrying off one of the hind quarters.

The two prominent features, in their moral character and social state, seem to be the indolence of the men and the degradation of the women; the latter being considered as perfect slaves, whose bodies are at the entire disposal of their fathers or husbands, and may be transferred by either of them, how and when they may please. The intriguing with a man's wife without his knowledge is however punished by a fine of two slaves; and if the adulterer cannot pay, the husband seems to be authorised to murder him.

Both men and women rise at daylight, and after washing their skins, those who pretend to gentility rub their shoulders and bodies to the waist with palm oil, which, though it keeps their skins smooth, gives even to the women, who otherwise have not the same natural effluvia as the men, a most disagreeable smell.

There are much fewer mulattoes among them than might be expected from their intercourse with Europeans, two only having yet been seen by us.

The mode of salutation is by gently clapping the hands, and an inferior at the same time goes on his knees and kisses the bracelet on the superior's ancle.

They have no other manner of reckoning or keeping an account of time than by moons; so that beyond half a dozen moons not one of them can tell the lapse of time since any event may have happened. The day they divide into morning or breakfast, noon or grand time, and evening or supper. The sea breeze was insufficient to bring the Congo up either this day or the 4th.

August 4. This forenoon I landed on the main land opposite Booka Embomma, and found it composed of very rugged hills, chiefly granite, with very little wood. An Adansonia here measured 42 feet in girth at the ground, and carried nearly the same circumference to the height of 30 feet. Where the boat anchored we found a regular tide, the rise and fall being 13 inches, and the current little or nothing during the rise.

August 5. Got the Congo up to a good anchorage on the south shore, opposite Chesalla island, where, finding we should be much retarded by persevering in the attempt to get higher, from the precariousness of the sea breezes, I ordered her to be moored, and directed Mr. Eyre, the purser, to remain in charge of her; together with the surgeon, a master's mate, and 15 men.

CHAPTER IV.

Progress up the River as far as Yellalla, or the Cataract.

CHAPTER IV.

ABOUT the middle of the day we proceeded up the river in the double boats, the transport's long boat, two gigs, and one of the punts, having with me the lieutenant, master, one master's mate, the four scientific gentlemen, and Mr. Galwey.

We found the river running between two high ridges of barren rocky hills, chiefly mica slate, with masses of quartz rising above the surface ; the slate running out in points, and the rocks under water forming strong ripplings and little eddies. In some spots, where the current has been turned aside by the rocky points, the river has deposited its mud, and formed little strips of soil covered with reedy grass, and some few little spots of Indian corn. Off these places anchorage is always found on a good clay bottom in from four to eight fathoms, a boat's length from the grass ; besides these narrow strips we this day counted several little vallies between the hills, forming the mouths of the ravines, the largest of which is named Vinda le Zally, and extends two miles along shore. In those vallies were some corn and manioc plantations, and many palm trees. The

S

two rocks named by the natives Sandy, or Zonda, are of slate; that named Oscar, by Maxwell, has a very large tree upon it, the other only brush wood; they are separated by a space of about 50 yards. On the north shore, nearly opposite these rocks, is a hanging precipice, to which may be given the name of Lover's Leap, though in a sense different from that of Leucadia, this being the place of execution of the adulterous wives of the king of Bomma, and their paramours, who are precipitated from the summit into the river.

The rocks to which Maxwell has given the little appropriate name of Scylla, lie close to the north shore, and form two masses of slate above water; about 20 yards beyond them the rippling denotes another mass under water, but on which there is six fathoms depth.

At seven o'clock the breeze failing, we anchored on the east entrance of the creek off the Gombac islands, close to the grass, in six fathoms.

Aug. 7. In the morning, it being calm, I went in the gig through the creek of Gombac, and found, though extremely narrow, that it had a depth of five to ten fathoms. There are but two islands, the separation marked by Maxwell between the two western ones being only a cove; they are mere rocks of slate with a good many trees. From them I landed on the main, and ascended the hills that form the Fidler's Elbow of Maxwell, which are also composed entirely of slate, with vast masses of quartz on the surface, and with only thinly scattered bushes of a shrubby tree, of which the natives make their spoons. These spoons are made with great neatness, and not inferior in any respect to the same utensil in many parts of Europe. Their knives, too, are not to be despised, but the blades are not always made by themselves; though they always prefer their own hafts and sheaths. These articles are here represented.

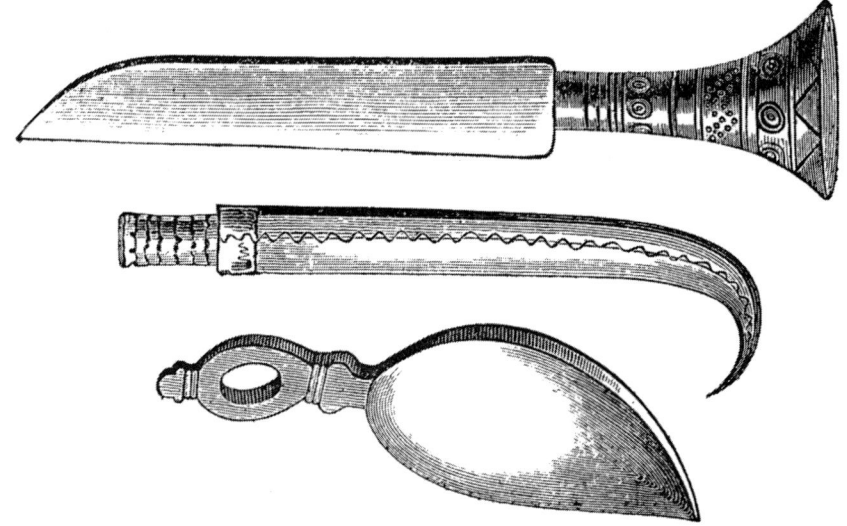

The ant hills were here extremely numerous, but now un-occupied, it appearing that these insects shift their habitations to the trees in this season; those on the ground have exactly the shape of a mushroom, consisting of a round co-lumn 18 inches high, surmounted by a domed head two feet in diameter.

We this morning observed a curious optical illusion, caused doubtless by the state of the atmosphere and the shadows of the high hills, the boats appearing to be placed on the pinnacle of an elevated mass of water, from which the descent was rapid on every side, so that in looking up the river the current seemed to be running up hill. By moonlight the reach we anchored in much resembled Loch Tay in Scotland. The rise of water in the wet season was here observed on the rocks, having worn two grooves, the first 8 feet above the present level, and the second and faintest $9\frac{1}{2}$ feet. The nature of the hills, as well as their appearance, prove that they do not absorb any of the water that falls in the rainy season, the whole of which is carried direct to the river by gullies and ravines, with which the hills are all over furrowed, and in which the only luxuriant vegetation is found.

We got a few very small shrimps from a fisherman, which he had just taken in a cotton scoop net, very well made.

At 11 we o'clock weighed with a light breeze at west, and crossed over to the south side of the river, to near the banza Sooka Congo; the Mafook of which sent his interpreter and gentlemen to ask for a bottle of brandy, which, not intending to stop near him, I did not think necessary to supply. We continued our course along shore until we reached the Diamond Rock of Maxwell, near to which, and to the south shore, we found the current too strong to be overcome with the sails and oars, and we anchored a little to the west of it. In the afternoon, however, the breezes freshened, we got through the channel, and at 7 anchored about 4 miles west of Condo Sono.

The rock called Boola Beca in Maxwell's plan, is by the natives named Blemba (the husband), and the rock named the Tinker, to the east of it, is an islet. The largest and westernmost of the three rocks named Weird Sisters, the natives call N'Casan (the wife); they lay nearest to the north shore, which, according to the natives, is all foul. The

Diamond Rock they name Salan Koonquotty, or the strong feather, alluding probably to the strength of the current, which is doubtless much encreased in the rains, it being now about $3\frac{1}{2}$ miles an hour, running directly on the rocks, and forming a strong upward eddy on its west side, where the ground is very foul and shoal, so that no vessel should attempt this channel without a breeze sufficient to ensure her going over the current.

On the summit of one of the hills which we passed close under, were upwards of twenty monkeys, which, had we not seen their tails, we should, from their great size and black faces, have taken for negroes.

Aug. 8. The hills which surround our last night's station are more barren than lower down, the strips of reedy grass and vallies less numerous, the palm trees are no longer seen, nor is there any cultivation whatever on the banks. Several persons came on board this morning from banza Noki, and from them we received the first coherent information respecting the obstruction in the river, higher up, by what they state to be a great cataract named Yellala, only one day's march from Noki.

At 10 weighed with a light breeze, and with the aid of the oars reached the spot named Condo Sono, where the European slave traders formerly transacted their business,

A Sketch near Aligator Pond.

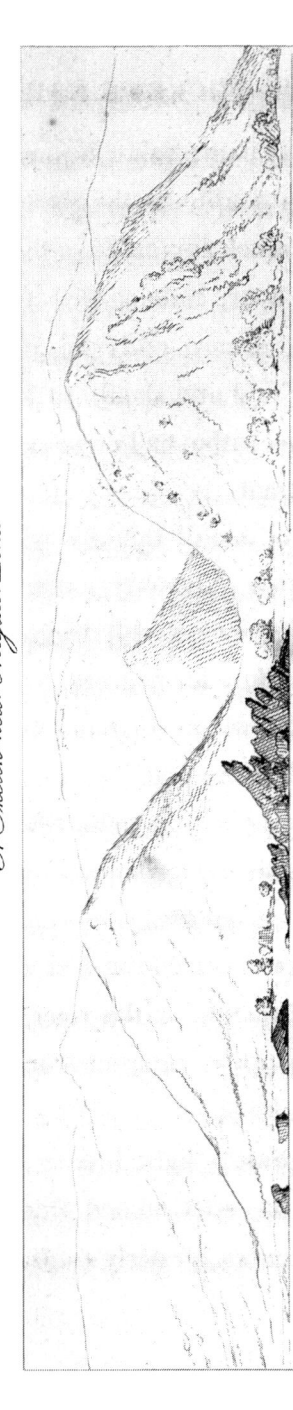

Sketch of the bank opposite the Diamond Rock.

J. Clarke sc.

Continuation of the same part of the coast.

CAPTAIN TUCKEY'S VOYAGE IN AFRICA.

From a sketch by Lieut. Hawkey.

but where there is now not a single hut. I immediately sent Simmons to the Chenoo of Noki, to request he would send me a person acquainted with the river higher up ; but on his return in the evening, I found that nothing could be done without my own presence, and the usual *dash* of a present of brandy.

August 9. I went this morning up the river in the gig, and found the difficulties encrease every mile, from the velocity of the current, and the ledges of rocks ; the barrenness of the hills also became greater, and the only trace of inhabitants was discovered in a few miserable fishermen, who take some small fish in scoop nets off the rocks and dry them.

August 10. There being every prospect that we should not be able to proceed much further in the boats, and finding that there were many Mandonzo men at Noki, whose country is situated very far up the river, and whose manners and language were described to me as totally differing from those of Congo, I this day paid a visit to the Chenoo of Noki, to endeavour to ascertain these facts, and to procure guides. We were led a two hour's most fatiguing march before we reached the banza ; sometimes scrambling up the sides of almost perpendicular hills, and over great masses of quartz and schistus, sometimes getting on pretty smoothly along their summits of hard clay, thinly scattered with brush wood, and sometimes descending into vallies covered with a rich soil and exuberant vegetation, the high and now withered grass choaking up the little plantations. In two of those vallies we found banzas, differing in nothing from that of Bomma, except that the roofs of the huts formed the segment of a circle instead of a triangle; close to them are two runs of water in ravines. At length we reached the banza, which is situated on the level summit of the highest hills amidst palm-trees, and plantations of vegetables, amongst which we were gratified with the sight of young cabbages in great perfection. In a few minutes I was ushered into the presence of the Chenoo, whom we found seated with two other Chenoos, in much more savage magnificence, but less of European manner, than the king of Bomma, the seats and ground being here covered with lions and leopard skins, the treading on which, by a subject of the highest rank, is a crime punished with slavery; and the care with which they stepped clear of them in passing to and fro, evinced that they never lost sight of the penalty. The Chenoo, besides his red cloak laced, had on his head an enormous high cap of the white feathers of the heron. One of the other kings was covered with an old hat, and the third was wrapped in a velvet mantle, and on his head a coronet, with a large

Sketch of Condo Sonio

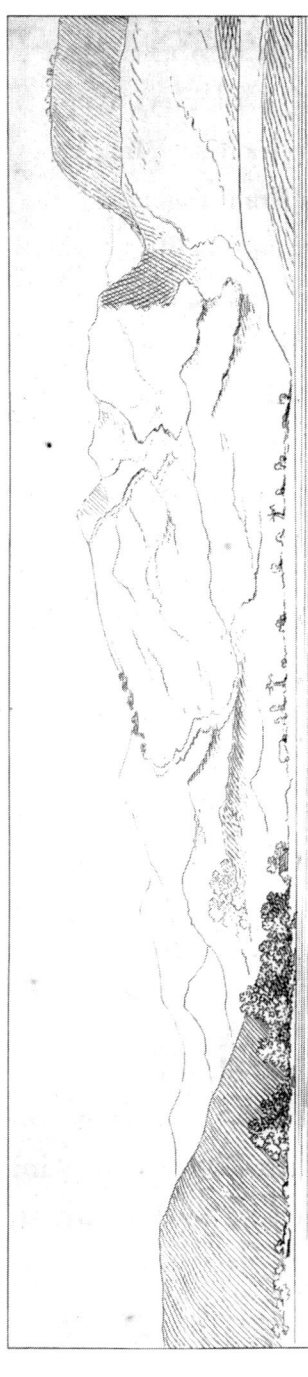

Sketch near Banza Noki.

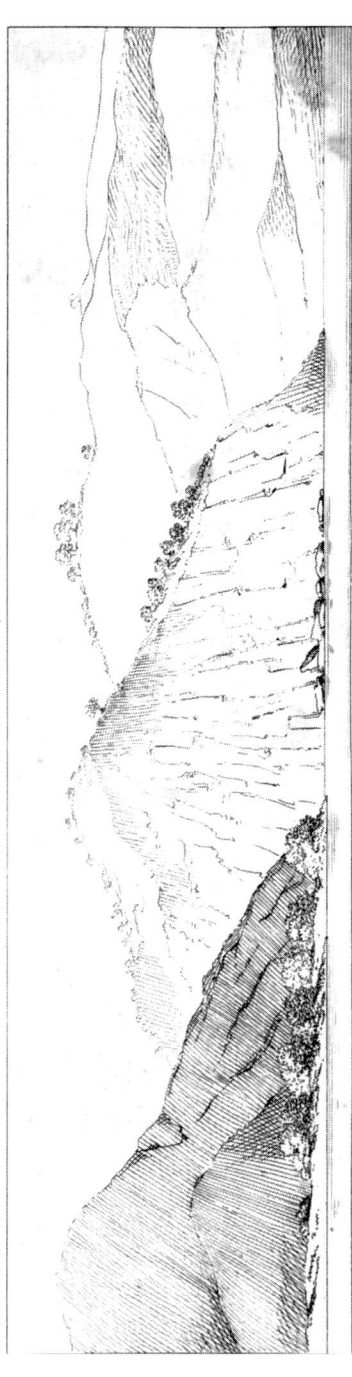

Continuation of the above.

CAPTAIN TUCKEY'S VOYAGE IN AFRICA.

Published by John Murray, Albemarle Street London, Nov¹.ˢᵗ 1817.

Page 236

button of coloured glass, which had evidently been procured from a theatre. The assembly was composed of about fifty persons squatted in the sand. Simmons having explained my wishes and the motives of the expedition, the Chenoo, with less deliberation or questioning than I had been plagued with at Bomma, granted two guides to go as far as the cataract, beyond which the country was to them a terra incognita, not a single person of the banza having ever been beyond it. The palaver being over, the keg of brandy I had brought was opened, and a greater scramble than even at Bomma took place for a sup of the precious liquor; and, towards the conclusion, one having been unable to catch a share, his neighbour, who had been more fortunate, and who had kept it, as long as he could hold in his breath, (as they always do), very generously spat a portion of this mouthful into the other's mouth! The Chenoo apologized for having nothing drest to offer us to eat, but directed a small pig to be carried to the boat, which on killing we found to be measly and unfit to eat.

We saw no women during this audience, but a considerable portion of the assembly was composed of boys of all ages down to four or five, and those young urchins were observed to pay the utmost attention to the discourse of the men, and to express their approbation by clapping their hands.

T

On our return we were conducted by a slave merchant of Simmons' acquaintance, by a road at first much more pleasant than that we had come, being along the summits of hills which are highly fertile, and in great part cultivated, but in the most careless manner. The vegetables we saw were manioc, Indian corn, a species of shrubby holcus, French beans, cabbages or greens, ground nuts in great quantity, and bird pepper. The fruit consisted of limes, papaws, and plantains, all at present immature.

Our conductor led us to his town (for every man of property calls his residence his town), where we were agreeably surprised to find a repast prepared, consisting of a stewed fowl, a dish of stewed beans, and cassava bread named Coanga. The stews were however so highly peppered that our gentlemen, not accustomed to such warmth of seasoning, could scarcely swallow them ; a bottle of spirits, in which some aniseed had been infused, was also set before us. The remains of our repast was served to the marines. The water brought to us issued from a rock, its temperature 73° ; the barometer at the banza fell two inches lower than at the river side, which, according to Leslie's scale, gives the elevation about 1300 feet.

While at our repast, the back ground of the court in which we were, was filled with women and girls, separated

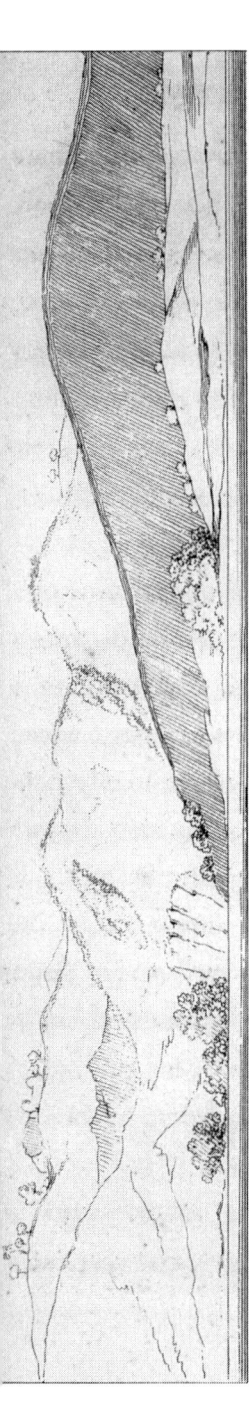

Slate Hills near Noki.

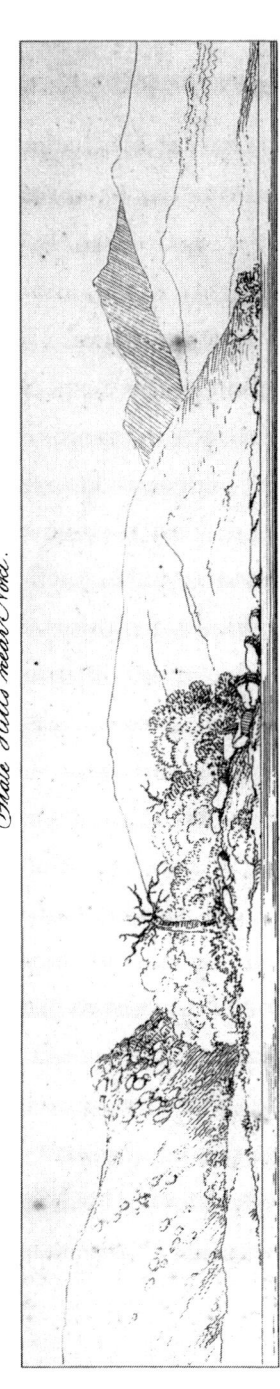

Sketch of Toobi Kakke Gonzo.

Sketch near Sandee River, bordered with masses of Slate.

CAPTAIN TUCKEY'S VOYAGE IN AFRICA.

From a sketch by Lieut Hawkey.

J.Clarke sc.

by a space from the men and boys. Though not one of them had ever seen a white man, they did not seem to feel any timidity, but on the contrary we had abundant opportunity of discovering that, as far as depended on themselves, they were perfect Otaheitans in their manners. One woman we saw spinning cotton for a fishing net exactly in the manner the French women do while tending their sheep.

The latter part of our journey, as we again approached the river, was even worse than our road going, being obliged to ascend and descend a succession of hills, of smooth rocks, so nearly perpendicular, that it required almost the legs of flies to crawl over them, and here the natives had greatly the advantage of us, the soles of their naked feet seizing hold of the rock, while our shoes slipped over them and threatened us every moment with a fall that would not have been without danger ; at length we reached the boats at 4 o'clock extremely fatigued, though the distance of the banza is not more than three miles in a direct line from the river, but by the circuitous route we took, could not be less than seven or eight.

The most striking features of the country we passed over are the extreme barrenness of the hills near the river, the whole being still composed of slate with masses of quartz, and sienite, the latter becoming the main formation, as we advanced to the S.E. with perpendicular fissures from three inches to $\frac{1}{4}$ inch in breadth, filled with quartz. The summits of the hills and the vallies are of stiff clay and vegetable mould extremely fertile. We did not see the smallest trace of any thing calcareous, nor the signs of any other metal than iron.

August 11. We had no visitors until near noon, when four women came to the river's side, opposite the boats, to make market, having a single fowl, half a dozen eggs, and a small basket of beans to sell; we were soon told that the oldest was a princess of the blood in her own right, and that consequently she enjoyed the privilege of choosing her husband and changing him as often as she liked, while he was confined to her alone, under penalty, if a private person, of being sold as a slave. This lady, after getting for her fowl and other articles twice their value, offered herself and her three companions (who, we were assured, were the Chenoo's daughters) for hire, to whoever would take them on board the boats, and seemed to be much disappointed at the apathy of white men, when they found their advances treated with neglect.

In the afternoon a couple of small sheep, a goat, and a few fowls were brought for sale ; but for one of the former the owner had the conscience to ask a full piece of blue baft and two caps, which cost 30 shillings in England, making the meat considerably above a shilling a pound ; so that we were obliged to confine ourselves to the purchase of a goat for four fathoms of printed cotton. Indeed, from the very little spare provisions the natives seem to have at this season, I do not think it would be possible to procure daily subsistence for fifty men in passing through the country. Towards evening two men were sent from the Chenoo as guides for Yellala, but one of them having evidently never been there, I sent him back. A Mandingo slave man was brought to me, bound neck and heels with small cords. His answers to the questions put to him were, " that he was three moons coming from his country, sometimes on rivers, sometimes by land ; that his own country was named M'intolo, on the banks of a river as broad as the Zaire, where we were at anchor, but so filled with rocks, that even canoes could not be used on it ; and that he had been taken when walking a short distance from his father's house, by a slave catcher, who had shot him in the neck with a ball, the cicatrice of the wound still remaining ; and that he had been about two years from his country." Although his reckoning of the lapse of time could not be depended on, he evidently had not been long caught, for he spoke the Congo language but very imperfectly ; nevertheless, as he understood enough of it to make himself intelligible, I thought he might be of use and purchased him, giving an order for his value to be paid on board the Congo, and taking care to explain to the natives my motives for buying him, as well as that I gave him his liberty on the instant, and only considered him as a servant; and finally, that if we arrived at his country, I would restore him to his friends.   When this was explained to him he expressed not the least mark of satisfaction, and permitted the people to take off the cords which had served to bind him with apparent apathy; indeed our people seemed to have more satisfaction in performing, than he felt in undergoing, this operation. In concluding this bargain, I had a specimen of the tedious manner of doing business amongst the native traders, the intervention of the Mafook, Mambouk, and a broker, being necessary between me and the seller; and each of these fellows expected two fathoms of baft, and as much brandy as they could drink. This roundabout way of trading, and the indecision of the sellers, must, I should suppose, have been a great draw back on the profits of the slave trade, by the time it kept them on the coast; for I am assured, that though fifty slaves may be brought to market in a day, not three are usually sold. The same huckstering is indeed visible in every branch of their trading with Europeans; the possessor of a single fowl, or a root of the manioc, examining the articles offered him fifty times, giving them back, taking them again, exchanging them for something else, and after putting patience to the test for an hour, often taking up his goods and marching off, because he could not get twice or thrice what he first asked.

Aug. 12. At nine this morning we weighed, and with the aid of the oars, and a track rope at times, got the boats up along the south shore, until we came to a large sand bank extending two-thirds across the river; here we crossed over to the other side, and ran along it as far as a little island named — — — — — — — Here we found the current so rapid, that with a strong breeze and the oars we could not pass it; besides, having observed when up here in the gig, that the north shore above this island was extremely foul, I crossed over, and after considerable difficulty succeeded in getting to an anchor in a fine little cove named Nomaza, entirely out of the current. In crossing the river we passed several whirlpools, which swept the sloop round and round in spite of her oars and sails, and not without some danger to so low and deep laden a boat. These vortices are formed in an instant, last but a few moments with considerable noise, and subside as quickly. The punt got into one of them and entirely disappeared in the hollows, so that the depression of the vortex must have been three or four feet. The schooner could not succeed in passing Zoonga Tooley Calavangoo, and anchored on the opposite shore; but a very strong sea breeze springing up in the evening she joined me.

In the afternoon I went on shore and ascended the highest hills under which we were anchored, and whose elevation might be 500 feet. From hence our upward view of the river was confined to a single short reach, the appearance of which, however, was sufficient to convince us, that there was little prospect of being able to get the double boats up much farther, and none at all of being able to transport them by land. Both sides of the river appeared to be lined by rocks above water, and the middle obstructed by whirl-pools, whose noise we heard in a constant roar, just where our view terminated by the closing in of the points. High breakers seemed to cross the river; and this place we learnt was called Casan Yellala, or Yellala's wife, and were told that no canoe ever attempted to pass it. The most distant hill, whose summit appeared above the rest at the distance of perhaps 7 or 8 leagues, we found was that of Yellala. The appearance of the river here was compared by Dr. Smith to the torrent rivers of Norway, and particularly the Glommen, the hills on each side being high, precipitous towards the river, totally barren, and separated by such deep ravines as to preclude the idea of conveying even a canoe

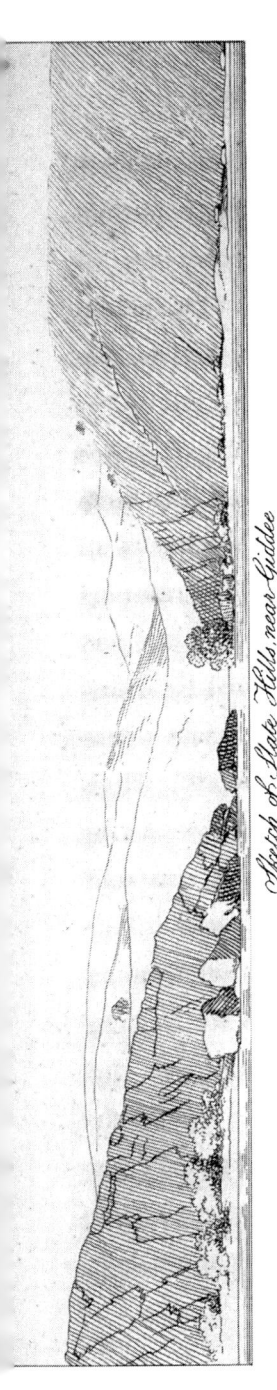

Sketch of Slate Hills, near Guddee

Continuation of the above

Continuation of the above

London, Published by J. Murray, Albemarle Street, 1877.

over them, without immense labour. Two tufts of trees on the summits of the northern hills, we understood from a fisherman, were the plantations round the banzas. The only other information we could get from him was, that Yellala was the residence of the evil spirit, and that whoever saw it once would never see it a second time.

This has been the only tolerable clear day since our entering the river ; the sun being visible both at rising and setting, and the thermometer at two o'clock at 80°. This heat produced a breeze in the evening stronger than any we before experienced, and which continued all night. On a little sand beach, off which the boats were anchored, there is a regular rise and fall of water of eight inches ; during the rise, the current is considerably slackened.

Aug. 13. This morning at daylight I went up the river with the master in the gigs, to ascertain the utility of carrying the boats any farther. By crossing over from shore to shore as the current was found slackest, we found no difficulty in getting up to Casan Yellala, which is about three miles above where the boats lay. We found it to be a ledge of rocks stretching across the north shore about two-thirds the breadth of the river (which here does not exceed half a mile), the current breaking furiously on it, but leaving a smooth channel near the south shore, where the velocity of

U

the current seems the only obstacle to the ascent of boats, and that I should consider as none to my progress with the boats, did there appear to be the smallest utility in getting them above it.   But as the shore on either side presents the most stupendous overhanging rocks, to whose crags alone the boats could be secured, while an impetuous current flows beneath; and as every information makes Yellala a cataract, of great perpendicular fall, to which the approach is far easiest from the place near which the boats are now anchored in perfect safety, I determined to visit this cataract by land, in order to determine on my future operations.

Accordingly at 8 o'clock on the morning of the 14th I landed on the north shore, in a cove with a fine sandy beach, covered with the dung of the hippopotamus, exactly resembling that of the horse.   My party consisted of Messrs. Smith, Tudor, Galwey, and Hodder, and 13 men, besides two Embomma interpreters (the Chenoo's sons), and a guide from Noki, with four days provisions.   Our route lay by narrow foot paths, at first over most difficult hills, and then along a level plateau of fertile land; in short, over a country resembling that between the river and Noki. Our course lay between E. N. E. and N. E.   At noon we reached Banza Cooloo, from whence we understood we should see Yellala.   Anxious to get a sight of it, I declined the Chenoo's invitation to visit him, until my return. On the farthest end of the banza we unexpectedly saw the fall almost under our feet, and were not less surprised than disappointed at finding, instead of a second Niagara, which the description of the natives, and their horror of it had given us reason to expect, a comparative brook bubbling over its stony bed. Halting the people, who complained of fatigue, I went with the gentlemen to examine it more closely, and found that what the road wanted in distance, which was not a mile from the banza, it abundantly made up in difficulty, having one enormous hill to descend and a lesser one to climb, to reach the precipice which overhangs the river.

The south side of the river is here a vast hill of bare rock (sienite), and the north a lower but more precipitous hill of the same substance, between which two the river has forced its course; but in the middle an island of slate still defies its power, and breaks the current into two narrow channels; that near the south side gives vent to the great mass of the river, but is obstructed by rocks above and under water, over which the torrent rushes with great fury and noise, as may easily be conceived. The channel on the north side is now nearly dry, and is composed of great masses of slate, with perpendicular fissures. The highest part of the island is 15 feet above the present level, but from the marks on it, the water in the rainy season must rise 12 feet, consequently covers the whole of the breadth of the channel, with the exception of the summit of the island; and with the encreased velocity, must then produce a fall somewhat more consonant to the description of the natives. In ascending two hills we observed the river both above and below the fall to be obstructed by rocks as far as we could see, which might be a distance of about four miles. Highly disappointed in our expectations of seeing a grand cataract, and equally vexed at finding that the progress of the boats would be stopped, we climbed back to our people, whom we reached at four o'clock totally exhausted.

The principal idea that the fall creates is, that the quantity of water which flows over it, is by no means equal to the volume of the river below it; and yet, as we know there is not at this season a single tributary stream sufficient to turn a mill, below the fall, we can hardly account for this volume, unless we suppose, as Dr. Smith suggests, the existence of subterranean communications, or caverns filled with water.

After having refreshed and rested ourselves, I waited on the Chenoo with a little brandy, and found less pomp and noise, but much more civility and hospitality, than from the richer kings I had visited. This old man seemed perfectly satisfied with our account of the motives of our visit, not asking a single question, treating us with a little palm wine, and sending me a present of six fowls without asking for any thing in return. In one of the courts of his tenement we had the disagreeable sight of two men slaves prepared for sale, one· having a long fork stick fastened to his neck, and the other with *European-made irons* on his legs ; on enquiry I found that there were 14 slaves in the banza for sale, who were going to Embomma.

The night was cool, the thermometer at one o'clock being at 60° ; in the early part the stars shone brightly, but towards morning it became very cloudy ; and at daylight we might easily have fancied ourselves amidst the blue misty hills of Morven.

August 15th. In the morning we were surrounded by all the women of the banza with fowls and eggs to exchange for beads. In the eggs we were however taken in, more than half we purchased having been taken from under the hens half hatched. Although the largest banza we have seen, we could not procure either a sheep, goat, or pig for the people.

Having engaged a guide to lead us above the falls (the hills close to the river being absolutely impassable by any thing but goats), we quitted the banza at seven o'clock, and after four hours most fatiguing march we again got sight of the river; but to my great vexation, instead of being 12 or 13 miles, as I expected, I found we were not above four miles from Yellala, our guide having persuaded me out of my own judgment, that the river wound round in a way that made the crescent we took necessary. Here we found the river still obstructed with rocks and islets sometimes quite across, but at one place leaving a clear space, which seems to be used as a ferry, as we found here a canoe with four men; no inducement we could offer them had however any effect in prevailing on them to attempt going up the stream, which I wished to do to examine the state of the river more exactly.

In this day's journey we crossed three deep ravines, the beds of torrents in the rainy season, but now quite dry, and but once found water at a very small spring. One antelope's skin was seen with the natives, and the dung of these animals occurred in many spots; several porcupines' quills were also picked up. On quitting the river I determined to cross the hills in a direction that I expected would again bring me to it considerably higher up, but the setting sun obliged us to halt on the side of a steep hill, at whose foot we fortunately found a fine spring, forming the only brook we had yet seen; and here we passed the night, which was much warmer than the preceding one, the thermometer not falling below 70°, the sky cloudy, but not the least dew. The constant dryness of the atmosphere is evinced in the quick drying of all objects exposed to it ; meat hung up a few hours loses all its juices, and resembles the jerked meat of South America; the plants collected by Dr. Smith were fit for packing in a day, while, towards the mouth of the river, he could scarcely get them sufficiently dry in a week. The oxidation of iron also entirely ceases here. The hygrometer at sun-rise usually marks 50°, at two o'clock in the afternoon in the shade 70°.

August 16. Finding Mr. Tudor and several of the people were unable to proceed farther, I sent them back to Banza Cooloo in the morning, and with the remainder proceeded onwards. Passing the brook and ascending the hill on the opposite side, we found ourselves on a level plateau at the banza Menzy Macooloo, where we again got a guide to lead us to the river. At noon we had a view of it between the openings of the hills, about two miles distant. Here the people, being extremely fatigued, were halted, while the natives went to the river for water, and I ascended the highest of the hills, which descends perpendicularly to the river: from its summit I had a view about five miles down the river, which presented the same appearance as yesterday, being filled with rocks in the middle over which the current foamed violently ; the shore on each side was also scattered with rocky barren islets. The river is here, judging by the eye, not more than $\frac{1}{4}$ mile broad, and I estimate the distance from Yellala at 12 or 14 miles. Upwards my view was stopped by the sudden turn of the river from north to S. E.; the concavity of the angle forming a large bay, apparently freer from obstructions than below. I descended a most precipitous path to the river side, where I found four women fishing with a scoop net; they had no canoe, and I learnt that persons wanting to cross the river are obliged to go from hence to the ferry above Yellala. Just where the river shuts in, in turning to the S. E., on a high plateau of the north shore, is the banza Inga, which we understood was two days march from Cooloo (though its direct distance is not above 20 miles,) and that it is out of the dominions of Congo. The only other information I could get here was, that the river, after a short reach to the S. E., turned again to the north; and the appearance of the hills seemed to corroborate this information; but as to the state of its navigation, or the possibility of getting canoes, I could not acquire the slightest notion.

On the return of the people with water from the river we dined where the men halted, and set out on our return for the boats; at eight o'clock we reached Cooloo (having this day walked ten hours), where we found Mr. Tudor in a violent fever.

It was by great persuasion we could get our guide to go on after sunset, through his fears of wild beasts, and his superstitious terrors of the night combined ; and every five minutes he sounded a whistle, which it seems had been *fetiched* by the Gangam kissey; and consequently both spirits and beasts fled at the sound. The only traces of animals we saw this day were the foot prints of buffaloes, who had been to the brook we crossed in the morning to drink ; and we were surprised how so bulky an animal could ascend the hills ; indeed the marks shewed that in descending they had sliden considerable distances on their hind legs. A wild hog also crossed us, making from thicket to thicket; but our men were not fortunate enough to hit him, though four of them fired.

Having crossed in a direct line from one of the fertile plateaus to the river, we had a good opportunity of seeing the formation and structure of the country near its banks ; of which I have attempted to give an idea below.

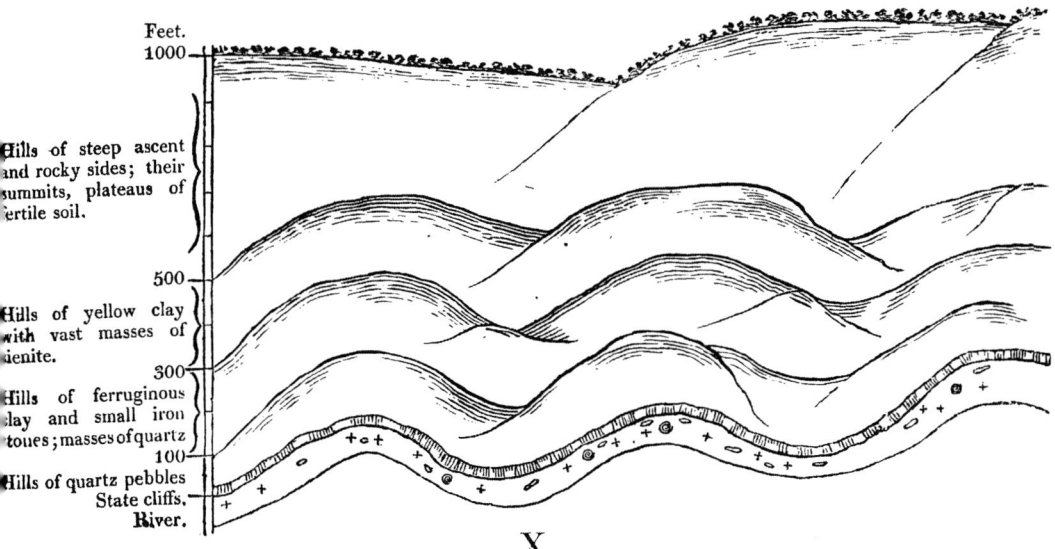

Feet.
1000

Hills of steep ascent and rocky sides; their summits, plateaus of fertile soil.

500

Hills of yellow clay with vast masses of sienite.

300

Hills of ferruginous clay and small iron stones; masses of quartz

100

Hills of quartz pebbles
Slate cliffs.
River.

X

The margin of the river and the rocks in it are of micaceous slate; next to which small hills of loose round
pebbles of quartz; then more elevated hills of ferruginous
clay with masses of quartz; next hills of yellow clay with
masses of sienite; and lastly, steep hills with rocky sides,
(sienite), but with long levels at the summit, covered with a
fertile vegetable soil, and on which most of the banzas are
situated. The hills of quartz pebbles have every appearance
of once having been the bed of the river, the pebbles having
evidently been rounded, and the hills received their forms,
by the long and violent action of water. If we assume, what
seems highly probable, that the present obstruction at Yellala was once a real cataract of equal elevation with the opposite shores, this surmise of the river having formed these
hills of pebbles will acquire a certainty; for in that case, the
water kept up by the cross ridge or cataract would have
covered these hills. A great quantity of quartz sand is
thrown up on the sides of the river among the rocks.

At night the hills appear to be in a continued blaze of fire,
from the hunters in the day setting fire to the long dry grass
to drive out the animals. The fire running to windward,
as is always the case, the hunters keep to leeward of the spot
fired, and the game, it would appear, being also aware of
the direction which the fire will take, endeavour to avoid it,
by also running to leeward, and consequently throw them-

selves within the hunter's reach. The guns we have seen among them are of Portuguese or French manufacture, and of a great length ; each has usually several fetiches fastened to it, to prevent it from doing the owner any injury.

Aug. 17. In the morning we prepared for our return to the boats, when Mr. Tudor, being too ill to move, was obliged to be carried in a litter made of one of the black men's blankets ; the other invalids, exceeding half the party, were sent forward with him under charge of Mr. Hodder, while I went to take leave of the Chenoo. Having performed this ceremony, and given him two fathoms of bafts, I quitted Cooloo, came up with the invalids in a short time, and reached the boats at noon, when a tent was immediately pitched on shore for Mr. Tudor, who was now in a most violent fever. The complaints of the other men were confined to fatigue and blistered feet.

During this excursion we seldom met with water to fill our canteens (that of the river excepted), more than once a day, and the springs were generally very small. The only provisions we could procure were a few fowls and eggs, some cassada root, green plantains, and beans ; but all in quantities totally insufficient to supply the daily expenditure of 20 men. We did not see a single sheep or goat, and not above two or three pigs. Palm wine, which we found infinitely more refreshing and grateful to the stomach, when heated and fatigued, than either wine or spirits and water, is at this season extremely scarce, owing to the long drought; so that, though every banza and gentleman's town is surrounded by these trees (from 20 to 200 at each), we often could not procure it even in exchange for brandy. It appears that the rainy season, for these last two years, has been very moderate, and the lighter rains, that usually happen in June, have been entirely wanting this year, which accounts for the burnt-up appearance of the country, and the very little water. It is however expected by the natives, that the ensuing rainy season will be proportionally violent; and they are now preparing for it, by fresh covering and repairing their huts. They say that every third or fourth year the river rises considerably higher than in the intermediate ones; and this accounts for the different elevation of the marks on the rocks.

Thus far the banks of the river do not afford a single timber tree capable of making a beam or timber for a sloop of war. The only trees that grow to a large size are the Adansonia and the Bombax, (or wild cotton), and the wood of both is spongy and useless; several varieties of evergreens, highly ornamental in their growth and foliage, are however met with in the vallies.

The only appearance of metals is in the ferruginous clay and stones near the river, which the natives grind, and of these form their pots for boiling (their only cooking utensil); these small stones rounded to the size of a pea, serve them for small shot. Small particles of copper were observed by Dr. Smith in some of the specimens of minerals he collected.

During my absence the seine was hauled, but not a single fish was taken. The only implements of fishing seen with the natives were the scoop-net, already spoken of, and a kind of fish pot of reeds. The fish we could procure from them are all very small, with the exception of one which I bought, and whose skin Mr. Cranch preserved. It appears to be of the genus *Murena*. Testaceous fish are extremely few; a single shell *(Helix)* found on the summit of one of the hills near a fishing hut, and an oyster taken up by the dredge, are all we have been able to procure.

On the banks we have found snares for taking beach birds, nearly resembling those used in Ireland to catch snipes, being an elastic twig with a bit of line and noose, which catches the bird's neck.

The higher we proceed the fewer European articles the natives possess; the country grass-cloth generally forms the sole clothing of the mass of the people, and gourds are substitutes for glass bottles, or earthen mugs. The women too approach nearer to a state of nudity ; their sole clothing being a narrow apron (the breadth of the hand and 18 inches long) before and behind, so that the hips on each side are uncovered. From every town near to which we passed they flocked out to look at the white men, *(moudele)* and without any marks of timidity came and shook hands with us. To the best looking and the best drest I distributed some beads. The price paid here, by a native, for a wife of the first class, the Chenoo's daughter for instance, is four pieces of baft, one piece of guinea, and a certain quantity of palm wine. We in no instance since we left Embomma found the men *allant en avant* in their offer of their women ; but this our Bomma men told us was from their little intercourse with Europeans, for that any of them would think himself honoured by surrendering his wife or daughter to a white man.

The population seems to be extremely thin, and (leaving out the insignificant number of fishermen who remain on the rocks at the river side) is collected into banzas and gentle-mens' towns ; the largest of the former (Cooloo) does not contain above 300 souls, of whom two-thirds are women and children. The extent of fertile land is, however, capable, with very moderate industry, of supporting a great increase of population, not the one-hundredth part of that we have passed over being made any use of whatever. The plateaus appear to be well adapted for wheat, and certainly all the garden vegetables of Europe might be produced here in perfection, as well as potatoes.

According to our informers, the dominions of Congo extend from below Malemba, cutting the coast and river to Banza N'Inga; but how far they extend to the south of the river's mouth, or up it on the south side, we could not learn, but it seems to be considerably higher up the river than Inga. The paramount sovereign is named Lindy, or Blindy N'Congo, and resides at Banza Congo, six days journey in the interior from Tall Trees (on the south side of the river); it has no water communication with the Zaire. Here the Portuguese appear to have a fixed settlement, the natives speaking of their having soldiers and white women.

The opposite sides of the river form two vice royalties, that on the north being governed by the N'Sandy N'Congo, and the south by N'Cucula Congo, both of whom reside at banzas in the interior.

The Chenooships, improperly named kingdoms by Europeans, are hereditary fiefs, passing in the female line; that is, on the decease of the Chenoo the succession, instead of passing to his son, goes to his brother, or uterine uncle

or cousin.   On every demise a fresh investiture takes place
by the viceroy's sending a cap (here the mark of all dignity)
to the appointed successor ; but though it is necessary that
the succession should be continued in the family, the viceroy is not restricted to nearness of kin or primogeniture,
but as favour, corruption or intrigue operates strongest, the
investiture is given.   The Chenoo, in his turn, appoints several inferior officers by sending them caps, particularly the
Mafook, or custom master, who interferes in all trading
transactions.   The Mombella, Macaya and Mambom, are
officers whose respective powers I have not yet been able to
ascertain with any certainty.   Slavery is here of two kinds,
which may be denominated household or domestic, and
trading.   When a young man is of age to begin the world,
his father or guardian gives him the means of purchasing a
number of slaves of each sex, in proportion to his quality,
from whom he breeds his domestic slaves, and these (though
it does not appear that he is bound by any particular law)
he never sells or transfers, unless in cases of misbehaviour,
when he holds a palaver, at which they are tried and sentenced.
These domestic slaves are, however, sometimes pawned for
debt, but are always redeemed as soon as possible.   The
only restraint on the conduct of the owners, towards their
domestic slaves, seems to be the fear of their desertion;

for if one is badly treated, he runs off, and goes over to the territory of another Chenoo, where he is received by some proprietor of land, which inevitably produces a feud between the people of the two districts. The trading or marketable slaves are those purchased from the itinerant black slave merchants, and are either taken in war, kidnapped, or condemned for crimes ; the first two of these classes, however, evidently form the great mass of the exported slaves ; and it would seem that the kidnapped ones (or as the slave merchants who speak English call it " catching in the bush"), are by far the most numerous. This practice however is certainly unknown at present on the banks of this river as far as we have yet proceeded.

The property which a man dies possessed of devolves to his brothers or uterine uncles, but prescriptively, as it would appear, for the use of the family of the deceased ; for they are bound by custom (which is here tantamount to our written laws) to provide in a proper manner for the wives and children of the deceased ; and the wives they may make their own, as in the Mosaic dispensation.

Crimes are punished capitally by decapitation, by gradual amputation of the limbs, by burning and by drowning. The only capital crimes, however, seem to be poisoning, and adultery with the wives of the great men. This latter

Y

crime, it would appear, being punished in proportion to the rank of the husband. Thus a private man accepts two slaves from the aggressor; but the son of a Chenoo cannot thus compromise his dishonour, but is held bound to kill the aggressor; and if he escapes his pursuit, he may take the life of the first relation of the adulterer he meets; and the relatives of this latter, by a natural re-action, revenging this injustice on the other party, or one of his relations, is one of the grand causes of the constant animosities of the neighbouring villages. If a man poisons an equal, he is simply decapitated; but if an inferior commits this crime (the only kind of secret murder) on a superior, the whole of his male relations are put to death, even to the infants at the breast.

When a theft is discovered, the gangam kissey or priest, is applied to, and the whole of the persons suspected are brought before him. After throwing himself into violent contortions, which the spectators consider as the inspirations of the kissey or fetiche, he fixes on one of the party as the thief, and the latter is led away immediately to be sentenced by a palaver. Of course the judgment of the priest is guided either by chance, or by individual enmity; and though (as our informer assured us) the judgment was often found to be false, it derogates nothing from the credit of the gangam, who throws the whole blame on the kissey.

The frequency of the crime of putting poison in victuals, has established the custom of the master invariably making the person, who presents him with meat or drink, taste it first; and in offering either to a visitor, the host performs this ceremony himself; this the natives, who speak English, call " taking off the fetiche."

Both sexes paint themselves with red ochre; and, before a bride is conveyed to her husband, she is smeared with this substance from head to foot. The men also make marks on their foreheads and arms with both red and white clays; but the only answer we could get to our enquiries respecting these practices was, that they were done by order of the gangam kissey.

Besides a prevalent cutaneous disorder or itch, several cases of elephantiasis were observed, and two patients with gonorrhœa applied to our doctor; this, however, they told us was a present from the Portuguese.

The only game we have seen them play at was a kind of drafts named looela, the implements of which are a flat stone 18 inches square with 16 cavities grooved in it, and a small stone in each cavity, as in the annexed figure.

The impossibility of procuring information to be at all depended on from the natives, respecting the course of the river or the nature of the country, proceeds equally from their want of curiosity, extreme indolence, and constant state of war with each other. Hence, I have never been able to procure a guide farther than from banza to banza, or at the utmost a day's journey; for at every banza we were assured that, after passing the next, we should get into the Bushmen's country, where they would be in danger of being shot or kidnapped. Thus at Cooloo, it was only by the promise of handsome pay, and still more through the assurance of safety offered by our muskets, that I could prevail on a guide to promise to accompany me to banza Inga.

All my endeavours to find a slave trader who knew something of the river have been fruitless. One man at Cooloo presented himself, and said he had been a month's journey from that place, but always travelled by land, except in the passage of several rivers by canoes and fords ; the direction of his course appeared to be to the N. E. and the country, according to him, more mountainous than where we are. Indeed it appears that the people of Congo never go themselves for slaves, but that they are always brought to them by those they call bushmen, who, they say, have no towns nor acknowledge any government. All however agree in asserting that the country on the south is still more difficult than that on the north, which, together with there appearing to be no traces of the Portuguese missions on the latter,* as well as the river again taking a direction to the north, induces me to prefer this side for my farther progress.

Aug. 19. In pursuance of my intention to endeavour to

---

* At Noki the crucifixes left by the missionaries were strangely mixed with the native fetiches, and the people seemed by no means improved by this melange of Christian and Pagan idolatry.

get as far as possible by the north bank of the river, I sent this morning Lieut. Hawkey with eight men, to form a depôt of provisions at Cooloo.

Aug. 20. Sent Mr. Fitzmaurice with eight men, with a second proportion of provisions, to proceed to Cooloo, where I learnt Mr. Hawkey arrived late last night. I shall to morrow proceed with 14 men for the same place, from thence for banza Inga, sending the boats down to rejoin the Congo.

CHAPTER V.

***

Progress from the Cataract, or Cooloo, by Land chiefly, to the Termination of the Journey.

CHAPTER V.

HAVING arranged matters at the double-boats, I quitted the river with the remainder of my party at 11 o'clock, and reached Cooloo myself at two; but for want of hands was obliged to leave part of the burthens behind, until the people could be sent back for them. Visited the Chenoo, who gave me a fine sheep, and promised me a guide, and some men to assist in conveying our things to Inga.

Aug. 21. With the usual delay, I waited this whole day for the guide and bearers, without either making their appearance; and in the afternoon I received a visit from the Chenoo, when I found that the delay proceeded from my not having given presents to the Mambom, Macaya, and half a dozen other gentlemen. As the giving way to such pretensions would very soon exhaust my stock, I positively refused all their solicitations for the moment, promising, however, that when I returned I would treat them as their conduct to white men should have deserved. After a long palaver, in which the disappointed party was extremely violent, the more moderate remained masters of the field,

Z

and it was determined that the Chenoo's son should accompany me as a guide the next morning.

The night scene at this place requires the pencil to delineate it. In the foreground an immense Adansonia, under which our tents are pitched, with the fires of our people throwng a doubtful light over them ; before us the lofty and perpendicular hills that form the south side of Yellala, with its ravines (in which only vegetation is found) on fire, presenting the appearance of the most brilliantly illuminated ampitheatre ; and finally, the hoarse noise of the fall, contrasted with the perfect stillness of the night, except when broken by the cry of our centinels " all's well," continued to create a sensation to which even our sailors were not indifferent.

The conclusion of the night I however found not so pleasant; awaking extremely unwell, I directly swallowed five grains of calomel, and moved myself until I produced a strong perspiration.

Aug. 22. Though still very unwell, I had every thing prepared to be off at day-light with half the party for Inga, intending to leave the remainder at Cooloo, until I had tried the practicability of advancing. It was however ten o'clock before I could get the guide, and six men or bearers, to each of whom I was obliged to give two fathoms of baft and three strings of beads.   I now found that Prince Schi (alias Simmons) had deserted, and taken with him four of the best men I had brought from Embomma, as porters; and just as I was setting off, the Chenoo and all his possé came to me, to let me know that my interpreter had violated their customs, and his own word, having bargained with two of the head gentlemen for their wives (one, the first time I was at Cooloo, and the other the night preceding), for two fathoms a night, which having no means of paying, he had concealed himself, or ran off to Embomma.   Though sufficiently irritated, I could not forbear smiling at their manner of relating the circumstance, as well as at their expectations, that I would either pay to the husbands the stipulated price, or permit them to seize Simmons, and sell him as a slave.   In order to avoid either of these alternatives, I promised that, on my return, I would arrange the affair amicably; with which, after a long palaver, which cost me some brandy, they were obliged to be satisfied.

Having thus lost my interpreter, I was obliged to offer very high terms to the only person with me, who could supply his place; a man whom we had picked up at Embomma, and employed as one of the boats crew, but who, having been in England five years, spoke the language as well as Simmons, and his own much better; I therefore at once promised him the value of a slave and other *etceteras* on my return, if he would accompany me; to which he at last acceded, all his countrymen attempting to dete·him, by the idea of being killed and eaten by the bushmen.

We reached banza Manzy (about nine miles north of Cooloo), at noon, the whole road being along a plateau. Here I was obliged to give four fathoms of baft for a pig of 15lb. weight; and after he was paid for, the people pretending they could not catch him, I was obliged to direct him to be shot. At four came to a very deep ravine (*Sooloo en-vonzi*), the bed of a vast torrent, covered with rocks, slate and quartz, in the hollows of which a considerable quantity of excellent water still remained, apparently since the rains. The sides of the ravine were thinly cloathed with wood, among which were trees perfectly straight from 80 to 100 feet high, and 18 inches diameter, the wood of which was nearly the density, and had much the appearance, of oak; they were the only trees we had yet seen of any utility.

In this ravine we halted for the night, on learning that we could not reach Inga, and that there was no water between where we were and that place. The country we passed over this day would appear to be from 8 to 12 miles from the river, and is more hilly and barren than any we have yet gone over; but the same structure, on a larger scale, appears to prevail, as that of which I have given a representation.

Our night scene here, though entirely dissimilar from the preceding one, was perfectly theatrical ; the trees completely shading the ravine, and the reflection of the fires on the tent, and on the foliage, and on the rocks, with the mixture of black and white men, each cooking his supper, might have exercised the pen of Salvator Rosa, and would give no bad idea of the rendezvous of a horde of banditti.

In the morning we found we had pitched our tent over a nest of pismires ; but although we were covered with them, not a person was bitten, any more than by the musquetoes, who, from its shade and humidity, had chosen this as their head quarters. At day-light we were roused by the discordant concert of a legion of monkeys and parrots chattering, joined with that of a bird named by the natives *booliloso*, (a crested Toucan) having a scream between the bray of an ass and the bleat of a lamb ; another, with a note resembling the cuckoo, but much hoarser ; and another crying " whip poor will" (a species of goatsucker). We also found that several buffaloes had been to drink at one of the holes in the rocks, about 200 yards from one of our tents.

Aug. 23. At seven o'clock (having given the people their usual breakfast of cocoa), we set out, and crossed a most difficult tract of hills and ravines until 11, when we found ourselves just at the angle of the river, formed by its returning to the S. E. ; this last reach not appearing to be more than three or four miles, but entirely filled with rocks, and absolutely, as far as we could judge with our glasses, without the smallest passage or carrying place for a canoe. At noon we reached banza Inga, having turned off to the west considerably from the river, and found it situated on one of the usual plateaus.   The Chenoo, we learned, was blind, and that the government was in a kind of commission, composed of the Macaya, Mambom, &c. which portended me no good; a palaver being immediately assembled to know what white men came here for.   I now found it would be necessary to deviate from my former assertions of having nothing to do with trade, if I meant to get forward ; and accordingly I gave these gentlemen to understand, that I was only the fore-runner of other white men, who would bring them every thing they required, provided I should make a favourable report of their conduct on my return to my own country.   At length I was promised a guide to conduct me to the place where the river again became navigable for canoes, but on the express condition that I should pay a jar of brandy, and dress four gentlemen with two fathoms of baft each.   These terms I complied with, stipulating on my part that the guide should be furnished immediately, (as this part of the river was said to be not above half a day's journey from Inga), and he was accordingly brought forward. Depending on these assurances, I proposed, as soon as the people should have dined, to set off, but was now informed that I could not have a guide till the morning. Exasperated by this intolerable tergiversation, being unable to buy a single fowl, and having but three day's provisions, I remonstrated in the strongest manner, and deviated a little from my hitherto patient and conciliating manners, by telling them, that if they did not furnish a guide, I should proceed in spite of them, ordering at the same time the ten men with me to fall in under arms ; at the sight of which the palaver broke up, and it was *sauve qui peut*. The women and children, who had flocked to see white men for the first time, disappeared, and the banza became a desert ; on enquiring for the men who had come with me from Cooloo, I also found that they had vanished with their masters ; in short, I was left sole occupier of the banza. Finding that this would not at all facilitate my progress, I sent my interpreter with a conciliating message to the Macaya, whose tenement was outside the banza, and which shortly produced the re-appearance of some men, but skulking behind the huts with their musquets. After an hours delay, the regency again appeared, attended by about fifty men, of whom fourteen had musquets. The Mambom, or war minister first got up, and made a long speech, appealing every now and then to the other (common) people who were seated, and who all answered by a kind of howl. During this speech he held in his hand the war kissey, composed of buffalo's hair, and dirty rags; and which (as we afterwards understood) he occasionally invoked to break the locks, and wet the powder of our muskets. As I had no intention of carrying the affair to any extremity, I went from the place where I was seated, opposite to the palaver, and familiarly seating myself along side the Macaya, shook him by the hand, and explained, that though he might see I had the power to do him a great deal of harm, I had little to fear from his rusty musquets; and that though I had great reason to be displeased with their conduct and breach of promise, I would pass it over, provided I was assured of having a guide at day-light; which was promised, on condition that the gentlemen should receive eight fathoms of cloth.

The people here had never before seen a white man, and the European commodities we saw were reduced to a little stone jug and some rags of cloathing. The language is a dialect of that of Embomma, but considerably differing. The Chenoo receives his cap from the Benzy N'Congo, who resides ten days journey to the N. W. and not on the river.

We purchased half a dozen fowls, but were obliged to pay for water, at the rate of three beads for a canteen. There is here a good deal of lignum vitæ, the largest seen about four inches in diameter.

Aug. 24. Though the guide was promised at day-light, I found that the people of the banza wished to throw every obstacle in the way of our proceeding, assuring us, that the people further on would shoot us from the bushes, &c. &c. which produced the effect of making the men that had brought our things from Cooloo refuse to proceed any further. At length I was under the necessity of secretly promising one of the gentlemen a piece of baft for his good offices; when he immediately offered himself as a guide, and five of his boys to carry our provisions. Leaving therefore every thing but these and our water, under the care of the Cooloo men, we at last set off, at eleven o'clock. At the end of the banza we passed a blacksmith at work, fitting a hoe into a handle; his bellows was composed of two skin bags, and his anvil a large stone. The progress seemed very slow, the iron never being brought to a red heat. Our route lay chiefly along the winding bottom of a valley between two ridges of hills; the valley generally very fertile, but now without water, though furrowed by extremely deep beds of torrents. In the valley we found two towns, sur-

A a

rounded by plantations of manioc growing almost to the size of trees, A flock of 20 to 30 goats was a novel sight; but the master being absent, we could not purchase one. The women sold us some manioc, and gave us a jar of water. At the upper end of the valley we found a complete banza of ant hills, placed with more regularity than the native banzas; they were very large, and had the shape of a mushroom, but sometimes with double and treble domes, the latter evidently intended to carry off the water in the rains. At four o'clock we reached the river at Mavoonda Boaya, where we found it still lined with rocks and vast heaps of sand, but free from all obstruction in the middle, from two to three hundred yards wide; the current gentle (not above two miles an hour), and a strong counter current running up on the north shore; its direction N. W.

The Macaya of Mavoonda being told of our arrival, visited us in a few hours, and was very civil, and seemingly rejoiced to see white men; in return for his civility and his palm wine, I gave him a cotton umbrella. The information received here, of the upward course of the river, was more distinct than any we have yet had; all the persons we spoke to agreeing that, after ten days in a canoe, we should come to a large sandy island, which makes two channels, one to the N.W. and the other to the N.E.; that in the latter there is a fall, but that canoes are easily got above it; that twenty days above the island, the river issues by many small streams from a great marsh or lake of mud.

Having thus ascertained that the river again becomes navigable at the distance of about 20 to 24 miles above Yellala, I endeavoured to ascertain if I could procure canoes, and was assured I might purchase them at banza Mavoonda. During the night we had two smart showers of rain, which, as we were bivouacked in the open air, wetted us through. In the morning we returned to Inga, whence I have sent Lieut. Hawkey back to Cooloo, with 14 men of this place, to bring up the provisions and presents; intending immediately to procure two canoes and proceed up the river. Mr. Cranch being ill, takes this occasion of returning to the Congo.

The country declines considerably in proceeding to the north.

Aug. 25. All the European articles procured from Embomma, which is the emporium of the Congo empire, and might once be considered as the university for teaching the English language, and breeding up factors for the slave trade, are now brought thither by the Portuguese, and consist chiefly of the coarsest of English cottons, aqua ardente of Brazil, and iron bars. Beads are only taken in exchange for fowls, eggs, manioc and fruits, which seem all to belong to the women, the men never disposing of them without first consulting their wives, to whom the beads are given. The fashion varies from day to day, but the mock coral, and black and white, seem to be most constantly in request.

Each village has a grand kissey or presiding divinity named Mevonga. It is the figure of a man, the body stuck with bits of iron, feathers, old rags, &c. and resembles nothing so much as one of our scare crows. Each house has its dii penates, male and female, who are invoked on all occasions.

A slave from Ben's country, (Soondy) was this day brought for sale ; it appeared that he had been pledged for debt, and not being redeemed, was accordingly to be sold. His information respecting his country, which appears to be that of Ben, and a large district, is, that it is a long way up the river; that he came down to Inga, sometimes by water, sometimes by land, and was in the whole 25 days on his journey ; but as he passed through the hands of a great many traders, and their days journey are very short, it is probable that it might be done in ten days. As Ben however had forgotten the name of his town, he could gain no information ; and indeed his knowledge of the language is so imperfect that he has never been of the smallest use as an interpreter.

From a sketch by Lieut Hawkey.

W.Finden sc

CAPTAIN TUCKEY'S VOYAGE IN AFRICA.

Fishermen inhabiting the Rocks of the Lovers Leap.

Page 130.

Published by John Murray, Albemarle Street, London, Nov.ʳ 1.ˢᵗ 1817.

Before marriage, the fathers or brothers of a girl prostitute her to every man that will pay two fathoms of cloth; nor does this derogate in any way from her character, or prevent her being afterwards married. The wives are however never trafficked in this manner except to white men of consideration.

The boys are taken from the mothers as soon as they can walk, and the father sits the whole day with them on a mat. The girls are entirely neglected by the father.

Whenever any thing brings a number of people together, the men immediately light a fire and squat themselves round it in the smoke; the men and boys together, the women remaining behind separate.

The *ficus religiosa* is planted in all the market places, and is considered here, as it is in the East, a sacred tree; for our people having piled their muskets against one, and some of the points of the bayonets sticking into the bark, a great clamour was raised until they were removed.

The hoe is their only instrument of husbandry, and is made out of a piece of flat bar iron beat out and stuck into a handle from one to two feet in length, as in the following figures

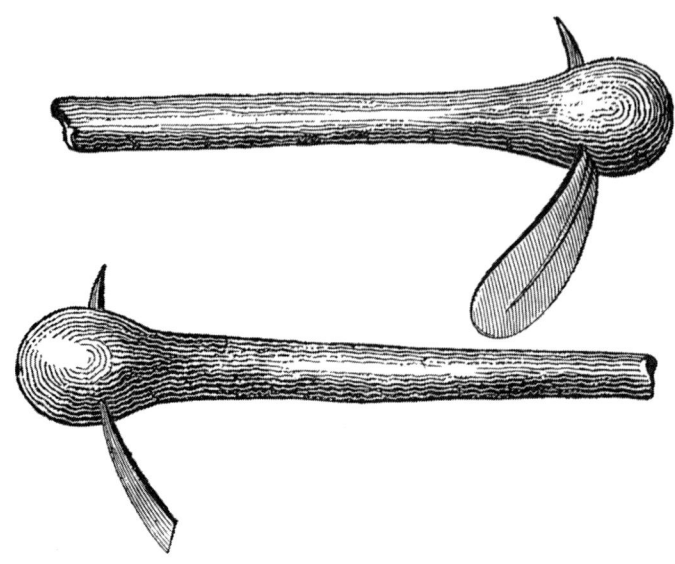

At Kincaya, in the valley of Bemba Macongo, we met with an Embomma slave merchant going into the interior for slaves. No information was to be got from him respecting the river, pretending that he had never been up it. It is evident these merchants do not wish Europeans to penetrate into the country, lest they should interfere with their business. Here the cicatrices or ornamental marks on the bodies of both men and women were much more raised than in the lower parts of the River. The women in particular had their chests and belly below the navel embossed in a manner that must have cost them infinite pain, the way of effecting it being to seize the skin between the fore finger and thumb, and scarify it longitudinally with a sharp knife; and when this is done so deep as to draw the blood, the juice of a plant is applied as a styptic, and the deeper the cut is, the more raised is the cicatrice.

When the natives first saw the new moon, they hailed it as the precursor of the showers of rain which they expect soon after its close (that is to say, when the sun is on the equator) : they then begin to sow their Indian corn and ground nuts. The heavy rains do not set in for six weeks later. The river begins to rise a month before the rains, that is immediately after the season of showers has set in. I find it useless to attempt talking about business until the palaver is finished ; that is to say, until the palm wine they have brought, and the brandy they expect to receive, are all expended, by which time they are usually half drunk, and their inclination to palaver, to lie, to chicane, seems to encrease in proportion to the quantity they may have swallowed.

This morning I received a visit from a Foomoo with a large calabash of palm wine, who, after sharing it out among *his own people*, expected I would, in return of the compliment, give him a jar of brandy; but as his sole business seemed to be curiosity, and I had little brandy left, I did not choose to comply; and in order to prevent further solicitation, I told him I had none, with which, though much discontented, he was obliged to be satisfied. At noon I returned to Inga.

August 26. While Lieutenant Hawkey was gone to Cooloo to bring up the luggage, I visited the river from Inga, being about a mile distant. The reach here runs east and west for three or four miles, and in that distance had three successive falls, the highest Songa Yellala. On my return I found a present of four chickens from the blind Chenoo, with a request that I would give him four fathoms of baft in return ; but as I deemed half that quantity to be four times the value of his present, I sent him two fathoms, which were soon after brought back to me, with a violent complaint that I had given four fathoms each to the Mambok and Macanga, and that the king expected the same; as the former was however an exaction before I could procure guides, I did not think it necessary to comply with the Chenoo's demand, but taking back the two fathoms sent him nothing, and in a very short time another message came from him requesting he might have the two fathoms, which I again delivered to be given to him.

A gangam kissey passed through the banza attended by his clerk or drum-beater, with all the instruments of his profession, viz. a big drum, a parcel of calabashes filled with small stones, a piece of tree, and a dozen stinking fetiches. We found that he had been sent for to a neighbouring village to discover the cause of a man's death.

I this day visited the valley of Kincaya, where I was told a man had canoes to sell; but he was gone into the country. The structure of the valley we found to consist of a vast mass of slate, the strata dipping 45° to the S.W. The hills on each side were also composed of slate, with masses of quartz. Here I purchased some smoked fish. Here also I again met the Embomma slave trader. The manner of conducting the traffic in slaves, we found to be thus: The slave merchant quits Embomma with three men, each carrying a jar of brandy, and a piece of cloth; on a bargain for a slave being concluded, a jar of brandy is then drunk, and a proportion of the cloth is given to the Chenoo and other great men as presents; the seller then sends one of his own men back with the trader to receive the price of the slave at Embomma, or at any intermediate place that the feuds or other impediments to tranquillity may render expedient.

August 27. This morning the gangam kissey returned, and we learnt that he had denounced three men of another village as the poisoners of the man that died, and that the accused were immediately to undergo the ordeal of chewing poisonous bark, which, if they were guilty, they would retain in the stomach, and thus it would occasion their death; but if innocent, they would vomit it up again

B b

immediately. On enquiry if the gangam did not sometimes undergo the revenge of the persons thus falsely denounced, I was told that such a thing was unknown, for that the accused considered the kissey to be only to blame; and that moreover the gangam could not be hurt, the kissey always forewarning him of danger. Even my interpreter, who had been baptised and lived five years in England, expressed his firm belief in these notions; which, by the way, are not more ridiculous than the augury of the Romans, or the inspirations and beatific visions of certain christians. The gangams do not appear to be numerous, the one abovementioned having come from a considerable distance. Each gangam has usually a novice with him, whom he initiates into the mysteries of the profession, to which he succeeds on his death. The manner of initiation is kept, as may be supposed, a profound secret from the people. Their pay consists of the country money, of which this fellow had received a large bundle.

The impediments to communication from the nature of the country, and the want of rivers, appear to be the great obstacles to the civilization of Africa. The abolition of the slave trade, though it will produce little or no effect on the state of domestic slavery, (which is not incompatible with a high degree of civilization,) must in the end tend greatly to improve Africa, by rendering the communication between different parts of the country free from the danger of being kidnapped, which now represses all curiosity, or all desire of the people of one banza to go beyond the neighbouring one. Every man I have conversed with indeed acknowledges, that if white men did not come for slaves, the practice of kidnapping would no longer exist, and the wars, which nine times out of ten result from the European slave trade, would be proportionally less frequent. The people at large most assuredly desire the cessation of a trade, in which, on the contrary, all the great men deriving a large portion of their revenue from the presents it produces, as well as the slave merchants, who however are not numerous, are interested in the continuance. It is not however to be expected that the effects of the abolition will be immediately perceptible; on the contrary, it will probably require more than one generation to become apparent : for effects, which have been the consequence of a practice of three centuries, will certainly continue long after the cause is removed ; and in fact, if we mean to accelerate the progress of civilization, it can only be done by colonization, and certainly there could not be a better point to commence at than the banks of the Zaire.

CHAPTER VI.

Excursion from Inga, and from thence to the Termination of the Journey.

CHAPTER VI.

LIEUTENANT HAWKEY having returned with a part of the presents for the chiefs, and provisions for the journey, I this morning (28th) determined to set out for Mavoonda to bargain for some canoes; and was just on the point of departure, when the Macayo paid me a visit, and informed me, that if I purchased any canoes, I would find myself taken in, for that at a day's journey above Mavoonda the river was again obstructed by a fall named Sangalla, over which I should not be able to get the canoes : this information at once throwing me again into as great uncertainty as ever, I determined, instead of going to purchase canoes, to visit this Sangalla without delay. It was however some time before I could procure a guide, when the old gentleman, who had led me to Mavoonda, again offered himself for half a piece, whom I was forced to accept. As he assured me we should be back at night, I took only Mr. Galwey and four men, with a very short day's provisions, not wishing to open a fresh case of preserved meat (which is now our sole resource for the people). After passing through the valley of Bemba, we ascended the hills that line the river, and which are more fatiguing than any we had yet met with, being very steep, and totally composed of broken pieces of quartz, resembling a newly made limestone road. At four o'clock we came in sight of the river, between the hills; and instead of getting back at night, I found it would be nearly dark before we could reach Sangalla, and, as I expected, we only reached it at sunset. Including the windings of the river, I suppose it about ten miles above Mavoonda; the intermediate reach running due north, studded with several islands, but the stream not very strong. At Sangalla the river is crossed by a great ledge of slate rocks, leaving only a passage close to the foot of the hill on the left bank about fifty yards wide, through which the stream runs at least eight miles an hour, forming whirlpools in the middle, whose vortices occupy at least half the breadth of the channel, and must be fatal to any canoe that should get into them. About two miles lower down the river the stream breaks quite across over a sunken ledge of rocks. Above, the river forms a wide expanse east and west, but filled with rocky islets; the great breadth however reduces the velocity of the stream, so that canoes easily pass. About two miles above the commencement of the narrow channel there is a ferry.

Having examined every thing here, and being told by our guide that there was a banza not much higher up, where we might get some victuals, I proceeded towards it, scrambling over the rocks with infinite fatigue for an hour, and then penetrating through a close wood (the first we have seen) near it, until it became quite dark. This seemed to be the haunt of buffaloes, whose dung fresh dropped, still smoked. At length about eight o'clock we reached the landing place; and by the light of the moon proceeded over new hills towards the spot where our guide supposed the banza to be situated. I soon however discovered that he had lost his way; and seeing a fire on the side of a hill, and hearing human voices, I desired him to enquire; but the people were afraid to come to us, and we could not find any path to get to them. After some time spent in halloing, we understood from them that the banza was deserted.

As we had neither victuals nor water, and nearly choaked with thirst, it was necessary we should endeavour if possible to procure the latter; and after an hour's walk one of the men came down from the hill, and conducted us through thick underwood, where we were almost obliged to crawl, and through grass twice our own height, to a spot clear of wood on the side of the hill; and finding it useless

C c

to go any further, we made a fire to dry our cloaks, which were literally soaked with perspiration. A little water brought us by the wives of these bushmen, for they had no hut, was our supper, and the broken granite stones our bed. The water was a strong chalybeate. The night was however fine, though cold, so that our bivouac, for want of our coats, which, on the expectation of being back the same evening, we had not brought, was not over comfortable; and at five o'clock in the morning of the 29th I quitted it to take a view of the river. One of the bushmen informed us, that after a short reach to the eastward it again ran to the south, and then turned back to the north, pointing out the hills and a banza, named Yonga, round which it turned; and according to his account, after two days journey in a canoe higher up, another Sangalla occurred, worse than the first. We also learnt that the banza, which we intended to have gone to the evening before, had been deserted for some time; the people, it seems, had robbed some slave merchants returning from Embomma with their goods, and fearing the consequences, had all taken to the bush. After a small portion of roasted manioc and a draught of water for breakfast, we proceeded on our return to Inga; and, having climbed a tremendous hill which hangs over the river, we came to three or four huts, where a woman had the conscience to ask us a fathom of cloth for a small fowl. We had however the good fortune to procure a calabash of palm wine, a little further on, without which we should scarcely have been able to continue on our march ; the sun, after nine o'clock, becoming extremely powerful; even with this, it was with the utmost difficulty I could prevail on the people to push on, the road being absolutely impracticable for a man with any burden ; and it was four o'clock before we reached Kincaya. Here I found the greatest difficulty in getting any thing to eat ; at last, however, we procured an old hen and some manioc, which, stewed up together, gave us a scanty repast; and after an hour's rest we set off for Inga, which we reached at seven o'clock, equally to our own satisfaction and that of our companions, who, expecting us back the first evening, had feared lest some accident had befallen us. On both days we saw great numbers of deer of two difent species, one evidently an antelope ; the other a large animal of the deer kind, of which was a herd consisting of thirty or forty. They seemed not to be very shy, but were too far off for our shots to take effect. In two ravines which we passed, we observed rather more rapid streams. The country to the eastward was low.

This excursion convinced us of the total impractibility of

penetrating with any number of men by land, along the sides of the river, both from the nature of the country, and impossibility of procuring provisions.

On the 30th, I sent Lieutenant Hawkey to Voonda to endeavour to hire canoes, to enable us to go up to the first Sangalla, being determined to make an attempt by water, though with little hopes of success.

Where there are neither written annals, legends, nor ancient national songs, nor chronology beyond a month, the history of a nation must be very vague and confined. The only idea I have been able to obtain of the Congoese history, is, that Congo once formed a mighty empire, the chief of which had three sons, between whom he divided his dominions at his death, giving to one the upper part of the river on both sides as far as Sangalla; to a second, the left bank of the river (the Blandy N'Congo), and to the third, the right bank, Banzey N'Yonga.

The Congoese are evidently a mixed nation, having no national physiognomy, and many of them perfectly south European in their features. This, one would naturally conjecture, arises from the Portuguese having mixed with them; and yet there are very few Mulattoes among them.

The creeping plants serve for cordage; some of which are not less then six inches in diameter. Fleas and bugs swarm in all the huts. A great scarcity of wood fit for building prevails in this country. The stony hills about this part are thinly clad with scrubby trees, which are fit only for fuel; in many places they resemble an old apple orchard.

The mornings are calm. The breeze sets in from the westward at noon, and is proportionably strong to the heat of the day, and when the sun has been very hot, continues strong during the night; the days and nights however are both very cloudy, so that it is impossible to get any observation even in three or four days.

The hoop by which they ascend the palm trees is formed of a moist supple twig.

The idea of civilizing Africa by the sending out a few Negroes educated in England, appears to be utterly useless; the little knowledge acquired by such persons having the same effect on the universal ignorance and barbarism of their countrymen, that a drop of fresh water would have in the ocean.

The scarcity of food at this time is extreme. The sole subsistence of the people being manioc, either raw, roasted, or made into coongo, and of this they have by no means an abundance; and a very few green plantains. A bitter root, which requires four days boiling to deprive it of its pernicious quality, is also much eaten.

The indolence of the men is so great, that if a man gets a few beads of different colours, he stops at home (while his wife is in the field picking up wood, &c.), to string them, placing the different colours in every kind of way till they suit his fancy.

They have songs on various subjects, love, war, palm wine, &c.

They have no other arms except knives and a few musquets; no shot, but small rounded stones: a piece of quartz makes a good flint. They take fish by poisoning them with a species of narcotic herb. They make good lines with grass.

They amuse themselves with a game which is played on a piece of board, having twenty-eight circular hollows on its surface; but I could not learn the principles of the game.

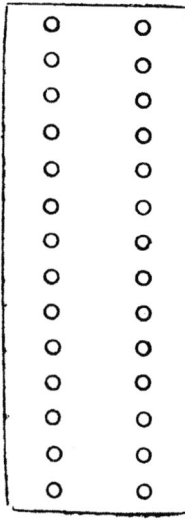

The palm wine is obtained by making an orifice a little above the insertion of the lowest branch or leaf, into which a tube is inserted for conducting the liquor into a calabash. These calabashes are extremely useful for all manner of household purposes, and are of all sizes ; the small ones are used for holding their snuff, or dust of dried tobacco leaves, and are generally ornamented with figures, sometimes cut in high and sometimes in low relief, as under.

The canoes are made far up in the country ; and it is said to occupy one man three months, in the making of one ; they are not however very dear.

Aug. 31. Lieut. Hawkey returned, having been unable to

procure a single canoe; indeed it did not appear that there is more than one or two, for crossing the river at Voonda. I sent some natives to Cooloo to bring up the remainder of the provisions, and with an order to Mr. Fitzmaurice to return to the Congo sloop with fifteen of the men, being unable to feed them at Cooloo. With the party thus reduced I am about to proceed to Bamba Yanzy, three days journey, where, according to all accounts, the river ceases to be obstructed with rocks or cataracts.

[*From this time the Journal consists only of short notices; but the dates and progress are regularly inserted, and accompanied with a Sketch of the River, the direction, length, and width of its several reaches, and strength of its current, as far as the extreme point to which they proceeded.*]

Sept. 1. Great astonishment of the people at seeing the pocket compass and the watch; especially the pocket compass always pointing to the same spot of the river.

Their houses would be convenient if a little larger, and if raised from the ground on posts to keep out the fleas and bugs.

It appears that the bark used in the ordeal is from a species of cassia; and the bitter root used as food is a kind of yam. Some Guinea fowls were killed, and some partridges of a very large kind. Plenty of bees in the rainy season, when the numerous flowers are in blossom. The natives eat wild honey whenever they find it, but are never at the trouble of searching for it; nor do they know the value of wax. Sweet-scented flowers smelling like the jasmine and jonquils.

The end of the rainy season is unquestionably the best time for a botanist.

This day for the first time observed that the river had begun to rise.

The names of the seasons are as under. The dry season, or winter, is called *Gondy assivoo*; it is from April to September inclusive.

The first rains are called *Mallola mantity;* they fall in light showers once or twice in 24 hours; commence the latter end of September, and continue to the middle of October. At this time the women prepare the ground with the hoe for planting Indian corn, beans, manioc, &c.

The second rains are called *Voolaza mansanzy;* commence in November and end in January; they are represented as being very heavy, attended with great heat, but with few tornados. They now plant Indian corn, &c. which ripens in three months.

The third rains are named *Voolaza chintomba,* and continue during the months of February and March. The rain still

D d

very heavy, with violent tornados, and storms of thunder, lightning, fiery weteors, and wind.

The species of *dolichos*, named voando, is ripe enough for gathering in the month of October. The *bozea* or *saphoo* also ripens in October. The *chichila* (papaw) plantain.

Macaya and Mambouk bought six fathoms of chintz and a bottle of brandy

Sept. 2. Mr. Galwey was to day very ill; sent Mr. Fitzmaurice orders to wait for him at Cooloo.

At eight o'clock, with much difficulty, got nine bearers for three fathoms each and three strings of beads. Set off with eight white men, Dr. Smith, Mr. Galwey, and three black men. No other road, it seems, but the horrible one we went before. At one P.M. reached a stream, the only one seen that was large enough to turn a mill; there we dined. Pratt was knocked up, and obliged to be assisted forward by four men.

I had some dilemma with the bearers, from the impatience of our people: the only way to manage them seems to be that of threatening to stop their pay; no use in personal intimidation, as they know they can run off; nor in promises, for, like children, they are only afraid of being deprived of what they have in actual possession, or think themselves sure of.

At five P. M. passed a second brook, but the bearers telling us there was another further on, we continued our march till six. Pitched the tent, and sent the bearers for water; but they returned with empty vessels, saying the brook was dried up; and most provokingly seated themselves with the utmost indifference round the fire; and tho' we had not a drop of water, they would not move until I put the above method in practice, when six of them went back to the brook we had left.

Some showers of rain now fell. In the middle of the night we found ourselves attacked by a swarm of ants, which fairly beat us out of the tent, as we preferred the wet grass to the torment they occasioned.

Sep. 3. Pratt being unable to proceed from his fever, I sent him back to Inga with two of the bearers; and at eight A. M. set off with the rest; the most horrible road, through narrow gullies not 18 inches wide and six feet deep, which had been formed by the rains excavating the rocks.

On the summit of one of the hills we met a caravan of slave traders going to Embomma, consisting of 30 men (eight with muskets), the rest loaded with cassava and ground nuts, some of which they kindly gave us. One man and four boy slaves were from the Soonda country: all said they were taken in the bushes. One of the boys made the most violent screams on seeing us : even the children of seven or eight years old held fast by the hand of their owners, while we were present.

We dined at a brook, named Sooloo Loo Anzaza ; and at three P. M. arrived on the bank of the river, a little to the eastward of the upper Sangalla, which is formed by a ledge of rocks running quite across the river, part of it near the right bank being now dry, and the stream close to the other side, forming another and a larger Yellala, or cataract. The direction of the river is here N. E. ; and at the distance of about a mile above this Sangalla it greatly expands, forming an elbow, and running into a creek called Condo Yonga, and then it turns S.E. ; the current about three miles an hour, with a strong eddy on one side. Some rocks are still seen in the river, but the appearance is much less rugged than below, the land on each side being lower, tho' still hilly and very barren, with no trees whatever. The formation also changes at this spot ; the pebbly quartz hills having almost entirely given way to clay and ferruginous earth ; and the rocks which line the river are now a perfect clay slate.

The constant setting fire to the grass must prevent the growth of trees, and render them scrubby by destroying their leaves.

Many hippopotami were visible close to our tents at

Condo Yanga, where we were obliged to halt, and to wait some time for a canoe to pass. No use firing at these animals in the water; the only way is to wait till they come on shore to feed at night. During the night, they kept a continual grunting like so many hogs, but none of the mcame on shore, though we had a constant watch on the beach.

This point of the river is the place of all others to set out from on an expedition to explore the course of the river; the creek offering a very fine place for boats, and the strand being an excellent spot for an ecampment.

Sept. 4. After much difficulty we obtained two canoes to ferry us over the creek, for which service I paid four fathoms and six strings of beads. As soon as they were loaded, the Inga men desired to go back, on pretence of being afraid to proceed; but as they had engaged to go to Bamba Yonga, the fear of loosing their wages at last induced them to pass over it. They had not however walked above a mile on the other side of the creek before they laid down their loads, and again refused to go on; and in this manner they plagued me until noon, putting down their loads every ten minutes, walking back fifty or sixty yards as if to return, taking them up again, and so on, with a palaver of half an hour between each stoppage. Finding I could get no good of them, I finally halted on the eminence that bounds the river, about eight miles above our last night's station, where I hope to procure a couple of canoes to proceed up the river, as far as my means will permit.

The river here expands to three miles in width; the banks have in some places low strips of soil and sand, with cliffs of clay slate, large masses of which are also scattered in the river, but do not in the least impede its navigation. In other parts low hills of gentle descent come down to the margin of the river; their summits clay, entirely bare of trees. Here we also found considerable masses of fine blue limestone; and a quantity of chalk was brought to us by one of the natives, which we were told was procured from hills on the opposite side of the river.

The population is here more considerable than what we have seen lower down. The *gentlemen's* towns forming a continued chain of buildings from the spot where we landed. Provisions seem to be also more plentiful. Several goats, pigs and fowls, with manioc, ground-nuts, and peas, having been brought to us for sale, as were also mats.

Amongst the croud that surrounded our tent, two or three of the foomoos or gentlemen only had any European clothes. The women were here the most horribly dirty looking wretches that can be conceived; equalling the New Hollanders in filth, and nothing superior to them except in the mere article of covering their nakedness. The huts were still of the same formation as below. The palm trees were more abundant. Fish very plentiful, and of several species unknown, excepting that one was a small bream; they take them with pots, having neither nets, hooks, nor lines.

The great encrease in the number of Paria dogs denotes an encreased stock of provisions; though it would appear, from their being half starved, like those of an Indian bazar, that they were not well fed: they never bark, but howl like a jackal; they have pricked ears.

Sept. 5 I discovered to-day that the Inga men were determined to stay here for my return, being, as they asserted, afraid to go back themselves.

Finding it impossible to get canoes without the interference of the Chenoo of Yonga, I sent forwards one of the black men (the guide of Inga,) with a piece of chintz, as a present to be divided among his great men. Having given to the interpreter and to my other Embomma man a dress of chintz each, they amused us by performing *Songa*, which is a kind of war dance, and a hunting dance, a pantomime, and a love dance. In the war dance, the performer, with a sword, looks about from side to side as if expecting the enemy; at last he sees them, flourishes his sword half a dozen times towards the quarter in which they are supposed to appear; advances; his eyes glowing fire; returns triumphant; while the spectators are clapping their hands and striking their breasts in turns; he then squatted down.

The only presents made by Europeans concerned in the slave trade, were brandy, musquets, and powder, all promoters of civilization and encouragers of population.

A foomoo waited on me with a present of a goat, for which I gave two fathoms of chintz and a necklace, and he promised to procure me two canoes at day-light the following mornining.

Dr. Smith made an excursion along the bank of the river to the northward, limestone and clay slate alternating.

The women braid their hair, and in this state it looks well; but when the braiding is undone, horridly frightful, like a black mop.

Alligators are so plentiful hereabouts, and so frequently carry off the women, who at daylight go down to the river for water, that while they are filling their calabashes, one of the party is usually employed in throwing large stones into the river outside of them. Here I learn that they have two crops of Indian corn a year.

Several instances have occurred which shew the minute divisions of property : three or four people have usually a share in a goat; and even a fowl is seldom without two

owners. The division of the price usually creates a dispute, if there should happen to be an odd bead.

Sept. 6th. After a constant battle with the natives from
daylight, and after using every possible means, by threats,
persuasions, and promises, I at last, about two o'clock,
got the canoes, which I was to have had at daylight, under
way, having first paid eight fathoms, and given a present of
two fathoms of chintz each to four different gentlemen.

As the canoes, though promised to carry twenty men
each, would barely hold eight, I was obliged to walk along
shore with a part of the people. About three miles from
the place of departure we passed two small rapids, but the
other side of the river was clear.

We came to a bay in which were ten hippopotami ; as
the canoes could not venture to come on until these huge
creatures were dispersed, we were obliged to fire vollies at
them from the shore, and although many shot evidently
told, not one of the beasts seemed in the least to be affected by them. The noise however, together with the
exorcism of our old guide, drove them away.

The river now, for the first time, bore a majestic appearance, having the land on each side moderately elevated,
with little hills of lime-stone further back, but still almost
without wood.

E e

At six P. M., after having rounded a rocky peninsula, we anchored in a fine bay named Covinda, formed by this and another rocky peninsula a little higher up. The night presented a beautiful picture of repose ; fine scenery, the moon, the stillness of the water. Alligators most abundant.

Sept. 7th. The morning set in with light rain. Here we observed the river to have risen three inches in the night, the total or highest rise by the rocks sixteen inches.

Some women brought us a large quantity of lime. Bought a goat for two and a half fathoms of chintz. The people very civil. Set off at eight A.M., rounded the marble peninsula, and opened out beautiful scenery, equal to any thing on the banks of the Thames ; the bare stone rock in many places resembled ruined castles over-hanging the river ; several fine grassy coves. Purchased a large bunch of plaintains for $1\frac{1}{2}$ fathoms.

The teeth of both men and women were notched thus, ⴸⴸⴸ. Saw a large flock of ducks, the precursors of rain.

Many marble promontories now jutted out on this side, round some of which the current sets three miles and half an hour. As it is necessary to take a long circuit round the bays by land, I hired another canoe where we stopped to dine at one o'clock, for four fathoms, to carry the rest of the people. We were however tired of this mode of conveyance, the inconvenience of sitting being hostile, besides the tedious rate, the paddles giving three strokes, and then waiting until the canoe had lost her way, so that they did not go two miles an hour, nor had our persuasions any effect in making them get on a bit faster.

During our dinner a boxing match took place between two of the canoe men about a little salt, at which they both handled their fists with much science; and after drubbing each other heartily, the others interfered, and the business being made up, both the combatants performed Songa.

At four, reached one of the rocky promontories, round which the current set so strong, that the canoe men refused to attempt passing it, neither would they cross the river to get out of the current, pretending they were at war with the people on the other side. I was therefore under the necessity of attempting to haul the canoes up the stream by the rocks with our own people; and had succeeded in getting one of them past the obstruction into still water, when, by the neglect of one of the men, the stern of the second canoe stuck fast in the rocks, and the current taking her on the broadside, broke her right in two, and several of the articles that were in her sunk, and others were swept away, among which were two swivels, my sword, a bundle of umbrellas, and all our cooking utensils, &c.

All was now confusion among the canoe men, who first ran off, and then, after a long delay, came back again, but nothing could induce them to go forward. I was therefore by this unlucky accident obliged to bring back the canoe which I had got up past the rocks; and to hire another to enable us to cross the water; which I effected without any assistance from the canoe men, who could not be prevailed upon to touch a paddle; though, when I landed on the opposite side, I found they were perfect friends with the people of this shore; consequently the story of enemies was nothing more than a pretence of the canoe rascals to go no further. Here I paid them their wages, and we en-camped in a beautiful grassy cove, from which both ends of the reach being shut in by land, presented the appearance of a mountain lake. During the night hippopotami were heard in great abundance.

Sept. 8th. In the morning the Zanzy canoe man would neither go up himself with his canoes, nor would he let the people of this side of the river hire me their canoes, until he was paid for the one that was broken, threatening them with war if they afforded me any assistance; and as I had but little merchandize left, I wished to pay him on our return at Inga. I was however at length, after three hours palavering, obliged to give him a piece and two fathoms, with which he set off; but the canoe men of this side had now gone off, and after three hours more palavering I at last procured six men to carry our baggage by land this day, and until to-morrow night, for which I paid them a piece of check and a bunch of beads. At length we set off at eleven o'clock, and at noon reached Masoondy, where the bearers lived, and where they laid down their loads, and set off to get their dinners; nor could I assemble them again for two hours. We now passed over a very hilly country, with some fertile spots both on the summits, on the sides, and in the vallies, but the general feature is still that of extreme barenness, and a total want of trees, except in the ravines and round the great towns, which are here very numerous. The barren hills are composed of clay slate, lime-stone in different states, with some little quartz; red ochrey hills. Half past three, we reached the river about ten miles from our last night's halting place; the reach running S. S. E. and forming a fine expanse or sheet of water from four to five miles in breadth, free from all rocks, and having a counter-current on this side. The north shore rises gently from the river till it terminates in bare hills; it is lined by a fine sandy beach, behind which, and at the foot of the rise, is a line of trees. The southern shore is steep and rocky.

Horrible face with the leprosy.

Natives extremely abstemious, a little raw manioc and water and their pipe, for a day : devour all the meat you will give them.

The wind always from the westward, inclining up the reaches, so that there is always either a free or leading wind ; and yet the natives have not the least notion of applying sails to their canoes : indeed the wars of neighbouring tribes render the water intercourse as limited as that by land.

Size of their canoes.

Their distribution of time consists of a week of four days, or a *sona* ; the first day of which is named *Sona*, and on this day they refrain from working in the plantations, under the superstitious notion that the crop would fail ; they however perform any other kind of work. The second day is named *Candoo*, the third *Ocoonga*, and the fourth *Cainga*. The month, or *Gonda*, is thirty days ; the year, *M'Voo*, consists of the rainy and dry season, that is to say several *Gondas*.

They believe in a good and evil principle, the former they call - - - - - and the latter *Codian Penba*, both supposed to reside in the sky ; the former, they say, sends them rain, and the latter withdraws it; however, they invoke their favour in the dry season, but it does not seem that they consider them as in any other manner influencing human affairs ; nor do they offer them any kind of worship. Their ideas of a future state seem not to admit of any retribution for their conduct in this world ; good and bad going equally after death to the sky, where they enjoy a sort of Mahommedan paradise.

Sept. 9th. In the morning some rain. Set off at eight, A. M. through a country more fertile, and having more land prepared for cultivation. The manner of preparing it is first by cutting down the long reedy grass, and placing it in small heaps, which are then covered with a layer of mold, and then set fire to ; in the little spots of ashes thus formed, the Indian corn and peas are planted, and the manioc in the unmanured places between them ; these ashes are the only kind of manure made use of. They never burn down the long reedy grass until it has shed its seed, so that the next rains bring a stronger crop than before. - - - - - - very soon blossomed, as well as the long reed grass twelve feet high, and the thorny sensitive plant.

Passed some fine lime-stone springs, the first we had seen.

At ten, reached the gentleman's town of Vooky filou, where our bearers had a long palaver for palm wine, which I was obliged to purchase at a high rate before they would stir a step further. Unfortunately I happened to spill some of it at the foot of our gentleman guide, on which he set off in the most violent rage, taking all his men along with him ; and I now learned that, next to pointing a musquet, the spilling of palm wine was the grossest insult that could be offered to a gentleman. It was full two hours, and after making an ample apology for my accidental impoliteness, and a present of three bunches of beads, that he deigned to shake hands with me, and call his men back.

The houses here are larger than below

Two gentlemen with heavy iron chains and rings on their legs and necks.

At two P.M. we reached the head of a deep reach named Soondy N'Sanga, running W. by N. and E. by S.; here we stopped to dine. After dinner I wished to proceed, but our bearers refused, asserting that they had already walked two days.

Finding all persuasions useless, I was obliged to pitch the tent at this place, and with Dr. Smith and Lieutenant Hawkey walked to the summit of a hill, where we perceived the river winding again to the S. E. but our view did not extend above three miles of the reach : the water clear of rocks, and, according to the information of all the people, there is no impediment whatever, as far as they know, above this place.

And here we were under the necessity of turning our back on the river, which we did with great regret, but with the consciousness of having done all that we possibly could.

On my return to the tent, I found that the bearers said they would not carry back the luggage unless they got another piece ; and I was at last obliged to give to each of the two head men two fathoms of chintz and beads.

Plenty of bees : the natives eat the wild honey when it comes in their way.

The rise of the river was marked at $3\frac{1}{10}$ inches.

The natives appear to be less civilized as we go on, totally ignorant of the relative value of European goods ; exorbitant in their demands.

Employed Peter to buy palm wine and fowls with the beads which I gave him.

Sept. 10th. At eight A. M. set out on our return ; our stock, two pieces of baft, two pieces of chintz, a few beads ; no brandy, no sugar, nor any thing else.

Dawson taken ill, and reaching the top of the first hill,

was obliged to be carried; got a canoe for him for four fathoms; and put all the baggage into her, though so small; another palaver about sharing the fathoms; at last, after three hours waiting in a burning heat on the beach, (therm. 86°) we got the canoe off with one man besides Dawson. Set off ourselves to return by land to - - - - - -. Went a different road from that we came, through the ravines of barren hills; lost three hours over the only fine place we have seen for an European town, being a plateau of ten miles in extent, descending gradually into a plain of double that size, all fertile. Pine apple plants growing close to the path; which descends gently through a small grove to the river side, at -- - - - -, which we reached at - - - - - -, and found the canoe but just arrived, having been nearly swamped by her smallness. Here we were therefore obliged to change her for a larger. The gentleman guide was in a great passion at Hawkey for touching his kissey, and also at our people for carrying the fowls with their heads downwards. Gave one umbrella for six fowls, and another for a small goat.

Inhospitality of the fellow with palm wine: obliged to give my handkerchief, baft, beads, &c. Tantalizing state with the palm wine, and not allowed to touch it for half an hour

Sept. 11th   In the morning got two canoes for six fathoms to go down to Yanza, where we had left the Inga
men.   Half way down we found one of the hippopotami
dead, lying on a bank, in a putrid state.   The people of
this part of the river having been *fetished* from eating it.
At Embomma a good hippopotamus is worth a - - - - ;   its
flesh is sold in the markets.

Here our boatmen wanted to stop, pretending they were
unacquainted with the river below.   I punished them by
carrying them six miles lower down.

The rapids we had before been obliged to haul the canoes
over were now smooth, the river rising about six inches
a day.   The velocity must be greatly increased in the rainy
season, but still the canoes are said to work on it.   Total
rise as marked by the rocks eleven feet.   The clouds
charged, and the barometer falling;   temperature of the
river decreasing ;   at Condo Yango it was 77°, and now but
74°;   the lime stone springs 73°

At one, stopped to procure men to carry our things to
Inga, the Inga men having returned thither without waiting
for us.

Sept. 12th   With great difficulty got a foomoo and four
of his boys to go down for two fathoms each, paid before
hand, and a canoe to ferry us across the creek to Condo

Yango.   A long palaver about a pig detained us till nine o'clock.   Discovered that the barometer was stolen.   Purchased ten fowls for empty bottles.   Found the river so greatly risen that the creeks we had crossed in our way upwards, were now filled, and we were obliged to go high up and cross them on fallen trees.

At noon we dined at the brook Sooloo Looanzaza, and at three encnmped at Cainga to wait for Dawson, who was obliged to be supported by two men.   I now found that besides the barometer we had lost our silver spoons, great coat, remnant of cloth, &c.   In the night we were driven out of the tent by ants.

Sept. 13th.   This morning we found that our bearers had gone off during the night, and left us in the lurch. Got a foomoo and four men of Cainga to go on, for two fathoms each; ascended the Mango Enzooma hill, the highest yet passed, covered with fern; and the transition from it between the mica and clay slate.

At eleven reached the brook Looloo: at three got to Keilinga, where we could procure nothing by purchase. Here we found the Mafook of Inga, who informed us that a goat which we were carrying was fetiched at Inga, and that we must not carry it there on any account, dead or alive, or even a bit of its skin; we therefore exchanged it it for two fowls, which we left for Dawson's use, and pushed on.

At five we arrived at Inga, where the Inga men had reported that one half of us had been drowned in canoes, and the rest killed by black bushmen. Greeted with Izacalla moudela by the people, but greatly shocked on learning the deaths and sickness on board the Congo.

Sep. 14. Sent off Mr. Hawkey with ten men and as many loads of baggage: though ill myself, I intended to proceed; but Dr. Smith and two of our people are too ill to be moved; remained therefore this day, and passed it most miserably.

After dark, the corporal of marines arrived with intelligence of Galwey's death. I passed a miserable and sleepless night, and at day-light mustered the boys with the intention of proceeding; but after paying them two fathoms each, the usual price, they refused to go without receiving three. Gave them three; being very weak myself and wishing to get on before the sun became too hot, I set off with Dr. Smith, leaving Mr. Hawkey behind to bring on the people; at noon he joined me, and from him I learnt that he had a terrible business to get the people off.

Four of the bearers of the sick men ran away and carried off a canteen of brandy and a case of preserved meat: a squabble for salt. Could not get a single fowl for eight bunches of beads. Terrible march; worse to us than the retreat from Moscow.

Arrived at Cooloo at five P. M. Hospitality of these people. Got a goat from the Chenoo, fowls and eggs; all ran cheerfully to assist us; brought us grass for our beds; water; wood for our fire.

At dark Dawson arrived; Inga men left him on the return of the people. Butler did not come in. Passed a good night; it rained hard, but the tent kept it out.

Sept. 15. At day-light sent two men to wait for Butler. Dr. Smith very ill; Dawson better.

Having arranged every thing for the men, I set off at eight, A. M. leaving Mr. Hawkey to bring up the sick. Reached the river at eleven. Thank God for his great mercies in bringing me on thus far!

Found no canoes; waited till two o'clock, when I learnt that the Chenoo of Bibbi, in whose district the landing place is, had forbidden his people to furnish any more canoes; according to some, on account of the commanding officer on board the Congo having ill treated one of the canoe men, who went down with the sick; while others said it was because he was not paid his customs for using his landing place; and I was told I must send my interpreter to him with a present before any canoes would be given. As I knew this would occupy the whole of the next day in palavering, and as Dr. Smith, Dawson, and Butler, were so very ill, that an hours delay in reaching the ship might prove fatal, I seized all the canoes, and a foomoo.

Sept. 16. Unable at day-light to procure any canoe men, I set off with our own people, and at 3 P.M. reached the Congo.

Terrible report of the state on board : coffins. - - - - -

Sept. 17. At day-light sent off all the sick in double boats, as well as the people who had been up with me, to the transport; hired fifteen black men to assist in taking the Congo down the river below Fetish rocks. The river bordered by a level plain, four miles deep, to hills of little elevation. Good place for a settlement

Mangroves commence at the east end of Tall Trees island.

Muddiness of water and red colour begins at the - - - - land, which latter is a bar of sand covered with clay, and under water when the river is at its heighth.

Extraordinary quiet rise of the river shews it, I think, to issue chiefly from some lake, which had received almost the whole of its water from the north of the line.

Commencement of its rise was first observed above Yellala, on the 1st of September; on the 17th of September, at Tall Trees it had risen seven feet, but the velocity was not at all encreased.

Hypothesis confirmed. The water - - - - - - -

Mistaken idea of anchoring ships out of the current for any length of time; the current always creating a current of air.

Palm wine in the dry season only. Palm trees, when two years old, begin to give out wine.

Sept. 18 Reached the transport; found her people all in health; her decks crowded with goats, fowls, pigeons, pumpkins, plantains, flaskets of palm wine; in short, the greatest appearance of abundance.

The difference of atmosphere perceptible between this place and Embomma. Fresh sea breezes.

Mangrove trees fit for?

No fish but cat-fish. Few hippopotami below Yellala.

Quartz sand in vast quantities on the banks of the river; must come from a great distance.

Maucaya - - - - - - child, child-birth.

Different foods fetished. Children fetished for eating the food which their fathers had been forbidden to eat. Women fetished for eating meat the same day that it is killed—with the men. When a man applies to a Gangam for a domestic fetish, he is at the same time instructed from what foods he must abstain; some from fowls, others from plantains.

Lindy N'Congo resides at banza Congo, far inland to the south.

In war the Chenoo of Embomma musters 1000 muskets; fire into the enemies houses at night. Cut off the heads of the prisoners and burn the bodies. All the women sent away before a war is begun. Some Foomoo makes up the business, and each party keeps the trophies, and puts up with the losses. All the men of a Chenooship obliged to go to war. Commanded by the Macaya, next brother to the Chenoo, and civil magistrate; Mambouk, relative of the Chenoo, war minister.

Chenoo of Inga dead, blind man substituted.

Dress. Old men, long thin aprons; young men, cat-skins; tyger cat most valuable; each costs a piece; common cat skins at Embomma for six fathoms

Canoes made of *camba fuma* (bombax)

Spoons, and mouth pieces of pipes made of *lemanzao*, and *pacabanda*.

Flocks of flamingos going to the south denote the approach of the rains.

THE END.

G g

PROFESSOR SMITH'S JOURNAL.

SECTION I.

Notices from England to our departure from St. Jago.

ON the 22d of February, I set out from Charing-cross for Sheerness, to which port the vessels belonging to the expedition had proceeded some days ago. Paul Hafgaard, with several others, accompanied me till I got into the coach. People who were nearly strangers to me, here bade me farewell with tears in their eyes, and looks that expressed their doubt of seeing me any more. Von Buch accompanied me as far as Sheerness. It was for the first time I saw this part of the Thames. Its beautiful windings, and the varying scenery on and around its waters, afforded us great pleasure. The view from Shooter's hill is of an immense extent. At Rochester, the Medway displays its greatest beauty. During our whole journey, those hills of Kent, which terminate in the chalk hills of Dover, were seen by us. On the other side of Rochester the country becomes more level, barren, and uniform. Orchards of cherry trees, for which this county is celebrated, were seen on all sides in great number. The rich but distant treasures I was going to enjoy occupied my thoughts.

On the 23d, in the morning, I took leave of Von Buch, and went on board. I had been expected for some days. Captain Tuckey came on board a few hours afterwards; on the 25th we weighed anchor. An unsteady wind carried us to the Downs, where we anchored, and remained there the 26th and 27th. On the 28th we passed through the Downs, and round Beachy Head. On the 29th we were cruising with variable winds. The S.W. wind at last prevailing, we were obliged, on the 2d of March, to anchor in the harbour of Plymouth, on the identical spot where Napoleon Bonaparte had been in the Northumberland. Here we remained to the 6th. On the 3d, Galwey, Tudor, and myself went on shore to see the beautiful country-seat of Lord Mount Edgecombe. The prospect from the highest hill has greater variety, and is more beautiful, than any I had seen in England. The great number of limestone-cliffs projecting in groups into the sea; the numerous bays; the two rivers that empty themselves into the harbour; the three different towns of Plymouth, Stonehouse, and Dock; the great number of ships of commerce and of war; the hills of Mount Edgecombe covered with wood,—and all this, comprehended in one view, forms a most striking, lively, and picturesque landscape. The immense Break-water, which is constructing across the harbour, is now nearly finished.

On the 5th, Lieutenant Hawkey accompanied us again on shore to see the men of war in the harbour. We went on board the three-decker St. Vincent, which had lately left the dock-yard, and is considered as the most perfect

vessel that ever was built. She carries 128 guns. Such
an immense machine, for the first time seen floating on the
water, makes a striking impression. The cabins are as
large as ball-rooms, with galleries as spacious as balconies.
The construction of the pump-works (qu. chain-pump ?),
and of the powder-room is remarkable. I was refused
entrance into the dock-yards as being a foreigner, though
in the English service. In the house of Lieutenant Taylor
I waited patiently for the return of the other gentlemen,
and amused myself in the mean time with reading the
travels of Bruce. We passed through Stonehouse town,
where they are breaking down a rock of limestone in order
to level the ground for the convenience of the increasing
towns in the vicinity of Plymouth. We visited the
castle, &c. On returning, we dined in Stonehouse. I was
somewhat tempted to remain on shore during the night,
but gave it up, and returned on board late in the evening.
On the 5th we put to sea again, and the wind proving
contrary, we anchored in the evening of the 6th in
the harbour of Falmouth. The 7th, all went on shore.
Tudor and myself hired horses with the intention to
travel about in the country, to see mines, &c. Passing through Penryn we arrived in Truro, which is one
of the mining towns in Cornwall. On the way we saw
a furnace for melting tin, the construction of which was
very simple . . . . . . without bellows. I dined in company with two Dutch ordnance officers, who had made
all the campaigns of Napoleon, and having been at last
taken prisoners at Beresina, had been sent to Archangel,

where they were detained a year and a half. They were now ordered out to Batavia in a rotten frigate, which is here repairing.

From Truro we proceeded to the mines in its neighbourhood, by the way of Sherwater. I descended one of the copper-mines to the depth of about 56 fathoms. The shafts, as also the adits are very narrow. Only a few men were at work in each of them, making constantly use of the pick-axe. The steam engines are in high perfection, and indispensable on account of the vast quantity of water in the mine. These engines have therefore been constructed in several places. The appearance of the country is that of a vast heath. The formation seems to be principally mica and clay-slate, with large veins of quartz, in which the ores of tin and copper occur between and at the side of each other. Long ranges of stones heaped upon one another are to be seen. The friends of Tudor in Truro detained us till late in the evening.

On the 8th, in the evening, we made an excursion to St. Maws, situated in the neighbourhood. It is an inconsiderable fishing-town ; but it is probable, that, before the growing importance of Falmouth, it was a considerable borough, because it has the right of sending two members to Parliament, who however at present may be said to be chosen by the Marquis of Buckingham. The manners of the inhabitants seemed to indicate that the spot is not much frequented.

March 9th. The wind veering towards N.W. we weighed anchor with hopes that the S.W. wind has now left us.

March 10th. We are off Scilly. The wind changed again to S.W. We made the signal in vain for pilots, they having already returned to Falmouth. In the morning of the 12th we anchored, for the second time, between Pendennis and St. Maws.

March 19th. All this time we have been detained in the harbour of Falmouth, partly from the stormy weather, and partly in order to take in water and provisions, and to fit the Congo with new rigging. We weighed anchor in the morning. I have written twice to Von Buch and to Hafgaard, to whom I have given a commission to send all my letters to St. Helena. A strong gale from N.E. has to-day (the 20th) carried us to the 49° 17' latitude, and 60° longitude. The weather is now almost calm. The Congo sails almost as well as the transport.

March 25th. For the first five days we have been favoured by a steady N.E. wind, which has carried us to the latitude of Cape Finisterre.

April 1st. The wind having been very favourable, brought us yesterday within sight of Madeira, which we passed on its eastern side ; and to-day I expect to see once more my old friends Palma, and the Peak of Teneriffe. The weather, as is usual in the vicinity of these islands, has continued for some days very squally. Unfortunately we are to pass, as it would seem, the Canaries, without stopping before we arrive at St. Jago. Being again so near to a place, where the last year I enjoyed so many delightful hours, which I never must forget, they were now as forcibly recalled to my memory as when present, but with a feeling of regret that they never more can be renewed. I had reckoned for certain on being able to fill up several deficiences in the observations that have been made on these islands. For the last few days we have seen but little that could be considered as re-markable on the ocean. A number of porpoises tumbling about the vessel; two large birds, the species of which, at the distance we observed them yesterday, could not be determined; some Medusæ, probably *Medusa pellucida*, but of which we have not been able to catch any, were all of the animal creation we got sight of. On board the Congo I saw a small whale, also a small turtle covered with two or three species of *Lepas*, which we dissected, and a small species of Cancer, probably the *Cancer fulgens* of Sir Joseph Banks.

On the 3d, at day-break, the majestic ridge of rock that skirts the shores of Palma was discovered. We passed its western side, at a distance of from two to three leagues from Punte-de-fuen-zabiente. The two highest summits, particularly that of Roche de la Mustachos, were covered with snow, and irradiated by the reflected beams of the sun a glittering light from the upper region of the air. In the course of the afternoon we came in sight of Tino, though it was inveloped in a dense foggy air. I considered its greatest height to be a little more than 4000 feet. We had not the pleasure of seeing the Peak of Teneriffe again. The weather is now very agreeable; the temperature between 15° and 20° of Reaumur. We have a favourable wind, which, however, has not yet changed to the tro-pical or trade wind, but is more inclined to the N.E. The night was star-light. Galwey and myself amused ourselves by studying the constellations, of which the northern ones will soon leave us. The sea begins to exhibit a greater abundance of animals. Cranch is at length preparing to fish up whatever he can catch. In the last few days we have constantly seen and caught a great number of Portuguese men-of war *(Holothuria physalis)* ; also a small eatable *Velilla* (which I sketched) ; a *Salpa?* which emits light, and a *Medusa*, with four tentacula. On board the Congo I saw a *Loligo vulgaris*, and a fragment of a small *Nautilus*, covered by a species of *Lepas*.

The Congo has for some days been sailing at a slow rate, and is now again taken in tow. Cranch, I fear, by his absurd conduct, will diminish the liberality of the Captain towards us. He is like a pointed arrow to the company.

April 5th. To-day the deep azure colour of the water has suddenly changed into a dark sea-green colour Last night we crossed the Tropic of Cancer. The latitude to-day at noon is 22°.

April 7th. The water of the sea has again resumed its usual colour. The anomalous temperatare of the sea, its green colour, and the great number of *Mollusca* that were seen, seemed to indicate that we were in soundings, but a line of one hundred fathoms was tried without success.

April 8th. We got in sight of the island of Bonavista, and passed its S. E. side at the distance of . . . . leagues. Its appearance resembled that of the eastern Canaries. Its height can scarcely be greater than theirs, but none of its higher hills had a volcanic form.

April 9. In the morning we were under the island of Mayo, which appeared to be lower than Bonavista; and soon after we got sight of St. Jago.   It is only a few days since the Captain informed us that we were to touch at Porto Praya, though we had reason to suppose this to be the plan long ago.   All eyes were therefore with double eagerness bent towards the island as we approached it. After having for so long time seen nothing but the heaven and the ocean, even the barren rocks, which were the first objects that met our scrutinizing eyes, conveyed an agreeable impression: though, in looking through the telescope, it was with difficulty I could discover the least vegetation, and scarcely expected on these rocks a richer harvest than in the deserts of Lonzocolo last year.   This island appeared much higher than either of the two we had just passed.   In its general formation it resembles the Canaries.   It is surrounded by a perpendicular wall of rock without any strand, and it rises gradually up to the summit of the hills.   The highest mountain is about the middle of the island.   Its shape is that of an oblique cone, the top of which rears its head above the clouds, appearing at a distance like a black spot.   The harbour soon opened out between the rocks, in the form of an half circle.   On both sides batteries have been erected, but merely for the sake of appearance.   Guns without carriages, negro-soldiers having muskets without locks, and the barrels tied to the stocks with twine, constitute the defence of the harbour. At the bottom of this circular inlet is situated the principal sea-port town of the island, which is nearly in the same

condition as the batteries. It is built on a flat rock, with
the third and principal battery in front of it, surrounded
by a continuation of the valley la Trinidad, which on each
side of the town forms a level plain, overgrown with groups
of the date-palm; and terminating in a sandy beach. A
rock situated towards the western battery, forms the west
side of the entrance. In the harbour were two vessels,
one of which having English colours, made us hope for a
conveyance of letters to England; but we were afterwards
informed that this vessel was to be very much delayed in
its passage, being previously destined to touch at Lisbon.
At 12 o'clock the anchor was dropt, and we were immediately ready to go on shore. It had been signified to us,
that the expedition would have to remain here two days
only; we were therefore anxious to make the best use we
could of our limited time. Most of the officers, Tudor,
the gardener and myself, went on shore. Our plan was to
ramble about in the neighbourhood the whole of the afternoon, afterwards to join Fitzmaurice, Galwey and Cranch,
who had set out on a fishing-party in a small bay situated
to the eastward; then to return with them on board; and,
before day-break the next morning, to make an excursion
to the higher mountains of the island, and not to return till
the last moment of our allotted time. Unfortunately, in
consequence of this plan, we had not provided ourselves
with more apparatus than would be necessary for an excursion of a few hours only. Captain Tuckey and the rest
waited upon the Governor. Lockhart and myself soon
found of course objects enough to attract our attention;

and we began to form a more favourable idea of the vegetation of the island, than its first appearance from the
vessel seemed to promise. The shore was covered with
Convolvolus soldanella, resembling *Ipomæa ;* and half a
dozen unknown plants, which were successively found
growing among the palms, made us soon regret, in our joy,
that we had so short a time to stay here. We took a view
of the structure of the rocks, which on this side the town
form a perpendicular precipice, and every where along the
coast consist of five or six strata in the following order.
1st. Conglomerate. 2d. Tufa of pumice-stone. 3d. Pumice-stone. 4th. A stratum consisting of an uniform and
somewhat blistered substance. 5th. Basalt. 6th. The
uppermost stratum, which is similar to the 4th, and by its
concentric formation, globules &c. made it evident that it
had been in a fluid state. It resembled the same corresponding stratum in the Canaries. We went into the town,
which consists of a few rows of miserable huts, situated on
a flat rock, about an English mile in circumference, and
surrounded by the lower part of the valley of Trinidad.
Most of the inhabitants here, as indeed on the whole island,
are negroes. Although we discovered nothing but naked
rocks and withered fields, wherever we turned our eyes,
yet many different sorts of fruit, poultry, &c. were offered
to our view, which proves, that the interior of the country
must have a different appearance. We descended and
crossed the other valley of palms, travelling over level and
scorched plains, in order to join our comrades of the fishing party. The night was fast approaching. We made

signals in vain for a boat from the east battery, under which the ships were lying at anchor. We ran back to the town and down to the other place of landing, hoping to find them there, but were now informed, to our great astonishment, that the boat, which had brought us on shore, had been upset in the surf when going out again, with all the officers. Captain Tuckey succeeded in reaching the shore by swimming, and escaped with the loss of his sword. Lieut. Hawkey's foot was entangled in a rope, and he was with much difficulty and almost lifeless brought on shore by Captain Tuckey and a negro. The fat purser wisely saved himself on the bottom of the boat. Thus they all got pretty well off, but their fright had made them forget us entirely. We were therefore once more obliged to return to the town, and take up our lodgings at a kind of public-house, kept by a fat negro woman, to whom we left the care of providing for our dinner, of washing our linen, &c. We were better accommodated than we expected. Before day break we were upon the alert, looking out for the landing of the gentlemen from the ship, with instruments, books, &c. for our excursion to the interior of the island. We waited impatiently till ten o'clock, but were disappointed. In the mean while we took a walk round the town, and descended into the plain to the east side. Here are several wells, one of which in particular supplies the town with water. It is about three fathoms deep, and its water was the temperature of 76° of Fahrenheit.

A great number of half-naked negro men and women flocked down to fetch water from this well. Tudor, in his journal, speaking of the latter, observes, that their whole figure, their hanging breasts, and other personal accomplishments, made our young gentlemen of the party dream of the Venuses we should have to admire in the kingdom of Congo.

No boat was yet to be seen. We had hired a negro corporal for our guide, and that we might not loose more time we determined to set out on our inland journey, though without instruments, and though our boxes, which were not calculated for a long tour, were already filled with plants. Our plan was, if possible, to proceed this evening to the foot of Pico Antonio (the highest mountain in the island), and having reposed ourselves in some hut during the night, to ascend the summit the following morning; then to cross the mountain in a S. W. direction, down the valley Publico Grande and Cividad; and lastly, to return by the road leading to El Matheo. That part of the island, through which we had to pass, is unfortunately the most level. From the cliffs near the sea, rising perpendicularly some scores of feet, large tracts of land, scorched by the sun, extend themselves upwards. They are intersected by level barrancoes, of which the valley la Trinidad is the largest, running from two and three leagues from the harbour up to the mountains in a W. N. W. direction. Through this valley, as being the most fertile, we took our way. Its lower end is horizontal, and at first sandy and naked. Some solitary trees of a *Mimosa* were the first pleasing objects that occurred; and soon afterwards we observed, for the first time, a huge Adansonia, divided into three large trunks, thick in proportion to their height, with bended branches, and the fruit (Calbufera?) hanging on long stalks. Though destitute of leaves, the sight of such a tree made an agreeable impression, and the fruit 1 found to be refreshing. In the upper part of the valley luxuriant thickets of *Jatropha curcas*, and thorny *Zizyphi*, covered with ripe and somewhat acid fruit, were found growing. After these followed lofty fig trees, *Annonæ*, with ripe fruit, &c. Our first stay was at a hut on the border of the valley, where we refreshed ourselves with delicious goat's milk. The scenery became more and more delightful. Shadowy mimosæ, oranges and fig-trees increase in number. A multitude of small *Fringillæ*, an *Alcedo* with tropical plumage, and many other birds fluttered about us. The valley continued to expand itself. New plants occurred in quick succession, and different kinds of cultivated trees became more and more frequent. For the first time we discovered a whole thicket of *Cassia fistula* with ripe fruit, and for the first time had Tudor the striking sight of *Bananas, Cactus opuntia*, &c. Of all the plants I have hitherto seen, none appeared more interesting to me than *Asclepias procera*, which here grows half wild like the shrubs in a thicket.

All at once we were called from out of the valley by Tudor. It seemed that our guide, John Corea, had no great inclination of venturing too far, without partaking of some refreshment, and that he had discovered a convenient spot for us to eat our dinner. A fowl, with the root of cassava and some eggs, supplied us with a very good meal. While it was preparing, we took a ramble up the rising ground to the westward ; afterwards, having finished our meal, we proceeded on our journey in the valley, which, from the eminences, was seen most to its advantage. A part of it, which was wider than usual, was planted with indigo, cassava, and sugar-canes of light green colour, and interspersed with groups of *Pisang*, oranges, *Annonæ*, &c. mixed with solitary tall cocoa-palms. Farther on, the valley winded up the mountains, and lost itself on their eminences. We had soon descended and pursued our way between the gardens along a rivulet, that ran from the upper end of the valley, and after having watered and fertilized this beautiful spot, here lost itself. We found the inhabitants very hospitable and good-natured. A rich farmer invited us into his house to drink tamarind lemonade; on arriving at it, we soon observed that it indicated a wealthy owner. He was governor, it seemed, of this district, and possessed a great part of the valley and the above-mentioned plantations. Here we saw the only *vineyard* which had yet occurred, and a large *Ailanthus* in flower.

The *barranco* now became narrower. We proceeded along its eastern bank across its lower part, and upwards on a gently rising and naked ground. It grew dark before we had advanced further than three leagues. The numerous objects in our way, and the tardiness of my fellow travellers, had much delayed us. We came to another *barranco* well watered and fertile, and soon arrived at a group of houses, where we were well received, entertained as usual with goat's milk and cheese, and had mats for our sleeping upon. The name of this place is *Faaru*.

Close to the houses was a steep rock, upon which I found several interesting plants, and among them a beautiful new *Lavendula*, and several others met with in the Canaries. Below was a clear spring, overshadowed by Pisangs and cocoa-trees. Its temperature was one degree higher than the well at Porto, though we had ascended to the height of about 1000 feet.

At day-break we heard a shot from the harbour, which made us doubtful whether we should proceed on our journey, but not perceiving, on looking through the telescope, any blue flag hoisted, we continued leizurely to walk upwards. We had not advanced far when the appearance of the country became entirely changed. After having for some time seen nothing, on the other side of the cultivated ground, but tracts of land scorched by the sun, and in some places overgrown with *Spermacoce verticillata* and a few *Sidæ*, it was an unexpected sight to perceive the hills covered with grass, from one to two feet high, being a species of *Panisetum* whose tropical nature was discovered by its ramifications. Innumerable herds of goats, sheep, and cattle were feeding all around. It had struck me that of the whole family of the *Euphorbiaceæ*, which are peculiar to a great part of the African countries, from the Canaries to the Cape of Good Hope, the *Jatropha* only is here to be met with, and this too is a foreign importation. In the small level valleys on the sides of the grassy mountains, I perceived groups of a shrub, which had something new in its appearance, and on approaching it, I found at last an *Euphorbia*, that bore so near a resemblance to *piscatoria*, as scarcely to be distinguished from it. I met successively with several old acquaintances on the hills, as for instance, a *Bupthalmum sericeum*; most of them however, in the Canaries, are growing in the lowest region.

Another beautiful view opened to the east. The valley of St. Domingo lay under our feet, between perpendicular rocks. South American and tropical fruit trees, plantations of sugar and other vegetables, in various places, and at the bottom a rivulet, formed by several streams springing from the steep rocks around, afforded a most delightful view of contrasting objects. We had now reached the ridge of the mountains, and followed it for some time over valleys and hills partly covered with high grass, and interspersed *Euphorbiæ, Jatropha curcas*, and some solitary *Mimosæ.*

The day was already far advanced, and on seeing the Peak again before us, we found the ascending it would take the remaining part of our time, and that the way we had followed, though it was the most commodious, was at the same time the longest. We resolved, therefore, to limit our farther journey to the ascent of one of the conical hills that surrounded us, in order to take as extensive a view of the island as possible, and then to return. Corea was dispatched to the nearest shepherd's hut to procure us some milk. We gave up our first plan with less regret, by considering that we had brought no barometer with us; but in return we missed many interesting plants, the number of which continued increasing as we walked on. We had soon reached the summit of the nearest hill to the left; and to the south-west, the

level land through which we had travelled, comprehending the whole barranco of Trinidad down to Puerto, was
spreading itself under our feet. Farther on to the west
some hills were seen, between which another barranco was
winding its course down to *Publico grande*. The ridge
of mountains runs in the direction of the longest diameter
of the island from S. E. to N. W., but it runs nearer the
sea and with steeper declivities at the N. E. than at the
opposite side. The valley of St. Domingo is one of the
deep barrancos to this side, and there are probably more
of them farther on. Their steep side towards the ridge
of the mountains, as also the Pico Antonio itself have a
complete basaltic appearance. No trace of real vol-
canoes were to be seen. The hills to the W. S. W. have a
more volcanic form, and it is in this direction that the
high peak of the island of Fogo is situated, but this we
unfortunately could not discover through the clouds.
Pico Antonio is very steep at its western side. At the
opposite side it would be easy to ascend it, the highest
summit perhaps, only excepted. Its perpendicular height
is scarcely more than 5000 feet, and from the place
where we stood, about 3000 feet. Pico occupies about
the middle of the ridge of the mountains, which is continued to the N. and N. W. by mountains more rounded, but not much lower. Its geological features, to conclude from the structure of the mountains, do not in any
essential point deviate from those of the grand Canaria.
All the specimens of minerals which I collected, are principally the same as those found in the latter island.

According to the notices which are to be found on some maps, there can be no doubt that the four islands to the N.W.N. are of the same submarine volcanic nature. Mayo and Bonavista may be compared with the lower part of St. Jago, and Fogo is in all likelihood the only volcanic one. The climate of the island is delightful, and considering its situation within the tropics, it is re- markably temperate. Of this the nature of the vegetation gave evident proofs. There were found very few tropical plants in proportion to the number of those which are common in temperate countries. The temperature which has been given for its wells probably does not much differ from the mean temperature *(isothermos)* of the island It was with difficulty I could draw the information from the inhabitants that it is now about a month or six weeks since the rainy season commenced. The *Adansonia, Jatropha,* and *Ziziphus* were already stripped of their leaves, and the *Mimosa* was producing new ones. Almost all annual plants were decayed by drought. The rainy season was said to last from five to six months, and to continue to the end of September. The atmosphere, after being heated in traversing the continent of Africa, is after- wards fully saturated in passing over the sea, and arrives at this island in a humid state, so that the fog comes down at the slightest degree of cooling. The mountains, even those of less height, are almost constantly covered with clouds. At the height of 1400 to 1500 feet, the ap- pearance of the country is completely changed. The hills are covered with grass of a tropical form and magnitude. A number of small streams spring up in the *bar-rancos* and water the valleys. The cultivation is extended with success to the tops of the hills. This sudden transition was, as usual, very striking in the plants, which induced me to represent in a table their physical and geographical distribution.* Whether the still higher parts of Pico Antonio might possess a third physical diversity we could not determine, but this is hardly probable. The whole number of the different plants collected did not exceed eighty, among which about a dozen are new species, and perhaps one new genus. We did not see any of the green monkeys that inhabit the steep hills in great number, but many of them were brought on board by the inhabitants. It was near noon. We waited a long while for Corea, and sent Lockart to fetch him. Corea returned by another road, and we waited again a long while for Lockhart, but resolved at last to leave a direction for him to follow, and to hasten back on our return. At *Faaru* we hired two jack-asses in order to save time, and galloped off by a shorter way across the plain. But I almost swore never more to make use of these animals, because by being obliged to beat them continually, our arms were as much fatigued as our feet would have been by running. Lockhart arrived nearly as soon as ourselves. The officers and the marines were on shore; and on being informed that the vessels were not to put to sea until the wind was fair, we walked quietly back again to the town, to remain there during the night. It was on the eve of Maunday Thursday.

* See *Table* at the end of the Section.

A procession with wretched music moved round the market-place. We understood from the officers, that several curious circumstances had taken place. The governor had been washing the feet of some of the poor inhabitants. Judas Iscariot was hanged in effigy, and had received some hard thumps by way of chastisement, and so forth. The governor had invited Eyres and Galwey to dinner, an hospitality which proved to be rather interested, he himself and the noble lady, his consort, begging as presents for every thing they saw or could imagine to be on board the ship. All the provisions that are brought to market, pass in a manner through the hands of the governor, and their price is enhanced by the duty, which is applied to the defraying the expenses for maintaining the garrison and the civil government. The colony probably does not afford any revenue to the government, nor charge it with any expenses; for which reason the communication with the mother-country is very little, or none at all. Almost all the inhabitants of the lower classes are black people. Eyres came the following morning ashore, in order to fetch the remainder of the provisions, consisting of goats, sheep, fruits, &c. to which we added a quantity of oranges, peaches, &c. Captain Tuckey, Hawkey, and Galwey had been at the end of La Trinidad, and in another small *barranco* which is the country-seat of the governor. Cranch had been rambling about the plain, and shot a number of birds. Fitzmaurice and Galwey, on the first morning, had been very successful in fishing in the bay, but the boat was upset, by which they lost the fishes.

The last of these misfortunes happened to Fitzmaurice and Galwey in the morning of the day of our departure. As they were going on shore with a view of making trigonometrical obsesvations, the boat was again upset. The instruments nearly escaped being lost, and the gentlemen saved themselves at the expense of some bruises only.

———

DISPOSITIO GEOGRAPHICA *plantarum quas legi in insula Sti. Jacobi die x*^{mo} *et xi*^{mo} *Calend. Aprilis; circa portum Prayæ in convalle Trinidad et montibus Pico St. Antonio confinibus ad altitudinem circiter* 3000 *pedum.*

A. *Regio inferior :* arida, 1500 ped. circiter alta.
  1. *Plantæ tropicæ.*
    a. *Propriæ.*

Mimosa glandulosa. - - -	MS.
Convolvolus jacobæus. - -	do.
——— affinis eriospermo. - -	do.
Boerhavia suberosa. sp. nov. - -	do.
——— depressa. ditto - -	do.
Glycina punctala. - - -	do.
Smilacina anomala genus forté novum.	do.

    b. *Senegalenses.*

Adansonia digitata.	
Achyranthes tomentosa. - .-	MS.
Spermacoce verticillata; *etiam in Jamaica.*	
Momordica Senegalensis.	
Cardiospermum hirsutum.	
Sonchus goreénsis	

K k

c. *Introductæ Americanæ num quasi indigenæ propartes tropicas.*

    Jatropha curcas.

    Anona tripetala.

    Tribulus cistoïdes.

    Argemone mexicana.

    Solanum furiosum?

    Datura metel.

    Cassia occidentalis.

    Ipomea pilosa.

    Eclipta erecta b.

    Malva ciliata?   -   -   -   -   MS.

    Sida polycarpa?   -   -   -   do.

    —— repens?   -   -   -   do.

    —— micans?   -   -   -   do.

d. *Introductæ Asiaticæ num quasi indigenæ.*

    Justicia malabarica.

    Calotropis procera.

    Abrus precatorius.

    Plumbago.

2. *Plantæ Zonæ temperatæ.*

  a. *Propriæ.*

    Herniaria illicebroides. sp. nov.   -   MS.

    Zygophyllum stellulatum, sp. nov.   do.

    Lotus jacobæus.

    Zyzyphus insularis.   -   -   -   MS.

    Antirhinum molle.   -   -   -   do

    Borago gruina.   -   -   -   do.

    Lavendula apiifolia, sp. nov.   -   do.

    Polycarpia glauca, do.   -   -   do.

b. *Canarienses.*
    Sideritis punctata?
    Heliotropium plebeium.   *Banks. Herb.*
    Lotus glaucus.
    Eranthemum salsoloïdes.
    Sacharum Tenerifæ.
    Physalis somnifera.
    Polygonum salicifolium.
    Sida Canariensis?

c. *Boreali—Africanæ, quæ simul Canarienses.*
    Cucumis colocynthis.
    Aloe perfoliata.
    Tamarix gallica, *var. canariensis.*
    Phœnix dactylifera.
    Cenchrus ciliatus.
    Celsia betonicæfolia.
    Comelina africana.
    Achyranthus argentea.
    Corchorus trilocularis.

d. *Capenses.*
    Sarcostemma nudum.
    Forskohlea candida.

B. *Regio superior:* *humida graminosa;* inter altis 1500 3002 ped. et forsan ad summa cacumina usque.

  a. *Propriæ.*
    Euphorbia arborescens, sp. nov.   -   MS.
    Pennisetum ramosum.   -   do.
    Campanula jacobæa sp. nov.   -   MS,
    Polygala?
    Lotus lanatus sp. nov.   -   -   MS.

Spermacoce? *divers. genus. videter.* MS.

Festuca?

b. *Canarienses.*

Bupthalmum sericeum.

Thymus therebinthinaceus.

Sideroxylon marmulana? (Madeira)

Festuca gracilis.

c. *Meridionali-Europeæ, quæ etiam in Canarien.*

Silene gallica.

Oxalis corniculata.

Sisymbrium nasturtium.

Centaureum autumnale.

Anagallis cerulea.

Radiola milligrana.

Gnaphalium?

d. *Capenses.*

Crotalaria procumbens?

Hedyotis capensis.

e. *Americanæ introductæ?*

Evolvolus lanatus.

Tagetes elongata,

Indeterminabiles absque flore et fructu.

Compositæ annuæ, (duo)

Liliacea.

Convolvolus.

Trutex. No. 90. - - MS.

Cenchrus.

Crypsis.

Bilabiata.

SECTION II.

From St. Jago to the Mouth of the River Zaire.

W E weighed anchor about noon of the 12th, the wind blowing fresh from off the island. At 5 o'clock, the fog at the horizon having disappeared, we got sight of the peak of Fogo, which reared itself above the skies. Though at a distance of 14 leagues, the sun, which was setting behind the island, afforded us a distinct and beautiful view of its form. The height of the peak probably is not less than 7000 feet, which is considerable for a spot of so small a circumference. The following days we proceeded at a moderate rate towards S. E. till the 19th, when the wind dying away, veered somewhat to the south. We had calm weather and squalls alternately. Thus we found ourselves in that miserable region which has been so much spoken of as exposed to an everlasting calm and violent rains. The wind that accompanied them generally blew from the south, which obliged us to turn the ship's head towards the coast of Africa. The horizon was constantly foggy. In the evening dark clouds arose from the east, and flashes of lightning gleamed all the night. The west was generally clear, sometimes with scattered dark clouds, which at the setting of the sun, being tinged with an infinite variety of colours, presented an interesting appearance never witnessed in the northern hemisphere without the tropic. The other parts of the canopy of heaven were mostly bright, except when they were overcast with a squall. The wind that preceded and followed the latter was seldom strong, but the rain fell in such torrents and in such large drops, as to keep the surface of the water smooth. We had not any opportunity of making observations on the depth of the sea. The polar-star was sinking low towards the horizon, and it was with difficulty we could discover it for the fog, in order to bid it farewell, for God knows how long a time. New constellations were making their appearance towards the south. The bright-glimmering stars of the Cross and the beautiful form of the Scorpion, were seen in the course of the night. The heat during the day was intense, but I did not feel greater inconvenience from it, than I had experienced upon former occasions. The evenings were cool and agreeable. During the last fourteen days I was occupied with examining the plants I had collected in St. Jago, and this employment being now finished, I am about to make some sketches of the physical objects of that island. On the evening of the 28th, Fitzmaurice and myself set out in a boat to make observations on the current. The sea was quite smooth, but a gentle swelling from the west, and an uncommonly dark cloud from that quarter, indicated an approaching change in the state of the weather. In the night I was awakened by the motion of the vessel, and on arising, I saw the Congo again taken in tow, and the vessels proceeding with a brisk and cool westerly wind. We are now at the distance of 14 to 15 leagues only from the coast of Africa. Should the wind continue as it is, we shall soon reach the latitude of Cape Palmas, and probably then be out of this hot and tedious *pacific* ocean, and approach the line, when, I suppose we shall have to go through the usual ceremony of being shaved by Neptune. Our whole party continues to be cheerful and agreeable. Poor Cranch is almost too much the object of jest. Galwey is the principal banterer.

April 30. To day we had the most violent squall we yet had witnessed, and the Captain himself owned that he had never seen a horizon so dark as that, towards which we were now sailing this afternoon. At the horizon flashes of lightning crossed each other in the deep darkness, which soon surrounded us. The rain, falling in torrents, and accompanied by flashes of lightning in rapid succession, approached us. We could no longer gaze at this dreadful phænomenon, but were obliged to retreat into the cabin. After somewhat more than the half of the squall had passed us, we ventured up again, and enjoyed the most sublime scene of the whole heaven beautifully illuminated by flashes of lightning. It generally darted forth from two points, now rising like spouts of water, now running in zigzag, and spreading itself into innumerable branching shapes. No lightning I had seen in Europe bore any resemblance to this.

From the third to the tenth of May we had a southerly wind, that carried us far into the Bay of Guinea. Innumerable shoals of fish of different kinds, but chiefly *Albicore* and *Bonitos*, were swimming in all directions. Every day some of them were caught. Flocks of birds belonging to the tropical regions and now and then some men-of-war birds were seen. From the 10th, the weather has been frequently calm, but the squalls have somewhat abated. To day a man died on board, who had been sick a long time. On opening him, a quantity of coagulated and extravasated blood was found in the pericardium. It was not without some emotion that I witnessed for the first time a burial at sea. Tuckey read the prayers, and two sailors in white shirts lowered him over the side. The common story that the sick recover, when the ship comes in soundings, was not in his case verified.

14th. In the afternoon we had the very amusing sight of a great multitude of fishes, flying over the surface of the water and chased by albicores, which hurried after them with the swiftness of an arrow. The chase took place close to the vessel.

15th A number of brown - - - indicated that we were not far from land; and on the 16th in the morning, we came in sight of Prince's island, at the distance of about ten leagues. This island had a singular appearance. Steep rocks of a cubical and conical form arose towards N. E. and E., some of which, on the side we passed, were perpendicular, and white as chalk. It is difficult to assign a reason for this white colour. Their forms prove that they are basaltic rocks, but no white substance, as far as I know, occurs in such large masses in this formation. Some of the

gentlemen supposed this colour to be owing to the multitudes of water-fowls which frequent these rocks; but the
number is scarcely conceivable that would be requisite to
supply matter for covering rocks of the height of 2000 to
3000 feet, and of such a substance as to make it appear at
so great a distance. The next two or three days we endeavoured to get to the westward, in order to pass St. Thomas
on the west side, that we might not make the coast of
Africa before we had passed the latitude of Cape Lopez.

May 18. This day we got sight of the island; and on
the 21st we passed its north-west side, at the distance of
two or three leagues, and were now arrived at the lowest
degree of the northern hemisphere. The island is uncommonly high, its ridge of mountains rising in a peak (St.
Anna), which cannot be of less height than 8000 feet: yet,
on looking through the telescope, we discovered the mountains covered with trees up to their very summits. What
an inviting sight! how many new and interesting objects
the natural historian has passed for centuries, without be-
stowing a moment's notice on them! We know little or
nothing of these islands, though they are situated as it were
in the midst of a track which has been for ages one of the
most frequented by vessels of all nations, while immense
collections of plants and animals have been brought together from Australia, China, the East Indies, and South-
America. For the last eight days we made a cruise, which
was perhaps unnecessary, in order to pass to the westward
of St. Thomas. How I could have wished to have spent
these days on the island!

L l

May 23. In the course of this morning we crossed the line. All the sons of Neptune were now busy in their preparations for a visit from his Tritons; who soon made their appearance with their hair dressed with hempen tails, their backs striped with tar, their heads covered with large caps, and gave notice to the Captain of the arrival of the God of the Sea, which was instantly announced by the sound of fifes and drums.

[*Here follows a long detailed account of the ridiculous ceremonies performed on those who for the first time cross the Equator, which, new and amusing as they were to Dr. Smith, do not deserve insertion here.*]

We continued to steer towards the west till the 26th, but the wind veering more and more to the eastward, it was resolved to try the other course along the coast. The sea is here uncommonly abundant in fish. The whole surface is often put in motion by the flying-fishes, when chased by others. Their number is immense. Shoals of them constantly surrounded the vessel, and at night they give out a white light, resembling that of the moon, when reflected by the sea. It was also chiefly at night that we were enabled to catch, with the net, the greatest number of mollusca and crustacea. Many different substances contribute to make the surface of the sea light. Some parts of the bodies of most of the crustacea have certain glittering points, and two or three species of crabs were perceived to give out the most brilliant light. The points, which are to be seen on the mollusca are larger, but less bright. But that luminous appearance which diffuses itself over the whole surface of the sea, arises from a dissolved slimy matter, which spreads its light like that proceeding from phosphorus. The most minute glittering particles, when highly magnified, had the appearance of small and solid spherical bodies.

May 28. We saw for the first time this day one of those floating islands, often mentioned, and which probably come out of one of the rivers of Africa. The Captain permitted us to put out a boat, in order to examine it. It was about 120 feet in length, and consisted of reeds, resembling the *Donax*, and a species of *Agrostis?* among which were still growing some branches of *Justicia*; and in the midst of these were seen a number of animals (*Sepiæ*)? For many days past the sea-water has been uncommonly cold. We were probably in soundings, though we could not reach the bottom with a line of 120 fathoms. I am often up at night fishing for marine animals, of some of which I make sketches.

June 2d. We this day got sight of the continent of Africa for the first time. The land was very low, but we did not approach it sufficiently near to be able to describe its appearance. The 3d and 4th we continued under the land with a southerly wind. The Captain resolved to stand out again to the westward, which course we followed, cruising till the 14th, but found that the wind was veering more and more against us, and becoming the steady tropical westerly (? easterly) wind. We stood in a second time for the coast, and got in sight of it on the 18th. In the morning we descried a large vessel. All on board believed it to be one of His Majesty's ships, which were known to be stationed here for the preventing of the slave trade. All were immediately busy with writing letters. On coming up, however, we felt no small disappointment on being informed that it was an East Indiaman going to St. Helena. I had indulged a hope of soothing my poor mother's anxiety on my account. We were now at a distance of two or three leagues from the coast, of which we had a distinct view. The shore consists of black perpendicular rocks, with yellowish streaks running in the same direction, and mixed with larger spots of the same colour ; and lower down resembling beds of pumice-stone. Behind these rocks a broad ridge arises covered with trees, some of which scattered here and there greatly over-topped the rest. In some places we observed plains of considerable extent, and of a yellowish colour, probably owing to the dry grass ; we observed also clouds of smoke, proving that the custom of setting dry grass on fire is even here prevalent. Several leaves and pieces of wood floated past on the water. About the parallel of Cape Yamba, and at some distance from the shore, is a conical hill, the height of which, though it is the highest part of an extensive ridge, did not appear to exceed 1000 feet. The longitude of this coast is very erroneously marked on the charts, so that according to the most recent, and we may suppose the best, it appeared from several good lunar observations, we should have been sailing inland to a considerable distance.

June 14th. A dead albatros (a bird rarely to be met

with so far to the north,) was fished up. On the 20th, a whale was swimming close to the vessel. For the last eight or ten days the weather has been humid and foggy. The Captain now communicated to us his instructions relative to our conduct in our future excursions. They are such as to afford a satisfactory proof of his liberal conduct towards us. We have two Congo negroes on board, both of whom speak English. One of them, named Ben, acts as my servant; but as Ben left his native country when twelve years of age, and as the other is but very little acquainted with the English language, we have not profited much by their information.

We have for some days past been proceeding at a regular but slow rate along the coast. The sea-breeze generally sets in at noon from S.W., and carries us somewhat forwards to the south. We now anchor in shallow water at the distance of two or three leagues from the shore. The country here is very low, and thickly covered with wood. The coast has a sandy beach, on which we can distinctly see the breakers, and hear them roar at night.

June 23. We anchored at a distance of three leagues from the shore, and had a beautiful view of it. The thick forests, which rise in two, and sometimes three successive ranges behind one another, varied by plains covered with grass of a light green colour, though they present an interesting appearance, nearly resembling that of the woody shores of the Danish islands, indicate at the same time, as far as may be judged at a distance, a great uniformity of the country itself, and of its natural productions. By looking through the telescope I did not discover any difference in the form or colour of the trees. Some of them scattered about towards the beach appeared higher than the rest, and were probably palms of the cocoa tribe. The pieces of wood which we have fished up, do not present any variety in their appearance. Ben asserts that the banks of the river Congo are perfectly similar to this shore. Whales (probably *Physeter* . . . .) are seen daily swimming near the vessel. We catch every day a number of *Sparus* resembling *Pagrus*. Its flesh, though dry, has a very good taste. On the 28th we had passed that vast tract of land, the appearance of which has been here described, extending from Cape Mayambo through Malambo and Guilango down to the bay of Loango. In the two last days the aspect of the country has changed. The trees do not form themselves into forests, but are scattered in groups only, or stand singly, having uncommonly large tops. These groups might be discovered from a great distance, even when the whole country presented itself only as a blue line, bordered by a gently rising ground apparently naked, with banks of a greyish white colour, which probably are banks of clay that have fallen down.

To-day we made several attempts before we could weigh the anchor. The current was very strong, and the bottom, which before consisted of a sandy clay, was here uncommonly uneven, with banks of coral rocks and mud alternately. Although the sea-breeze blew fresh, the vessel made not the least progress. On weighing the anchor for the fourth time, we found it now to be so difficult, that we were obliged at last to cut the cable. To-day, the 29th, in the afternoon, we are again under sail. We have ascertained that we already must have passed the southern point of the bay of Loango, although this point on all our charts is placed much farther to the southward. The weather is clearing up and the heat is again encreasing. The nights are resembling those we had in the bay of Guinea, the atmosphere being clear, except at the horizon, where it is foggy. The sea-breeze enables us to get to the southwards, and we shall soon see a new hemisphere, with new constellations appearing at night. The sea-breeze generally continues until midnight, but is not followed by any land-breeze at all, the weather continuing calm until the sea-breeze sets in again at noon, or somewhat later. This may be partly explained by supposing, that by the returning current of the air in the higher regions of the atmosphere, the eqilibrium is restored ; a supposition which is the more probable from the fact, that the fog, which had been driven together towards the shore, as soon as the calm comes on, again covers the heavens, which before were clear ; but the principal reason of the want of the land-breeze may probably be this, that the great current of air setting from the two coasts of this narrow part of Africa towards the interior, is deflected towards the north, where the continent is greatly extended, and where the heat is much more intense.

Some days ago the sea had a colour as of blood. Some of us supposed it to be owing to the whales, which at this time approach the coasts in order to bring forth their young. It is however a phenomenon which is generally known, has often been described, and is owing to myriads of infused animalculæ. I examined some of them taken in this blood-coloured water : when highly magnified, they do not appear larger than the head of a small pin. They were at first in a rapid motion, which however soon ceased, and at the same instant the whole animal separated into a number of small spherical particles. The sea has again assumed a reddish appearance ; but this is probably owing to mud, that has been dissolved. We have of late not had any sick on board. When we were in the bay of Guinea, several symptoms of a putrid fever were dis-covered ; but this disorder, as also a peripneumonia that frequently occurred, and sometimes was very violent, were easily cured,

July 1st. This morning we found ourselves near the coast at the large mouth of the river Loango Luisa, on the south side of it. This river is called Caconga in the chart ; and in the place where Loango Luisa is marked, no river exists. The coast before us had perpendicular cliffs towards the sea. Its banks consist of a reddish substance, which, as far as may be judged from a piece of about one foot in length, which was taken up with the anchor, is a hardened chalky clay or marl. Their upper part, which is flat, is over-grown with scattered groups of palms and other solitary trees. The bay of Malambo is situated lower down within the banks. The harbour of Malambo, as also that of Cabenda, which is next to it, were formerly the principal trading-places of the French on this coast.

At noon we went all upon deck on hearing the Captain hailing some canoes that were in sight. They soon came along-side. This circumstance, more than the aspect of the country, reminded us of the place in which we were. On looking at the hollow trunks of which their canoes were made, each pushed forwards by two or three naked negroes, who stood upright in them, the figures resembling those I had seen in South-sea voyages were brought to my mind. The canoes were from twelve to fourteen feet in length, and from one to one foot and a third in breadth, and about as much in depth; the upper part of the sides were somewhat bent outwards, the bottom was flat, and both ends pointed. The oars were made of a rounded flat piece of wood, fastened to a staff. A half-dressed negro addressed us in English, and appeared very much pleased by being answered in the same language, and invited on board. He called himself *Tom Liverpool*, and said he was the interpreter of the Mafook, whose visit he came to announce; but he appeared to be greatly astonished on being told that we were not come for the sake of trade. The Mafook came soon after in an European boat, bearing a small white flag, and he had an umbrella over his head. He welcomed us in the English language. They were all invited into the cabin, and the conversation became general, partly in broken English, and partly in somewhat better French. They were very much alarmed on hearing from us, that hereafter no other nation except the Portuguese, could carry on the slave-trade with them; and one of them, considering the King of England as the

M m

cause of it, broke out into a violent passion, abusing and calling him " the devil." The inhabitants of Malambo, who were formerly wealthy, since the abolition of the slave trade, have become very poor, because their town was little more than the general market-place for the disposal of slaves, having no other source of profit, than what those slaves from the interior and the trade of the harbour afforded them.

After having refreshed themselves with a glass of brandy, and with great appetite partaken of our biscuit, filling their pockets with it at the same time, it was resolved that the Mafook, with some of his gentlemen, should remain on board during the night, and that in the mean time the boat should be sent on shore, and return with refreshments the next morning. They now put on their court-dresses, which they had brought with them. One appeared in the coat of an American officer, another in a red waistcoat, a third in a sailor's red jacket, the Mafook himself in a red cloak. All of them had a piece of coloured stuff wrapped round their loins, and a skin worn as an apron. Their legs were naked. Those, who called themselves *gentlemen*, wore caps of several sorts, mostly red caps with tops, but the Mafook and two others wore round caps, that were made in the country itself, and neatly embroidered. They were presents from the king, who gives them on appointing any one to the office of Mafook. The Mafook *(Tamme Gomma)* had worn his cap six months only, but his predecessor, who was with him and retained the title, had worn his cap for ten years. On showing them the plate in the voyage of Grandpré, it was found that *Tatu Derponts*, at that time Mafook of Malambo, who is represented on this plate, was the uncle of Tamme Gomma. But the luxurious - - - - and polite manner with which Grandpré was received by the former, was very much contrasted with the poverty of the latter. Tamme Gomma was a man of the middle age, tall and well formed, with an interesting and noble countenance, which resembled more that of an Arab, than of a Negro. This was the case with several of his retinue. He wore over his shoulders a riband with a fetishe of some inches in length and breadth, representing two figures in a sitting posture, each of them holding a globe in their arms: they were tolerably well executed. Grandpré observes of these figures that they have European features, and the resemblance struck us immediately. They had high bare foreheads, aqualine noses, painted white, and bore some resemblance to the Egyptian, and in some parts to the Etruscan figures. Those of the better sort of people wore skins of a kind of tiger, but the rest wore simply skins of calf. All of them wore round their necks pieces of cord twisted from the hair of elephants' tails, and above the wrist a thick ring of iron or copper, with figures, the execution of which proved that they have some skill in working these metals. Most of them however wore rings of iron. They told us, that both these metals were abundant in the interior, but that the country produced no gold dust. We showed them samples of beads and small looking-glasses, &c. in order to be informed of what value they considered them to be, which, as merchants, they understood perfectly well.   A row of beads or a looking-glass, worth sixpence or thereabouts, they thought might be exchanged for a fowl, and twelve cowries they considered to be worth as much.   This price is considerably higher than what would be offered for these commodities on the coast of Guinea, or in the East Indies, where forty cowries are not worth more than a *penny*.   They were now called to dinner, and behaved themselves perfectly after the manner of the French.   The Mafook carved the meat.   They drank to the health of every one of us.   When the night came on, they all complained of cold, and were very much pleased to get our great coats to protect them. Tom Liverpool, having got my Norway cloak on, walked about with great gravity.   They had several times been half drunk, and went now to sleep between deck.   In the night we were boarded by a boat from Cabenda, which was not received.   The next morning, having waited in vain for the Mafook's boat, and the breeze coming on, we made sail towards Cabenda.   The people of Malambo probably thought it too far off to follow us thither.   In the course of the day two boats boarded the Congo, and by them the Captain took the opportunity of having our guests sent ashore, though they seemed not to be very good friends with the people of the boats.   These at first refused to take them at all, till the Captain threatened to sink them if they persisted in their refusal.   The person who had the command put on a red embroidered coat, being otherwise naked, and came on board, where he stayed a few minutes only.   He confirmed what the others had told us, that there were nine Portuguese vessels in Ca-benda. As this is the most northern trading-place that has ever been in the possession of the Portuguese, and which has been a matter of dispute with the French, it will also probably for the future, be declared to be the highest place to the northward at which they will be allowed to trade. It is not improbable, that the large vessel we met with some time ago, is also in this part.

With the assistance of Simons (the Congo negro) we cal-lected a great many words of the language of the people of Malambo. Many of them had been high up the river Congo, and told us that it is navigable for boats to a very considerable distance. Yamba Enzadi is, according to the explanation they gave of the word, a whirl or violent current, which may be passed by one side of it. We felt disappointed in not getting refreshments here, as we pro-bably, for some time, shall have no opportunity of pro-curing any. The Mafook, among other things, had ordered a goat and a pig as a present to the Captain, besides fowls, fruits, &c. We should have abundance of all this, they told us, if we would go on shore. They had been taught by Europeans, among other things to be offered, never to forget women, and were highly pleased when we were joking with them on that head. After all, we were very glad to get rid of them, because they made a terrible noise, particularly when their boats boarded us. They were all immediately quite as home, and ran to assist the sailors with the utmost willingness to weigh the anchor. Some of them danced with the lascivious gesticulations usual with them, while the pretended gentlemen encouraged them.

Fitzmaurice, accompanied by Tom Liverpool, had made an excursion in the morning, in order to examine a sand-bank to the southward of Cabenda, mentioned by Grandpre; but it was found to be so near the shore, that it was not thought to be of any consequence to us. About noon, we could discover some of the vessels at anchor under the high banks; and, on a projecting low point to the south-ward of the harbour, some huts were observed to be scattered among the trees. After our guests had left us Fitzmaurice was again sent out; Galwey and myself accompanied him, and Hawkey soon joined us. Though it was already growing dark, the thought of approaching for the first time so near the soil of Africa, afforded us great joy. From our anchorage, which was four miles distant from the shore, the depth was gradually diminishing. No breakers were observed until we came close under the land, the breakers running parallel with it to a great distance. We proceeded along the banks. The breakers were here so inconsiderable that a landing might have been effected without any difficulty; but as the officers did not participate in the eagerness I felt to visit the shore, we did not approach it nearer than at a distance of some hundred yards. It was a level tract of land, covered with wood, and in some places having small open plains, on which stood larger trees; from this plain we were separated by a white strand. A bright moon-light aiding the awakened enthusiasm of my mind, made me fancy that I beheld charming landscapes, which appeared to me to be en-livened by the loud chirping of the grass-hoppers. No birds or other animals were seen. On hearing the signal-gun we returned.

July 3. The Congo was ordered this day to proceed along the coast down to Red Point, which we had in sight, and which is the last point before entering the river ; but the current carrying her away, she was obliged to anchor, and to day she was brought up by the boats. These two last days we have had a gentle sea-breeze from S. S. E. (? land-breeze,) which sets in in the morning, and commonly carries with it some butterflies. Yesterday it was squally, to-day it is foggy and dark. With the drag we have fished up several kinds of shells and crustacea. An uncommonly large eel of a very good taste was caught on board the Congo. The temperature of the sea-water did not present any difference. To-day it is colder than we had yet found it in the southern hemisphere. We feel already the strong current of the river.

July 4. We remained all the day at anchor with an indistinct view of the land.

July 5. We went into a boat to sound the nature of the bottom, which was found to be rocky and uneven. Nothing was taken up but branches of an *Antipathes.* In the after-noon, after the calm and warm morning, a strong sea-breeze set in. We weighed anchor and sailed on briskly through the rapid current. In the evening we found ourselves all at once out of soundings. The wind was dying away, and we thought we should be under the necessity of driving back with the current; but at midnight we were again in soundings with a depth of 20 fathoms, and the anchor was dropt. We had unexpectedly passed to the other side of the channel, but we have got in sight of the long-wished for Cape Padron, and Fitzmaurice is gone to try the depth.

SECTION III.

―――――

Our Progress up the River as far as Cooloo, opposite the Cataract.

W E weighed anchor on the morning of the 6th, in the expectation of soon welcoming the land of promise, but were obliged to anchor again at seven o'clock under Shark Point, at a distance of a hundred yards from the shore. It is low, with a sandy strand terminating in a very steep bank. Along the beach and close to it is a thicket of shrubs ; above this another of palms extends along the whole shore (probably *Corypha*,), and higher up is a lofty wood. The vegetation has the appearance of being decayed by drought. To the right hand is Hippopotamus Cliff, with a low shore, running southward in a long straight line. The anchor broke at the heaving, and we were for a moment in a dangerous and critical situation, the Congo having run foul of us. For the first time in four months we now saw the sails taken in. A number of naked negroes immediately assembled on the shore. The land to the north (Mona Mazea,) is perfectly similar to Maxwell's representation of it. The mouth of the river is very large, perhaps fifteen miles wide. The naked hills of the interior consist probably of sand. Fathomless

N n

Point is just before us. At this point, towards the bank that runs down perpendicularly 100 fathoms, the wood appears as if it were cut off. It is at this point that the river first commences its proper breadth. Shark Point is a narrow and unsafe anchorage. The bottom is steep and very uneven, as we have a depth of 17 fathoms on one side the vessel, and a depth of 30 fathoms on the other; and a few fathoms father off is the whirl of the rapid current. The Congo had also her anchor broken by carelessness. We expect to be very busy tomorrow morning, but there is scarcely any hope of getting on shore. This is a tedious tantalizing busines. At ten o'clock this evening we have a bright moon-light. The dark forest and the white foaming surf below, present an interesting appearance. The negroes have made several fires on the Point. About the twilight the cries of parrots were heard from the vessel, and large eagles hovered over the forest. It is rather cold (69°), and the dew falls in large drops. The current runs at the anchorage about three knots an hour.

July 7. Early this morning the Mafook or governor came on board in two canoes, with his retinue. At first his pretensions were very lofty. He insisted upon being saluted with a discharge of cannon, and on observing us going to breakfast, declared that he expected to be placed at the same table with the Captain, and endeavoured to make his words sufficiently impressive by haughty gesticulations. Sitting on the quarter-deck in a chair covered with a flag, his dress consisting of a laced velvet cloak, a red cap, a piece of stuff round his waist, otherwise naked, with an umbrella over his head, though the weather was cold and cloudy, he represented the very best caricature I ever saw. He soon became more moderate on being informed that these vessels were not belonging to slave-merchants (who generally for the sake of their own profit grant these gentlemen every possible indulgence), but to the king of England, and that our object was not trade. In order however to give him a proof of our good will towards him, a gun was discharged, and a merchant flag hoisted. At table they ate and drank immoderately ; and in return promised to procure refreshments if we would send a boat on shore. In the meantime they were to remain on board. Fitzmaurice was accordingly ordered to go on shore in the boat. The Captain observing my extreme impatience to accompany him, said that if an excursion of a few minutes could be of any use to me he would readily give his permission, but that there was no place for more of us. I was not long in jumping into the boat. With a fair wind and with the assistance of oars we were scarcely able to double the point on account of the current. At last with great difficulty we reached the shore. Ali, our pilot, flung a rope to the negroes, who flocked down, and they drew us for some time through the surf along the coast. The vegetation was magnificent and extremely beautiful. Shrubs of a rich verdure, large gramineous plants, and thick groups of palms met the eye alternately. The country displayed the most beautiful forms, the most charming scenery. I found myself as in a new world, which was before known to me in imagination only, or by drawings. We had still to double several points before we could arrive at the village, but our orders did not permit us to go farther. We leaped therefore on shore with one of the sailors, each of us carrying a rifle barrelled gun. I ran a few paces to the left, where the thick and dark forest came down close to the strand; but my progress being obstructed by shrubs and grass so as to make it impossible to proceed, I turned to the opposite side. The ground was sandy. The strand was in a few places some feet broad, but in general the vegetation left no intermediate space. I met Fitzmaurice surrounded by negroes, and bargaining for a turtle of immense size and a singular form, being no doubt a new species. On going farther I was so much obstructed by thickets of shrubs, that I was obliged to step into the water up to the middle, which I found to be the only way of getting at the plants, and of taking a view of the outside of the trees. The most common shrub was a *Chrysobalanus*, bearing a strong resemblance to *icaco*. It was mingled with another, which, though without flowers and therefore hardly determinable, is probably a *Ximenia*, and the same I found at St. Jago, (whither I believed it to have been carried,) with a fruit resembling much a yellow . . . . ., which had a fragrant smell, and an acid but not disagreeable taste. The inhabitants higher up the river called it *Gangi*. The Portuguese missionaries tell long stories about its use in putrid fevers. *Chrysobalanus* has also a fruit called *Mafva*, that is *blind*. I saw also two large species of *Arundo*, three of *Cyperus*, one of which was the *papyrus*. It rather surprised me to find this last growing in low and inundated places. Various shrubs of *Hibiscus* with flowers in bunches, and growing near the water; a *Papilionaceous* plant with ripe pods, and thick groups of *Mangrove* were successively met with. The palms are probably the *Hyphæne*, with large *frondes*. A creeping *Jasmine* (the same I found at St. Jago,) was also discovered. The forest consists of a tree resembling *Cæsalpinia*, but I could not approach near enough to be able to examine it. Most of the plants here mentioned are commonly called South American, but they have probably been carried thither from Africa along with the negroes. The *Hyphæne* and *Cyperus papyrus* are the same as those of Egypt. I found those parts of the ground the most accessible, which had been set on fire by negroes. Near the Point a great quantity of fish was hanging exposed to the air to be dried, but no huts were any where to be seen, nor did any women appear, which shows that their habitations must be situated at some distance in the forest.

The current carried us back in a few moments. The large turtle was exchanged for a pair of knives, worth about one shilling. Almost all the negroes who came on board our ship are Christians. One of them is even a Catholic priest, ordained by the Capuchin monks at Loando. They were baptized by these monks two years ago at St. Antonio, situated seven days journey from hence. The monks have given to the priest a diploma. This barefooted black apostle, however, had no fewer than five wives, the number of whom is always in proportion to

their means and inclination, and which they never forget
to make known.   If the early missionaries had used a little
more indulgence on this point, their doctrine would certainly have met with a better reception.   A few crosses on
the necks of the negroes, some Portuguese prayers, and a
few lessons taught by heart, are the only fruits that remain
of the labours of three hundred years.   If the exertions of
the missionaries had been deliberately directed towards
civilizing the natives, what good might not have been
effected in the course of so long a period!

July 8. We had this morning a visit of another Mafook
from the point of Tall Trees, situated higher up the river.
His countenance wore the mark of great good-nature.
He brought with him as a present a beautiful little goat.
They tell us that there are eight small Portuguese vessels
at Embomma, which seems to be the general market-place
of all the surrounding nations.   Simons, who is an important personage in his capacity of interpreter, has already
got some intelligence of his relations, after an absence of
eleven years, and Ben is in hopes of meeting with some of
his countrymen.   They all agree in considering it impossible for the Dorothy to proceed as far as Embomma.
The rapidly advancing season will not allow us to stop until we have proceeded far up the river. The current under
the point runs from 1 to $3\frac{1}{2}$ knots, and in the channel of
the river from five and six to eleven and twelve knots.

On the setting in of the sea-breeze, in the afternoon, we
weighed anchor, and were nearly boarded a second time by
the Congo sloop. We had no sooner doubled the Cape, than

the vessel was whirled round by the force of the current. The wind however blowing fresh, enabled us to advance about a mile before we were again compelled to anchor. The Congo went on briskly, following Fitzmaurice, who was sounding ahead. She anchored at Sherwood's creek, near the shore, her people having thereby the satisfaction of having sailed before us up the river, though she had been towed all the day. As we anchored we saw a schooner steering towards Kakatoo Point. While we were at table this vessel discharged a gun, and hoisted the Swedish flag, which by mistake was reported to be Spanish. On seeing the British flag displayed, she fired another gun, which was said to have been loaded with shot, and that the ball passed very near our vessel. Every thing was now in a bustle, and all in arms; the guns were loaded, and every thing in readiness to answer such an unexpected salute. We were all courage, though aware of our inferiority in point of strength. The schooner anchored at a short distance, and sent a boat with a few men on board with the Captain's compliments, and an offer of his services. Captain Tuckey began to suspect that it was an American slave-vessel under Spanish colours. He returned for answer, that he would send a message on board their vessel, with thanks for the Captain's civility (the real object was to reconnoitre); adding, that he commanded an expedition, of which the principle vessel, a corvette, had proceeded before us. The sight of regimentals, and of marines, and the return of Fitzmaurice in a boat from the upper part of the river, confirmed the Captain's statement.

The schooner immediately turned back and put to sea, though her Captain had just before mentioned his intention to sail the next day for Embomma.

On leaving the point we had an interesting view before us. The sea-coast with its sandy beach and foaming surf were gradually disappearing; before us was the Cape of Kakatoo, of which the lower part is covered with shrubs and palms to a distance of about half a mile from the shore, and farther up is a dark lofty wood. On the other side of the point the shore is covered with grass and shrubs, and one point after another is seen projecting into the great bay. Making allowance for the tropical form of the vegetation, the scene bears some resemblance to the Danish lakes. In the forest of palms, some larger trees were to be seen; they were destitute of leaves, and *Maba* is the name given to them by the natives. I have no doubt but that *Adansonia* even here is to be found.

9th. We are still in the middle of the great bay. On one side is Cape Kakatoo, where a number of canoes are seen fishing. Farther on before us lies the Congo (schooner) under the land, at the mouth of a rapid river, the banks of which her people are exploring. The natives are still on board. The Mafook of Kakatoo cannot forget that he has been seated at table with the Captain of a King's ship.

10th. We have been driving back during the night to a considerable distance. To day we weighed anchor, and endeavoured again to sail up the river, but on perceiving that we were driving back in the mid-stream, we were obliged to steer out of it. The vessel was very difficult to manage in the current. We had soon passed the mid-channel, which is scarcely more than $1\frac{1}{2}$ mile broad, and reached the Mona Mazea bank, where the depth was only seven fathoms. The vessel was for a moment in danger, but by the exertion of all on board, she was brought to anchor again under Kakatoo, a little farther back, and in shallower water than the first time.

We remained here all the day waiting for the sea breeze. The Captain began to entertain doubts whether he should be able to bring our clumsy bark up to the Congo. To the joy of all of us it was resolved that a fishing-party should go ashore to morrow. Every one is permitted to be of the party, who may be inclined, provided he will return with Fitzmaurice at breakfast time. I thought this time too limited, and procured permission to remain on shore till dinner time, on condition that I would risk to be cut off from the ship, in case the re-embarking at that time should prove impossible ; which condition I eagerly accepted of, hoping that if such an event should happen, I might be able to get a canoe to convey me on board.

11th. We were all up before day-break and went into the boat. Most of us were completely wet, though the swell of the sea was not very high. After having dried our clothes at the fire made by the negroes, and waited until the first draught was made with the seine, when not a single fish was caught, the company, consisting of Cranch, Tudor, Lockhart, a marine, and myself, proceeded along the shore. This being the only opportunity we should

O o

probably have during the whole voyage, of examining the coast, I preferred to take this course, though less interesting than a walk into the thick forests. A great many strand-plants were growing in the gravel; some of them presented forms similar to those at the Cape of Good Hope, and detained us a great while. At last we thought it necessary to turn back and enter the forest. Thickets of thorny shrubs made every step difficult, but in return every step was repaid by the discovery of some new plants. Among them were the following; a *Jasmine* with large fragrant flowers; pyramids of *Flagellaria* of a light green colour, covering the trees up to their tops; a *Rhamnus* covered with flowers; and several shrubs, almost all with fruit, and for this reason undeterminable. We tried to follow a path leading into the wood, and soon met with Mangrove trees, which sometimes rise to a considerable height. Here we sunk to our knees in a morass. The first tropical *Felix* was seen here. I was now obliged to return. The whole peninsula seems to of the same structure. The surf, opposing the current of the river, has thrown up a high sand-bank along the shore; above this bank is a vast morass, covered with mangroves. The land at the river side is level and sandy, partly overgrown with lofty groups of *Hyphæne*. The ground is covered with an *Arundo*, and an *Andropogon* with broad leaves. They were each of them about twelve feet high, but had been set on fire in many places. Our way led through these *Hyphæne* palms, the fruit of which was hanging down in branched clusters. We met with some negroes, who informed us that our comrades were in the neighbourhood, near a pool of water like a fishing-pond, where Cranch had just shot an *Anhinga*. We went down to the river side, and to our great joy found here the whole company, who had just had an immense draught of large fishes of a species of *Sparus*. It was near dinner-time, and we thought it best to accompany them on board. We returned accordingly with a full cargo, more like wood-cutters than botanists. Cranch had not been very fortunate. We observed but few insects, and the birds were very shy. I saw a number of parrots, small parroquets, a black-bird on the wing, and two small *Moticillæ*, but did not fire a single shot. Galwey brought to me a beautiful violet-coloured *Robinia*, which I had not seen before. We had collected plants, the examination of which would require weeks, though our excursion had been so very short, not more than four hours. Lockhart and myself were occupied the whole afternoon and evening in laying in specimens of plants. I considered that the best way would be to preserve them immediately, and put a specimen of each in water. We found ourselves quite exhausted. The heat was from 6° to 8° of Fahrenheit greater inland, than on the sea-shore.

July 12. We examined plants all this day. The two double-boats had already been put into the water last night. Both are fitted up to-day, and to-morrow we shall leave the ship. A boat from Embomma is with us. The natives tell us that all the vessels which were at that place had betaken themselves to flight before we entered the river, having got over land from Cabenda intelligence of our approach, which makes it probable that they were Americans under Portuguese colours. Every thing yet seems to indicate that the descriptions of the great breadth of the river, of the length of its course, &c. have been exaggerated, and that the whole expedition will sooner terminate than any body suspected. The channel is very narrow, and the current never more than three knots, though six are marked on the chart. The gentlemen from the Congo came on board the transport in the afternoon, bringing with them specimens of several interesting plants collected during their excursions. Yesterday they had penetrated into Sherwood's Creek to a distance of about five miles, till they came nearly close under the high in-land country, which is more open, and accessible. They had seen traces of elephants and of numbers of antelopes, but the negroes inform us that these animals, as also the hippopotamus, are first to be met with in great numbers higher up the river.

July 13. All our baggage is sent on board the double-boats, and the Dorothy is now like a deserted village. In the afternoon all were on their appointed posts on board the small vessels, which make a kind of flottilla. Our two double-boats formed the van, and were soon found to be excellent sailing vessels. The double-boat and skiff of the Dorothy followed, and Captain Tuckey in his gig was in the rear. Galwey and myself were with him. Favoured by the sea-breeze we soon passed the bay, and in the evening found ourselves under the southern bank of

the river.   The weather, as usual, was clear and pleasant.
Small floating islands, sometimes overgrown with waving
shrubs of *Cyperus papyrus* passed by us.   The river-side
was thickly covered with a variety of plants of interesting
forms.   We heard the cries of parrots and the warbling
of some small singing birds.   The scene was solemn and
beautiful.   The night was approaching when we arrived
on board the Congo, where we found our room very close, as
we indeed had expected; but our agreeable prospects made
us forget this and every other inconvenience.   These
prospects were only clouded by the gloom which some of
us could not avoid betraying, when speaking of the exaggerated accounts of the river, and the forebodings they
were apt to entertain of the voyage being speedily terminated.

July 14.   In the morning all wished to go on shore,
which was at the distance of about two gun-shots only;
permission for this purpose was given after breakfast.
The landing-place was in a small bay just opposite the
ship, with an island covered with plants, and a *Rhizophora*
with bended branches.   At the bottom of the bay was a
sandy beach, with a small plain, in the middle of which
a large *Adansonia* was growing, and higher up a thick
lofty forest was discovered.   To the right and along the
shore the country was more open.   A number of shrubs
and trees, from their form and novelty, excited admiration
and surprise.   Farther on the surface had been set on
fire, which is the only method of making a way through
the impenetrable thickets.   After having walked about

in all directions as far as the thickets and mangroves permitted, hunting for birds, insects, and plants, we sat down near the river to rest ourselves. The strand being now broader and more accessible in consequence of the ebb, we had an opportunity of discovering many new and interesting objects. The evening was approaching when we gave the signal for our returning on board.

July 15. I was fully occupied all the day in laying plants in paper, having very little time for examination. I have already got three or four new genera, and two-thirds, I conceive, of all the plants that have been collected will probably turn out to be new species. The collection of birds and insects is small. Some of the gentlemen are again on shore for a short time.

July 16. We went into the boats on another excursion. Having passed along the shore up to Sherwood's Creek and entered several of the innumerable small inlets, we proceeded nearly to Alligator's Creek, and then returned with the current. The river-side promises us a large collection of plants, the vegetation being so luxuriant, that there are only a few places where a landing may be effected, either where the negroes coming down the river in their canoes encamp during the night, or where there lately has been fire in consequence of their encampment. The high thickets growing near the water consist chiefly of *Pterocarpus, Convolvulus,* a new species of *Hibiscus,* related to *tiliaceus,* with scattered red and yellow flowers, *Pandanus candelabrum* (a little farther up) an *Eugenia,* covered with flowers, and a vast number of shrubs and trees round

which are twisted a great variety of climbing plants, and
among which are *Quisqualis ebracteatus*, and a plant resembling *Schonsbœa*, with brilliant purple coloured flowers
in bunches. Farther on, in the bay, we met with more uniformity. *Rhizophora* covers the lower part of it, having
bended and branched roots, on which some of the officers
observed marks, that in their opinion determined the height
of the flood tide to be about $2\frac{1}{2}$ feet, but they are rather
to be considered as indicating the rising of the river in a
former season. A species of *Acrostichon* is the only plant
growing among this *Rhizophora*.

A complete calm and the deep shade of the forest give
to the contiguous places a dark and solemn appearance.
We saw traces of buffaloes on the sand of the shore. Of
birds, we saw two species of the eagle, an *Ardea alba,* and
Platus anhinga, three *Certhiœ*, two *Alcedines*, two *Fringillœ*,
a large grey parrot and a parroquet.—The negroes come
daily on board in great numbers, bringing refreshments,
consisting chiefly of goats. As yet we have not met with
any remarkable fruit, except that of *Rhaphia*, which I
have seen but once. A journey to Sognio is daily spoken
of, but the distance is about ten leagues, and all on board
being very busy, it is consequently put off. The Dorothy
has at last advanced a little higher up, but will scarcely
reach Embomma. We have a great many Negroes on
board, who come under various pretexts, being by their
own account all Mafooks, all good pilots, &c. The principal enticements, however, are our provisions and our
brandy.

July 17th. I remained again on board all the day. Lockhart and Tudor went on shore. Lockart had collected several new plants, which were all found within the limits of our small bay.

July 18th. Lockhart and myself went on shore in the bay, in order to discover what still might have been overlooked. We found a remarkable tree at the river-side (*Didyn. drupac. fol. 5-natis.*) We heard a shot from the Dorothy. The Mafook Senu, from Embomma, having been dismissed from the Congo, was received on board the Dorothy with a salute. The Mafook with his retinue have been constant visitors on board for a long time. His pretensions were not much attended to. He was, however, presented with a small boat, in which he might return, because the Negroes, it was said, would laugh at him, if he returned without some present. The Captain resolved to proceed to Embomma in the double-boat before the Congo, having there affairs to settle with the Mafook Senu, which will take several days.

July 19th. We prepared ourselves to accompany the Captain, but the weather continuing calm we were obliged to remain here all the day.

July 20th. We left Sherwood's Creek and the beautiful bay where the Congo had been at anchor, proceeding along the bank of the river, which presented the most charming and varying scenery, and, after having passed Knox's island, we anchored between several low sandy creeks and small islands.

July 21st. Early in the morning we were under Knox's

island. While the other gentlemen were filling the boats with fishes in a few draughts, we were walking about in the thickets. Our acquisitions this day were; a palm with two sorts of leaves; a *Rhamnus*, which being too weak at the root to stand upright, was supported by high shoots; an *Amomum*; a plant of the *Liliaceous* tribe; *Rhaphia*, and many other plants. A large monkey was seen on the shore from the Congo. The *Simio cephus* was frequently brought to us by the inhabitants.

July 22d. We made a short excursion on the shore nearest to our anchorage, where it wore a more varying aspect. A *Rhaphia* with brilliant flowers was seen here. In the evening the double-boat was ordered to make sail. We bade farewell to the Congo with three cheers. Our accommodations were in this boat much better in every respect. Tudor, Galwey, myself, and the Captain, were on board. We sailed on briskly along the shore, though the water was shallow. We frequently came in contact with the shrubs that were spreading their various branches over the water. The *wine-palm* waved its leaves above them. The parrots, flocking together like crows, leave this side of the river, where they have been seeking their food in the day, and retire to the northern shore, were they remain during the night. The wind was dying away, and we were obliged to cross over to the opposite side of the river. We anchored under the first island, which is one of the many banks formed in the middle of the river. The Captain offered me his boat for an excursion on shore the next morning.

P p

July 23d. being awaked early in the morning by the cries of the parrots, we went on shore on a low island (Mampenga.) I had not expected to find here any great variety in the vegetation. A number of new objects, however, presented themselves every moment as I walked on. The island is the point of an extensive sand bank formed in the river, and at the north side is almost joined to the main land. In the middle were low morasses, in which a number of different plants were growing, and among them two species of *Nymphœa*, a *Menyanthes indica*, and several others. They were almost all herbaceous. Tudor chased several water-fowl, but with little success. Traces of *Hippopotami* were seen every where in the sand. A singular species of *Sterna* and an *Alcedo* were shot. The river here abounds with canoes, which come to this side to fish. About thirty new plants collected here, will fully occupy my time until we shall weigh anchor in the evening. We proceed along the islands, several of which are overgrown with whole forests of the *Cyperus papyrus*. We are again near the main-land, which has a most beautiful appearance. . . . .

. . . . A great many *wine-palms* growing among the thickets, seem to indicate a populous country. A number of negroes walking about among the shrubs, call out to us as we go along. Behind the thicket is a village, but there is scarcely any penetrable landing-place on the shore. We continued our course along the north side, and anchored under Sangam Compenzi (Monkey's Island), and we are again preparing ourselves for another excursion on shore to morrow at an early hour, and to renew, as it were, the exquisite pleasures of this day.

July 24th. We landed on the second range of the low sandy islands in the river, or Monkey's Islands (which name the inhabitants give to these islands, not to those of Maxwell). At low ebb they formed two ranges of banks, of which those parts which are above the water at high flood were thickly overgrown with a tall grass, and with scattered thickets of shrubby plants. One of these shrubs was a *new genus*, and another a new species of *Limodorum*. These plants and some *Cyperoideæ* were the principal acquisitions of this day. I was running for a long time after several strand birds, but shot only a few of them. Galwey and Tudor were in another quarter. I waded to the island farthest to the south, and returned in a canoe. These islands are probably only inhabited periodically, and must be partly inundated in the rainy season. The fishing for oysters seems to be the main object of the negroes who live here. Large heaps of the shells of a *Mya* lay spread all over the shore, and a great quantity of dried and half-roasted oysters were hanging under the straw huts. We have not yet seen any regular built hut, but only thatched roofs, supported by four poles. A few women only were to be seen. We were proceeding to the northern woody side of the river, when we observed the fore-top sail hoisted on the boat. Innumerable canoes are continually passing upwards and downwards. The object of some, that of fishing; of others, that or drawing palm-wine from the trees.

The breze sets in to day somewhat early, or about eleven o'clock, but, as usual, is scanty It generally begins first to blow fresh after it is dark, when we are obliged of course to remain at anchor, We proceeded at a slow rate along the coast. Our black pilot steered between the two islands to the northward (where the boats touched the very bottom) and into the great channel along Monkey Island. The picturesque and varied scenes occasioned by the thickets and forests have now disappeared. We have now passed the mangrove country, and see only thickets of *Hibiscus* near the water, with some solitary trees. To the right the low land is thickly overgrown with high grass like a corn-field. No forests are now visible, unless that name should be given to vast ranges of the *Cyperus papyrus*, which with their lofty and waving tops present a singular appearance. Farther up some scattered *Hyphæne* palms are still visible. The appearance which these two plants give to the whole country strongly reminds me of the drawings of Egyptian landscapes. The palms as we proceed increase in number, forming groups, and higher up even whole forests. Great numbers of the natives make their appearance on the shore, walking about in the grass between the thickets. We are just by the village of Maliba, whose name is derived from its palms. When it grew dark we anchored a little higher up, a few fathoms from the shore. A black cloud for the first time made its appearance in the northern horizon. The negroes told us that it was the prognostic of the approaching rainy season. While I am writing this at our anchorage I hear the evening music of the grasshoppers.

July 25th. A gentle breeze carried us about two miles higher up, where we anchored near the shore in a small bay. The natives welcomed us. A water-snake about four feet in length was caught in the cabin of the Captain. He must have swam on board during the night. For the first time we saw here the bank of the river rising perpendicularly about four feet, and consisting of indurated clay The surface was thickly overgrown with grass ; but as the soil consisted of a hardened clay, it was not so high as usual. Groves and forests of *Hyphæne*, mingled with some *Adansoniæ*, were the only kinds of trees that occurred. The village *Condo Tjongo* was at a short distance. It consisted of huts covered with neat mats, made of grass or straw, and supported by poles. We walked about an hour in the high grass and between the trees, on which calabashes were hanging to receive the juice which constitutes the palm-wine. I shot a large species of *Corvus* and some small pigeons, of which a great number were cooing in the trees. We saw numerous traces of buffaloes, and were told by the natives that they had seen a whole herd of these animals early in the morning. The vegetation here is without variety. A gentle breeze carried us somewhat higher up the river. We crossed the channel, and passed to the shore on the right, near Farquhar's island.

I had now the opportunity of going on shore for a few minutes. Here at length, for the first time, some traces of cultivation were seen. A considerable extent of land was planted with maize, and some tobacco, both of which were now about one foot high. I collected two new plants, and

some specimens of a *singular tree with large leaves.* We
proceeded along the island, which seemed to be well cultivated and very populous. At a distance an *hippopotamus* was seen with his head above the water. The land on
both sides of the river is again quite level and sandy, overgrown with high grass, and in some places with *Cyperus
papyrus.* An old woman, a relation of Simons, came on
board, accompanied by her grand-daughter, a little black
Venus, the sight of whom kindled an amorous flame in the
breasts of several of our gentlemen. Some tempting offers
were made to her before hand, and Embomma appointed
as the place of meeting. A canoe, in which was the son of
Mafook Senu was upset. He gave a specimen of his skill
in swimming by diving from under it. We passed Rough
Point, and came within sight of the high country on both
sides the river. To the left were some rounded hills, and
behind them a ridge of mountains, with a pointed pillar
on the highest summit, which is called *Taddi Enzazzi,* or the
rock of lightning. These hills are almost naked and do
not therefore promise much in the way of botany, in comparison with the lower parts of the river side. To the
right we observed a large projecting point terminated by
a rock (Fetishe rock) resembling a mass of ruins, but not
high. We crossed the channel and anchored near the
shore to the right. On the north side and at a short distance is Coyman's Point, where three channels of the river
meet together. We shall probably reach Embomma to-
morrow. The whole appearance of the river, its numerous
sand-banks, low shore, inconsiderable current, narrow

channel, seem but little to justify its extravagant fame. Its sources cannot be farther inland than those of Senegal and Gambia.

Though we were at the distance of four or five miles from Fetishe rock, the Captain was desirous of sailing down to it. Our pilots entertained some fears of passing through the whirlpools, which, however, we found to be nothing more than the main stream of the river running with a strong current. We had no sooner landed on the rock than we observed two hippopotami, at which unluckily we were not prepared for firing. Tudor however fired a random shot near the head of one of them. The rock is steep, and difficult of ascent. Its lower part was overgrown with trees of various kinds. Its conical shape gave us an opportunity of discovering its structure. It consists of a coarse-grained granite. Large pieces of feld-spar occur in several places, and at the extremity of a low flat point some of them are not less than from one to two hundred feet in circumference. The towering cliffs of Fetishe rock, the new and varied vegetation on its sides, and the extensive view of the river which it commands, presented a magnificent scene. This rock is the terminating point of the high mountainous land which is seen to extend into the interior in undulating ranges of blue mountains, two or three ridges one behind the other. Beyond the great sandy and grassy islands in the river, scattered palms are seen in the horizon, appearing as if they were growing in the water. The high land rising from the banks of the river would form interesting landscapes, if it was not quite naked.   Some scattered *Adan-soniæ,* stripped of their leaves, were almost the only trees that were visible.   A few minutes only were allowed to me for examining the rich vegetation of the Fetishe rocks. The steep cliffs and the impenetrable thickets of shrubs, climbers, and lofty trees, limited my acquisitions to a most superb climber *(Polyan. monog.),* a *Limodorum,* and a *Liliaceous* plant, the thick fibrous stalks of which (without leaves) are used by the natives for making ropes.   The sea-breeze set in late as usual.   We passed several villages.

We are now under the rounded mountains, of which Taddi Enzazzi is the highest; perhaps it may be from a thousand to fifteen hundred feet.   They seem to be naked, with the exception of some solitary trees and dry grass. Groups of the *Maba palm* are seen in small valleys be-tween the hills, and shrubs of a *Mimosa* are common at the river side.   We arrived soon at the market-place of Embomma, where a number of negroes had assembled. The Captain's intention was to go to the banza this evening, but when the Mafook came on board and promised to accompany him to the King's residence the following morning, the journey was put off.

July 27.   The negroes intruded themselves upon us be-fore we had left our beds.   I went into a boat in order to visit Molineaux's island, which at the river-side consists of a steep rounded rock.   A patch which had been set on fire opened a way down to the more level parts.   Here I found several new and brilliant plants, and lofty trees with upright leaves, which I had not seen before.

Impenetrable thickets obstructed our farther progress. We made some attempts from the boat to ascend the steep hill. The climbers and creepers hanging down its sides resembled worked tapestry. The ground was covered with *Ipomœæ*, one with white and another with violet flowers. The rock consists of blocks of quartz. I was called back on board just when I was taking a view of the surrounding scenery. In the evening we arrived at the banza, and dropped anchor near the shore, which was level, and covered with grass. The residence of the king is on the other side of the hill. The river is already narrow, and is perhaps not larger than the bay of St. . . . . . . The land to the left rises gently, and forms a long ridge, better covered with wood than the lower parts, and behind it is seen a range of undulating mountains. A salute was fired and soon after a palanquin made its appearance, having been sent by the king to the Captain, who however declined going on shore till the following morning.

July 28. This day was fixed upon for the ceremonious procession to the royal residence. Early in the morning we were already teazed with visits from Mafooks and Princes by dozens. The Captain preferred to walk the greatest part of the way. Some marines followed us as a guard and we ourselves, dressed in borrowed coats, formed the van. Our way led over a grassy plain, varied by cultivated patches of maize, different kinds of pulse, and cassava, resembling asparagus as to . . . . . . Single trees were scattered around. The grass was almost dry. A path leads over the rising ground (which consists of clay)

Q q

up to a hill, which commands an excellent view of the
river, winding behind a high island, and re-appearing beyond it. The Captain having got into the palanquin,
which was constructed after the ancient Portuguese fashion,
and the party disposing itself into a sort of processional
order, and being put in motion, which, contrary to European
custom, is here a full gallop, we soon arrived in the royal
city, which did not appear to us a very large one. With
its situation however, and neatness, we were more satisfied.
On an eminence among palms and *Bombax*, *Adansonia*,
and *Ficus*, and several other trees ; the straw huts were
scattered about, all of them surrounded by fences, near
which young trees had been carefully planted. We halted
by a large old tree, under whose sacred branches the
meetings of the elders are held, and all public business
transacted. After some time had elapsed, we were invited to proceed to the residence of his Majesty, which is
composed of several huts ranged within a small enclosure.
The Captain was seated on an elevated place with a large
umbrella over his head, and we took our seats around him.
The King, dressed out as if going to a masquerade, in
drawers and a cloak of silk, and boots of marroquin,
with a large cap on his head, resembling that of a grenadier, adorned with flowers, made his appearance, and
took his seat near the Captain. Next behind us were his
counsellors, surrounded by a great assemblage of people.
Mr. Simons, in an ancient court-dress, with a sailor's
hat on his head, had here the weighty charge of being the
interpreter, and had told us the preceding evening that

great doubts were entertained as to the real object of our visit, and that, since we did not profess to trade, it could be attributed to nothing else than a hostile intention. The Captain ordered him briefly to explain the object of the expedition; but this they could not at all comprehend. When any thing occurred in the conversation that struck them, one of the nobles rose, and with attitudes resembling those of a fugle-man, gave the signal for the *sacala*; on which all smote themselves on their breasts with an expression of joy.

As a first present a cask of brandy was brought forward. This they comprehended well enough; and having poured it into a large washing bowl, they fought among themselves for the nectar. The king and his ministers soon after went away, and we were shortly invited into a spacious straw-hut, which, like all the rest, could not be said to be encumbered with too much furniture. We observed in it, however, a number of small fetishes. Here we were entertained with a dish of boiled fowl and with palm wine in mugs of English manufacture. The sailors dined after us. We took a walk in the town and were allowed to enter into every enclosure. Tudor and Galwey found here the daughter of the Macage of Loomba, which is a market-place, where they had been the day before, and made some bargains in favour of their other friends. The women are considered in the light of merchandise; and a husband generally takes care to make an European pay dear for his bargain. If any of his countrymen, however, should violate his wife without his knowledge, he is permitted to put him to death; but every one most readily will lend his wife to his neighbour for a very trifling consideration. Some of our gentlemen had thoughts of hiring a house; but the distance from the vessel was considered to be too great, and the Captain did not by any means approve of it.

A long deliberation had taken place among the King's advisers; after which we again assembled. The Captain repeated what he had said before. One of the sons of the King was the person who showed most uneasiness, distrust, and coldness during the whole deliberation. At length, an old man, uncle of the King, who commonly communicates the resolutions of the assembly, and who had voted with Simons in our favour (Simons had been called upon by them to take the oath), broke leaf, which is the symbol of peace and amity, with the Captain; implying, that they believed that he had come on friendly purposes, &c. The most interesting personage was the father of Simons. He had given his son in charge to a trading slave captain, of Liverpool, in order to be educated in England; but instead of this, the boy had been sold by this captain in the West Indies, where he had suffered many hardships: from which he escaped by getting on board a King's ship, from which he was by mere accident sent on board the Congo. The father, after waiting for his return eleven years, during which time he had made many inquiries for him in vain, had now unexpectedly got intelligence of his being with us, and came on board the first evening of our arrival near Loomba. His excessive joy, the ardour with which he hugged his son in his arms, proved that even among this people nature is awake to tender emotions. As a token of gratitude he offered to the Captain a present of nine slaves. Simons was here an important personage, was called Mafook, carried in palanquin, &c.

In the midst of this business I could only give a hasty glance at the treasures which surrounded me; among which, however, I observed a *Clerodendrum*, with flowers of a scarlet colour; but I hope that none of them will elude my researches, as soon as I shall be at liberty to dispose of my time. The evening was again noisy, as a number of black gentlemen had come on board. They have already nearly exhausted our store of spirits. To-morrow our pilots, whose names are *Gun, Brown*, and *Tati Maxwell*, are to bring the Congo up to this place.

July 29th. In the morning I set out for the small creek, accompanied by four boys and a servant: here I shot several water-birds. The shore was overgrown with a thick sod covered with a species of *Jussieua*, and higher up with a thick grass, chiefly consisting of *Ischæmum*. I made an excursion on the shore to the right, which is a peninsula almost surrounded by water, being connected with the main land only by a small chain of rocks in the middle. Cultivated patches are seen here and there, with free access to a considerable distance. Lofty *Hyphæne*, a *Bombax pentandr. trunco spinoso*, and a great many new shrubs and other plants, are scattered about. We penetrated through the thickets up to the mountain which occupies the middle of the peninsula. Near the summit we saw a superb tree with eatable fruit resembling dry . . . . . The inhabitants seemed to have no knowledge of its being an esculent fruit, but after I had given the example, they partook of it readily. A tree of *Pandanus* caught my attention.

The prospect from the summit was grand, comprehending the windings of the river, its islands, the points of its shores up to Taddi Moenga and Bamba, the eastern Boka de Embomma, the western Boka de Embomma which is overgrown with wood, Kæti, which is the third island above Loomba, with a rounded mountain in its middle. Boka Embomma consists of a half decompassed gneiss, which occurs in large projecting blocks.

The Captain had pitched his tent to day. Frank Clark came to inform him that his Majesty had sent one of his daughters to Clark's wife, and that he might send for her. I took a walk to the town, chiefly with an intention of seeing Galwey and Tudor in their habitations. On entering the enclosure of the King's residence we found him sitting amidst a number of children, and distributing food among them. They were all boys. Having stayed there a while, during which I was considered as the Embassador of the Captain, a young girl, naked, as all the others were, presented herself kneeling at the entrance. I took her hand, when unexpectedly N'Kenge, the Prince of Embomma, offered her in a very polite manner as a present to the Captain. She was very much alarmed; and on being brought into the Captain's tent, which was lighted by lamps, she ran under the bedstead. The Captain seemed not much pleased with this obtruded civility.

July 30th. Early in the morning I went out with Hodder, and proceeded towards the upper end of the creek; and here I shot a number of birds. After having been on shore on the island Boka, I pursued my course up the creek to a bank in the middle, and passing the villages Thimanga and Vinda, I followed the shore until I arrived at Tinyanga, a village belonging to the brother of the King. He was sitting on a mat, surrounded by his grandees and armed young men. The rock above the village is steep, with two or three springs, and covered with an interesting vegetation. We went to the village of Frank Clark, who accompanied us, and invited me into his house, One of the Captain's people had been sent to fetch a cow, which was promised him by the King. In assisting to catch this half-wild beast, my foot was entangled in the rope which was fastened to her, and with which she was running away. I was thus placed all at once in an awkward situation; it might have been still worse if the rope had not broken. I was brought to the King's house almost senseless, except to exquisite pain. From thence I was removed on board in a kind of litter. I was confined to the cabin till the afternoon of the following day, when, with the assistance of another person, I was enabled to limp on shore, and to collect such plants as I might have overlooked on my former excursions.

Gun brought us the agreeable news that the Congo had already passed Taddi Moenga. The Dorothy is at Tall Trees. They say they have killed an alligator. Frank, according to a promise given long ago, brought on board a weeping girl last night, who was soon followed by another. To day Tudor and Galwey have removed their light huts to the tent of the Captain, which is already called the Captain's village by the inhabitants. A village means with them a pater-familias and his private dependants. The village of the King (Banza Embomma) is the only village where several families have sufficient land capable of cultivation in the neighbourhood, and to enable them to live together. The land is cultivated in patches only, and the labour is performed by the women, whom we saw frequently in the fields, carrying with them their children and baskets of provisions, the daughters of the King as well as others. The only plants we observed to be cultivated were cassava in small quantity, and maize, planted chiefly along the river side, probably because the air there is more humid. The cotton shrub was growing wild in the plains. *Pisangs* were frequently brought from the market of Loomba, though none were seen in this neighbourhood. They were said to grow plentifully higher up the river on the opposite side, as also oranges and other fruits. The order of precedency is : Tjenu, or the King, the princes Malibere, Mambous, Macaya, Mafook. The inhabitants are addicted to some superstitions with respect to food ; as, for instance, to abstain from eating eggs and milk. A Fetishman is not allowed music at his meals, except when he has not partaken of *Leimba*?

July 31. My foot being much better, I was enabled to take a short walk on the plain. I caused a pit to be dug in order to ascertain the temperature of the earth, but the elay which here every where is predominant, and hard as stone, did not permit a deeper pit to be dug than three feet, which proved insufficient. The temperature however appeared to be 80°. The weather, which continues cloudy and unsettled, does not allow of any remarks on the dew, and in general makes all observations on the climate in this season less intreresting. The temperature to-day, before the setting in of the sea-breeze, is about 77°, and in the tent of the Captain, 88°. The grand-father of Simons is on board. Hodder is sent with orders for the Congo.

August 1. In company with Galwey and Tudor I visited our creek again in the morning. We passed through floating *Jussieuæ* and *Ipomææ*. We shot a number of birds of a species of . . . . ., which is here very common. We proceeded upwards to the sand-bank, which Galwey and Tudor yesterday had found to be passable in a flat-bottomed boat. It was overgrown with floating grass, through which we could not pass but with great difficulty and labour; but in return, the most beautiful scenes presented themselves on the other side. . . . .

The river runs between rounded mountains, the higher parts of which are naked, but the lower parts along the river side are partly covered with a luxuriant vegetation. The mountains incline towards the north, with their precipices-facing the south, which is indeed evident by the different appearance of both sides of the river. The declivity is in some places vertical. The Congo with the two double-boats arrived in the evening and anchored near the opposite shore. To-morrow we shall take leave of the Tjenu.

R r

August 2. We went to the banza with the officers to take our leave. In the meantime Lockhart made an excursion to the left between the plain and the hills, and I took the opposite side, where I again met with the superb *Brownea*. The visit was unceremonious, and not very hospitable. His Majesty did not so much as offer the Captain a mug of palm-wine. His dress was that usually worn by the natives. He was seated near his house, opposite the bench of the officers, surrounded by about fifty negroes. I visited their burying-place, where a very deep grave had been dug for a woman who had just died, and I passed by her house, where half a dozen women were howling terribly. They are said to repeat this mode of shewing their grief for several days, keeping up the same kind of concert for about an hour each day. I returned in company with my friend Frank Clark (who is the best of all the members of the Royal family), to the precipices, by the way of Tihenyanga. The rocks consist of a completely decomposed quartz, and granular red masses. In some places water is continually pouring forth as from springs. The temperature under the shade of the thickets is down to 70°, probably in consequence of the evaporation. I found several new plants, among which was a shrub bearing berries, of which I am uncertain whether it be cultivated, but I saw it growing wild all around. I met Lockhart, who had collected several beautiful plants. Hawkey and Mr. Kerrow returned to the village on a visit to the ladies. Tjenu offered another daughter to the Captain in addition to the first-mentioned, who was better looking than the former. I was occupied with laying in plants till late in the evening.

August 3. Yesterday the village of the Captain was removed in about the same space of time as that which had been requisite for its erection. To-day, at day-break, the schooner proceeded, with the assistance of oars, along Boka de Embomma, and anchored nearly opposite to the middle of it, where the rock was overgrown with thickets of shrubby plants. An alligator was seen swimming near the shore. Lofty trees over-shadow the steep sides of the rock, which rises to the height of four or five hundred feet. Its upper parts are almost naked. I was ordered on board the Congo, in consequence of some misunderstanding between Tudor and Cranch, relating to the ammunition belonging to the latter, and of some irregularity which had taken place with respect to the boats.

August 4. Accompanied the Captain to the rock Taddi Mansoni on the opposite side. Plains and deep vallies vary with hills, which sometimes terminate in peaks and grotesque cliffs. Groups of the *Mimosa spinosa* and some cultivated patches were scattered about. We saw in our way the skeleton of a hippopotamus. We measured a large *Adansonia*, which at two or three feet above the root was found to be forty-two feet in circumference. The river is now no longer divided into several channels, but continues to be free from islands to a considerable distance. Its breadth here I calculated to be about the same as that of Drammen (in Norway) at the bridge. I mounted the hill, whose base consists of a yellow mica-slate, which con-

continually decreases in approaching the summit, where the formation is almost entirely quartz. Among the trees we found lying on the ground several pieces of wood, which was called *lignum vitæ* by our carpenters. It is extremely hard, but I could not ascertain what it is. A *Maba palm* was cut down, which I examined. We found several curious insects. Our situation, as we are walking here under groups of trees of various kinds, would be envied by many of our friends. We saw monkeys running to and fro on the branches of the trees *(Simia cephus)*, and several birds, among which was a spotted *Alcedo*. Divine service was performed on deck, where we had an agreeable temperature. We observed on a small bank of mud, situated a few fathoms only from our anchorage, that the flood tide rose about ten inches.

August 5. In our flat-bottomed boats we arrived at Tchinsala, and run deep in the mud. We observed the Congo proceeding upwards. The Captain was just gone on board the Congo and brought her up to the height of Tchinsala on the opposite side of the river, where she anchored near the end of the island Kinyangala. The Captain finding that she only retarded our progress, resolved to leave her in this place. The evening was spent in arranging for our further progress in the double-boats. Fitzmaurice and Hawkey went on board the schooner. Cranch was at first resolved, after long deliberation, to remain with the Congo, but changed his mind on hearing that the third part of the apparatus should be delivered up to us. All left collections to be conveyed to the Dorothy. I left all my dry plants and a box full of seeds.

August 6. I had landed for a moment on Tchinsala, but a shot called me immediately back again. Our flotilla was already under sail at one o'clock. The longboat of the ship is to go as far as Benda. We crossed the river several times according to the strength of the current. The mountains come down to the river, and in many places contract its channel to scarcely half a mile. They are rounded, and commonly sloping hills with narrow and short vallies between them. They are overgrown with dry grass, but otherwise entirely naked. The southside had at first more level ground, with rising blocks of slate dispersed over its surface. At Vinda, a small plain, the mountain along the northern bank is very steep, the declivity about forty-five degrees, and sometimes almost vertical, out of which were projected several rocky points. It is only near the high grass on the banks of the river and in the vallies that trees of luxuriant growth are to be seen. On the small plains at the heads of creeks we observed some villages amidst palms of *Hyphæne*, which were also seen scattered about higher up the sides of the mountains. The banks, with their precipitous cliffs of slate, overgrown with a hanging green tapestry of climbers, and surmounted with plants and trees of various kinds, among which is a high and always naked *Adansonia*, present indeed a picturesque view, but nothing yet inspires the notion of an extraordinary grand river. We anchored above three islands (*Tanyanda*), on the northern bank, near where it is studded with high rocks. One of these was said to be used for the same purposes as the Tarpeian rock of old, from which criminals, as for instance, seducers of the wives of the King, &c. are sometimes precipitated.

August 7th. In the morning we found ourselves between Gamba islands. We saw several birds. The mountains consist every where of mica-slate. These picturesque islands are very much visited by hippopotami. At noon a gentle breeze carried us a little higher up. The river at Fidler's Elbow is again somewhat more expanded. In its middle are some rocks, with scattered trees. The river turns northward, and is contracted by the mountains. These are here somewhat higher up intersected by narrow valleys. They are all of the same mica-slate formation, inclining in all directions, but commonly towards the west. The inclination is at half an angle (qu. 45°). We observed half-a-dozen monkeys, from three to four feet in length, on a hill near the southern bank. The breeze seems to die away as we pass between the mountains. We anchored on the southern bank. The natives of Benda begin already to talk of the fall of Yallalla, of which, they tell us, we may hear the roar. At one time we could only approach this cataract by a journey by land, or by double-boats composed of canoes, &c. The river is here broad, resembling the Scotch lakes.

August 18th. I went on shore up a narrow valley, opposite to us, which, as is usual, forms at its lower end a small plain, covered with grass, and higher up a narrow, deep ravine, the sides of which are covered with trees. Galwey and myself ascended the hill just before us. It consists of mica-slate.

We had only a confined view of the river, the prospect being obstructed by the mountains, which appeared somewhat higher to the east. The river runs in a winding course between them. They form columns rounded at the top with fragments of quartz, which sometimes form veins and beds in the slate. We found here some scattered shrubs of *Eugenia,* and two or three species of grass. We rowed up opposite Congo Binda, which is situated at some distance inland, and high upon the platform of a mountain. We went on shore and followed a rugged ravine, whose sides consist of a compact mica-slate. . . . . . We observed the traces and excrements of several kinds of animals; chiefly, however, antelopes; but we did not get a sight of any of them. I walked over some flatly-rounded rugged hills, on which only a few shrubs were growing. I attempted in vain to get a view of the windings of the river as far up as the cataract. The mountains are of the same form, and are seen to a great distance, undulating with sloping declivities, and frequently intersected by deep ravines. The level parts are luxuriant, but the sides and tops of the mountains are naked. I descended the ravine through climbers and shrubby plants, almost all of which left me in uncertainty as to their genus, and regretting, as usual, our coming here so late in the year. I followed another ravine, intersected with numerous narrow but deep holes, which generally makes the ravines very difficult to walk in. I met Lockhart amidst a thicket, in which were several lofty trees, but almost all of them were in fruit. We discovered an *Arum foliis 3-nat. dichotomis;* the root of which we afterwards boiled, and found to be eatable. We returned in the dark quite loaded.

During our whole excursion we did not meet with any animal, a few pigeons and small birds excepted, but we saw a great many traces of hippopotami on the shore. Simons was sent to the Tjenu of Benda, in order to procure some men who were acquainted with the country higher up. His account of Benda would seem to prove that the inhabitants have some intercourse with the Portuguese. He was ceremoniously received.

August 9th. We sailed somewhat higher up. I passed over to the north side of the river. The vegetation is without variety, and the steep hills overgrown with grass, in which was only found growing an Euphorbia. I fired three shots at an alligator. In the evening it was resolved that we should go to Banza Nokki, the residence of a Tjenu, by the way of Condo Sonjo. The north side is generally called Benda.

August 10th. Simons was dispatched to announce the intended visit of the Captain. Early in the morning we set out on our journey, and proceeded over the hills and across the small plain, where we had been two days before. The slate formation here ceases. The rounded mountains on the other side rise to a greater height, with projecting cliffs of a more cubical form. The transition is formed by a compact mica-slate with a large proportion of feldspar. The . . . . of the mica mountains . . . . . . perhaps sienite. It is granular as granite, and composed of quartz, feldspar and a third metallic substance. Blocks of rock, which we have often met with before, were scattered about, but they occurred now in greater number, and in masses of a cubical form.   We passed two or three small villages, situated between the mountains.   A spot planted with *Pisangs* was seen from amidst the palms.   A few moments afterwards we arrived at a rivulet.   Springs, they tell us, are here very common.   This circumstance also proves a new formation of the mountains.   These small villages and cultivated patches, surrounded by mountains of more grotesque form than those hitherto seen, and the luxuriant vegetation, afford us a new and beautiful view.   The straw-huts have here rounded roofs. At eleven o'clock, after a march of six or seven hours, we reached the summit of those mountains, which appeared to be the highest within our view.   On a small plain the ground had been cleared and planted with *Ficus religiosa.*

Here is the market-place of the inhabitants of the banza, and close by is the village of the Tjenu.   In passing between the huts we observed poppies, cabbages and other vegetables.   Having observed the height of the mercury in the barometer, we found, according to the calculations of Leslie, that the elevation was about 1450 feet, which, generally speaking, may be considered as the greatest elevation of the highest ridges of mountains in these parts.

The manner in which we were received by the Tjenu had more of stateliness, but also more of savage manners, than that in which we were received at Embomma.   A silken hanging served to cover one side of the hut, in which the Tjenu was discovered ; he was dressed in a red

cloak, with a cap on his head resembling that of a grenadier, and adorned with feathers. His two ministers were
on each side of him, and seemed to be eunuchs. A little
palm-wine was the only refreshment he offered to us. The
conversation was short. The Captain only wished to have
some guides. A smile was the only answer. The ministers
ran several times to the fence and back again, crying out
some words expressive of the king's understanding us. Tatti,
a good-looking man, the father of Simons, invited and
entertained us. The temperature of a spring down in a
small plain we found to be 73°. Following a rugged path I
walked down the valley and passed over another rivulet.

August 11. The interesting scenes that now surround us
demanding a more close examination, I proceeded, in
company with Tudor and Lockhart, to the lower end of the
valley, over patches planted with cassava, following the
course of the rivulet. In our way we met with thickets composed of shrubs and trees, of which many were entirely new
to us. The women screamed out on seeing us, but the taste
of our brandy had soon made them less shy. Yesterday
we were constantly followed by a number of people, chiefly
boys. They said they had never seen white men before.
We ascended some of the high rocks, but were unable to
climb up the highest of them, which is of a conical form.
They are all of the same formation. We rested ourselves
at the source of the rivulet. Its temperature was 71°. Its
coldness is probably to be attributed to the evaporation.
I missed Lockhart, who came down a long while after me.

August 12. The forenoon was spent in laying in plants.

The Captain bought a slave of the Mandingo tribe, that is said to live high up inland. Their language is somewhat different from that of the Congo language. In the afternoon I followed the Captain up the hills towards the point, in order to observe the winding course of the river, which is now contracted within a narrower channel, by projecting points. The conical mountain of *Yallalla* was just before us. A projecting bank of rocks is seen on both sides the river, which is not passable even by canoes. The country appears equally barren higher up, and not likely to enable us to procure sufficient provisions on a journey by land, which must now be resolved upon, nor are any here to be bought. The population is scanty, only a few scattered small villages, situated on the hills, are discovered amidst some groups of trees. The natives are not willing to part with their commodities, except at extravagant prices. They crowd daily around the vessels with commodities for sale, but all we have been able to get are only a few fowls, some eggs, and a single sheep. Their knowledge of the country is very limited. They are pretty uniform in asserting that the cataract of Yallalla is a considerable one, and that above it the river is divided.

August 13. We remained in order to make some observations with the barometer, the oscillations of which nearly correspond with those observed at the Canaries. Last night an uncommonly strong breeze set in. In the morning we made a short excursion on shore. The Captain went in a boat towards the conical hill of Yallalla, but he found the river here to be impassable. A journey by land of four days to Yallalla was therefore resolved upon, in order that, after having examined this cataract, we might take such further measures as should appear to be advisable.

August 14. We set out to the number of about thirty, and passed across the river to a small sandy beach, termi-nated by sand hills, on which the hippopotami commonly dwell when on shore. A steep ascent led to a village. Having gained the platform, we found the declivity on the other side precipitous, but no trace of mica slate in the formation. - - - - - - - - The mountains here consist of compact feldspar, and resemble those at Nokki. On reaching the high land, the country assumed its usual appearance, which resembles very much that of old or-chards, composed of stunted Annonas, and three other sorts of trees. Having soon after reached the hill on which the banza stands, and passed through the lofty palms and Adansoniæ in which the village is buried, a view of the river opened all at once upon us, and we discovered the cele-brated fall of Yallalla, at a distance of about a mile and a half. But how much were we disappointed in our expec-tations on seeing a pond of water only, with a small fall of a few hundred yards! We descended the steep barren hill, and arrived at the fall. The rocks on both sides of the river were precipitous. The mica slate is slightly un-dulating, and abounds with veins of quartz and compact feldspar. This formation of the rocks may possibly have very much weakened the force of the waters of the river. The inclination is half an angle (qu. 45°) in the same direc-tion as the course of the water and the declivities are oppo-

site to it. I descended some of these declivities to the depth
where they are washed by the water of the the river in the
rainy season, and found a great number of rounded exca-
vations. In the middle of the fall is an islet at the distance
of about a short stone-throw from the shore. The river
above Yallalla winds between two projecting points in a
northerly direction. On both sides the river, rocky hills,
intersected by ravines, are visible to the distance of two
miles. They are all lower than the high platform of the
hills on both sides ; that on the west continues quite flat as
far as the horizon. Yallalla may in fact be considered as
placed in the line of the greatest elevation of the mountains.
Towards the east the country is more broken, and in some
places may be called mountainous, but the mountains are
searcely any where so high as at Nokki. The summit consists throughout of a hardened clay.

In the evening we made a visit to the Tjenu, who is a
plain good-natured man, who expressed his satisfaction on
seeing a few gallons of brandy, for which in return he presented us with some fowls. Scarcely any information can
be drawn from the natives of the state of the country higher
up the river. A slave merchant affirmed that he made a
journey of a month on the eastern bank, and found the
river, as he proceeded, expanding as wide as it is at Shark's
Point. The eastern side of it, he told us, was more populous and civilized, than the western side, which they pretend they do not venture to visit, for fear of the savage
disposition of the bush-men on that side. Some even asserted that the people on that side are canibals. Our tent

was pitched in the evening at the end of the village, from which we have a view of Yallalla. Below is a valley covered with wood, in which runs a small stream of water, supplying the inhabitants with that article. We made a circuitous route over the elevated plain, in order to come upon the river higher up. We passed Gongola, which is the residence of another Tjenu; but both are subject to the Suxum Congo, the province to the northward on the north-west side of the river. The south-side is called Kukulu Congo. Gongola (Concobella of the charts?) is said to be the last regular village in the dominions of Congo. After a long march on the summit we directed our course towards the river, following the common path down to its banks, where two small canoes, which are the only ones close above Yallalla, are used as a ferry for crossing the river. Notwithstanding the repeated remonstrances the Captain made to the guides that they should conduct us straight forwards up the river, we found that we had advanced a few miles only above the fall of Yallalla. At noon we halted. We made an excursion on the hills near the river, the banks of which now consisted of a white sand. The river above the cataract is full of rocks. We returned towards a village, where, out of humour with our guides, we took a more direct way through the grass. In the evening we were obliged to halt on a hill covered with grass. A small valley, with a rivulet winding its course through it, was seen below. The rivulet is said to be visited every morning by buffaloes and antelopes, of which some had been seen at noon. When near the villages we were constantly followed by a crowd of negroes, particularly women, who came out to enjoy the sight of white men, which was perfectly new to them. Though more shy, they appeared to us to be more good-natured than those we had seen before.

August 16. We went across the valley and the hills on the other side, (which were last night illuminated by fires,) and arrived at the village Monzi. Having here procured guides who were better acquainted with the country, we proceeded on our journey over the hills. Between them are small plains with a luxuriant vegetation of trees. Towards the north, the country (which is already called Mayamba country), is more level and more woody. Elephants are reported to be plentiful here. A wild boar rushed forth in a valley, and though it broke through the whole line, the sailors, from their hurry and want of skill, all fired amiss. Unfortunately I happened to be somewhat behind, and consequently had not the pleasure of seeing it. We continued our route over the steep hills, following the course of the river, which is here bent into the form of the arc of a segment of a circle. At a short distance above the bend, and on this side of the river is situated the village Jonga (Inga), which is the first village inhabited by bush-men. From the hills we had a distinct view of the river running again northwards to a considerable distance. I went down to the steep banks of the river. Thickets of shrubs and rocks of slate that are scattered about in the water, give to the line of the river a beautiful appearance. A young alligator was basking himself in the sun. 1 fired at him with small shot without success.

The river here is full of rapids and rocks. It is however navigable, though not without great difficulty, and no canoes are to be had ; and as it would take too much time to carry canoes over land by the way of Kullu, the plan of the Captain to make double-boats of them must therefore be given up. A journey by land over Jonga was resolved upon. We may reach it, we are informed, in a day, by proceeding from Kullu over the high plain. We returned in the evening by the shortest way to Kullu, where we did not arrive until late. Mr. Tudor and several men, who were already quite exhausted by fatigue, were sent back. Next morning Tudor was attacked by a violent fever. Unfortunately the greatest number appear to be men quite unfit for a long march. The few marines we have are of the veteran battalion.*

August 17. The sick were sent before our departure early in the morning. We made a visit to the Tjenu, who promised to send at noon twenty men to carry their baggage back again by the shortest way down the river.

* There is no such battalion ; the oldest marine was not 40, and all were stout healthy volunteers. Ed.

SECTION IV.

From Cooloo to the Extremity of the Journey.

August 20. THE boat which had been stationed at Nokki went down to the Congo sloop. We accompanied the Captain. Two days before he had dispatched Hawkey and Tudor, with most of the people and part of the baggage carried by black men, who now met us on their return in order to fetch the remainder. Our whole number was about thirty, but of whom there were but few fit for a long journey. After having reached the summits of the hills along the river side, and passed by a village, Tudor, in company with Lockhart, turned somewhat to the left through a small beautiful valley, where I found at last a *Myrianthus* in flower, and a new dried fruit. Passing by Areba, and descending a steep hill, we reached Kulloo at four o'clock. In the evening we paid a visit to the Chenoo. Here we received some farther information about the country; but the men who are to carry our baggage are so expensive, that they will soon have exhausted our whole stock. A depôt is to be left here in Kulloo. The views of the country are very limited.

August 21. In the morning I went to the other end of Yallala. The rocks here present a new structure, being

T t

mixed with quartz, and bearing some resemblance to sienite. A shrub with black fruit, which I had seen a long time before, was found to belong to *Pentandria trigynia*; and a tree, the fruit of which was sweetish and of a blackish colour, I discovered to be of *Hexandria monogynia*. Observations on the barometer were made during the day. In the evening I went down into a small valley on the other side, where I found a lofty tree of an uncertain genus, which I had seen before in the swamps. A - - - - - was seen at the village, with brilliant flowers and an *Apocynum (Nkennyewumba)*.

On the 20th, the thermometer in the morning at nine o'clock was at 75°; at two o'clock it had risen to 83°, and at five P. M. it was 73°. On the 21st at nine A. M. it was 71°, at two P. M. 81°, and at five P.M. 78°. A young *Adansonia* thirty feet high, and half a foot thick, was found by its annual circles, as well as by its branches, to be thirteen years of age; which would seem to prove the opinion, grounded on its very great size, of its long life, to be erroneous; and indeed this might already have been inferred from its spongy and succulent texture.

August 22d. After having, with much difficulty got some men to carry the baggage, we set out for Inga, leaving Fitzmaurice and Hawkey behind, and proceeded almost by the same way as on our former journey over Gongala to Mansi, across a valley covered with wood and well watered.

Sangala woo is a kind of reed (perhaps an *Amomum*) which is always kept fresh in the house. In time of war it is rolled between the hands in invoking the war fetish.

They chew it and spit it out on their enemies. Mazenga are fetishes used for the discovery of robberies. They are covered with . . . . We proceeded over rugged hills and small ravines thickly covered with wood in a direction somewhat more to the north than on our former journey. We had been told that we would not find any water until we reached Inga. We found ourselves all at once in a deep ravine about 120 feet in breadth, the bottom of which was level with holes; and consisted of sand and mica-slate. We found plenty of water. A dark thicket of trees, which, as usual in ravines, grow upright to an extraordinary height, gave to this spot a picturesque appearance. As the day was declining the Captain was prevailed on to pitch his tent there. A number of plants were found, among which a *Hillea hexandra, tubo coroll. longiss. deflexo*, and three species of ferns. The monkeys were seen coming down the hills, and we heard the noise of buffaloes, mingled with some singular cries of birds issuing from among the dark thickets.

August 23d. We continued our route, passing over high and more woody hills and ravines, round the bend of the river, where it again winded its course towards the east. We had a view of a fall that was nearly opposite to Inga, and appeared to be higher than that of Yallalla. We made our entry into the banza of Inga about noon. It is situated on a high plain, and, as usual, surrounded with palms and Adansoniæ. It was with difficulty we could procure a guide, by alternately using menaces and entreaties. The Chenoo had given his permission, but the Macages seemed to interfere, and appeared to be jealous of our intercourse with the neighbouring nations, whom they were anxious to describe to us as being exceedingly evil disposed. We were obliged to remain here all day, and meanwhile made a tour round the village. The tree Safu* grew here in abundance, *3-cal. 3-petal. 6-andr. c. gland. 6-altern.-monogyn.* The fruit, which was not yet ripe, is valued very highly.

August 24th. A stranger had undertaken to serve as guide. The greatest part of the baggage was left behind. We proceeded eastwards parallel with the river, and into a long valley called Vimba Macongo, which we found tolerably even in traversing. We then passed by the conical mountain Sansa Madungu Mongo, which had long been within our view. Though the composition of the rock is very mixed, it is, however, probably nothing else but mica-slate. We arrived at the villages of Condoalla and Kincaya. The horns of a large antelope were seen. I fired at a small species of this genus. We proceeded further through a valley covered with ant-hills, bearing a resemblance to a fungus. After a march of eight or ten miles we were again in the vicinity of the river, whose waters here moved gently along. We observed on the opposite side a large canoe at no great distance. . . . . . We encamped on an agreeable spot near the river, surrounded by green banks and trees. Several new plants occurred.

A Chenoo, who appeared much more polite than the

---

* In a detached note in pencil: Safu. fruct. edul. nigre tingens. (Kullu-M'galo Galo)

people of Inga, paid us a visit, and was presented by the Captain with an umbrella and some other articles. Several others came afterwards, particularly on the following morning; but our want of the all-powerful malava (presents) gave great dissatisfaction. They promised, however, to assist the Captain in purchasing canoes.

August 25th. The great number of traces observed all along the river induced me to go very early on an excursion to a sandy beach of the river, situated somewhat higher up at the end of a level ravine. Recent traces of buffaloes, antelopes, and turtles were seen, but no living animals, except two large wild ducks. We were much incommoded by numbers of people from the opposite shore and from Kullu. Some of them, however, informed us that the river higher up forms a great sandy island, and that it divides into several branches across low swampy lands. From this place set out on our journey back again over Kincaya and Condoallo to Inga, where we arrived in the afternoon. In the mean time I visited a ravine, and got a magnificent view of no less than four rapids, of which Sanga Jalala (Sangalla), situated above the others, and not far from our last encampment, appeared to be the most considerable of the whole of the falls, the number of which probably amounts to six.

August 26th. Hawkey was dispatched with fourteen negroes and some of our men to Kullu to fetch the baggage. We accompanied the Captain through the valley of Dimba to Kincaya with a view of buying canoes, but did not find the owners at home. The inhabitants appeared

to be of a benevolent disposition. At noon we arrived
at Condoalla, majestically situated in a lofty wood, and
surrounded by plantations of Pisang. The . . . . . tree
foliis terminal. pennat. grew here in abundance. Its fruit
was not yet ripe. I observed flower-buds of the shrub
Echinophora, which we saw in the district of Kullu. In
returning, I ascended the top of the high Madungo Mongo.
The rock in the valley consists chiefly of mica-slate,
stretching as usual towards N. E. and inclining under an
angle of forty-five degrees towards N. W. Undermost in the
valley, the slate is thickly mixed with a granular feldspar
and hornblende. The top part consists of a very loose
mica. The mountain here inclined gently in a long-ex-
tended slope. Its opposite side was covered with wood,
which was now on fire. I had a full view of the whole
valley, which from my station appeared very flat. The
river was seen behind in its whole course, forming several
larger islands in the line of Inga and my station. The
country behind rises into a platform of uniform elevation
with the plain on which Inga is situated. The summits of
the hills are somewhat higher in the back-ground. Farther
still; at the extremity of the view, the river runs round an
eminence, but afterwards probably turns somewhat more
to the northward, where the country is considerably
lower. No eminence appeared more elevated than the
summits of the hills bounded by the horizon. The general
direction of the valleys and the plains is nearly N. and S.
I climbed up a lofty tree bearing fruit, some of which I
had previously brought down with my musket. I shot

some pigeons; and being induced to pursue a flock of large crested toucans, I did not reach Inga till it was quite dark, where they had been in some uneasiness about me.

Last night, when encamping in the open air near the river, we had the first shower of rain since our arrival in Africa. The negroes on seeing the new moon expressed great joy, because during the next moon there would be abundant showers of rain, and the time would then be arrived for planting their grain. The river would then also begin to swell, and in two months the rainy season, properly so called, would be set in. Lockhart had made a tour to the river, where he had found another *Epidendron*, and an *Euphorbia* ten feet high. *Caudice 3-ang. spinis marginalib. binis, ovalibus crassis, petalis oblongo-peltatis.*

August 27. Hawkey returned from his journey already before noon, having proceeded beyond Mansæa, where he had found the fruit of a singular water-plant, *calcare longo*. Accompanied by Galwey I descended into a deep ravine, situated N. E. from the village. The plants grew here thicker than we had observed in any place before, but the greater number of them bore only leaves. We now ascended a hill, which afforded us a view of four of the falls. That part of the river, where they occur, is contracted between rocky hills, forming nearly a continued line of rapids. Sanga Jalala is the uppermost fall but one, and the highest. The lowest is under the village of Inga, a little before the river turns towards the south, and the fall of Yellala begins at the last turning of the river towards the general direction, or south-east. The length of the rapids is perhaps twenty miles. In the evening I went in company with Lockhart to the ravine situated northward, where I found the superb tree *Musanga* in flower.

August 28. Captain Tuckey and Galwey proceeded through the valley to Sangala, the uppermost rapid, in order to ascertain whether canoes might pass over it. Accompanied by Lockhart, I descended the hill, in order to cut down the Musanga-tree, round the top of which several climbers, *Dioscorea*, and a singular species of *Aggregata* were twisted. The flowers were withered. A man died in the village. Gangam Kissy was busy all day in order to discover the cause of his death, and found out at last that he had been poisoned by three men, among whom was the Macaya . . . . Macasso is a nut chewed by great people only. It is rather bitter, and perhaps is the Cola nut. Casa is a purgative legumen. Jandu, a *Dioscorea* growing wild. Its root is used for food, but it has a bitter taste, even after having been boiled a whole day.

August 29. Captain Tuckey returned in the evening after a long and fatiguing tour. Above Sangala occurs another fall, but beyond this the river is said to run quietly, and that canoes with two pair of double oars are to be found there. A number of antelopes had made their appearance. A village had been deserted from fear of vengeance for a crime committed in a neighbouring territory.

August 30. Hawkey was ordered to go to Walla to hire canoes, by means of which it was intended to endeavour to convey the baggage up Sangala. I accompanied him on this journey. We followed the usual way through the valley of Dimba and its villages. From the eminence over-looking the valley we observed a herd of antelopes. On our arrival at Wallu we were informed that there was but one single canoe at the ferry, and that it was used chiefly on market days.

August 31. Before day-break Hawkey and I set out for the small sandy creek. Here we again fired at some of the large wild ducks, and observed recent traces of hippopotami. On ascending some hills we saw several flocks of Guinea-fowls, of which we shot one, and when engaged in the pursuit of these birds, we came in sight of four antelopes, two of which were of a blackish colour on the back, the rest of the body brown, with large spiral horns. We met with but few new plants.

Sept. 1st. Lockhart this day discovered the female tree of Musanga. Another tree which had been seen with fruit a long time before was now found with flowers. The bark and leaves of the Casa tree, which the Gangam Kissey made use of as an ordeal, were brought to us. They are said to be poisonous. Of some climbers I got only imperfect specimens.

Sept. 2d. We set out for our final tour in company with the Captain and Hawkey; Fitzmaurice was ordered to go back to the ships; Hodder was sent up to Inga, and Galwey was carried back again sick. Accompanied by twelve men, and the negroes carrying our baggage, we proceeded by way of Wallu and along the valley under the foot of the hills. Some antelopes were seen. We halted at a rivulet

U u

called Lullu, where I found a single specimen of an *Angi-osperma* and an *Euphorbiacea*. Our way went over rug-ged hills till we approached the high land towards the evening; we then crossed some rivulets, near the last of which was a luxuriant growth of trees and thick grass. A *Clerodendrum* was found here, but I lost the flowers.

Sept. 3. We passed over the high mountain at Man-goama Gomma, where I found a *Limodorum*. At noon near a rivulet was found a *Labiata herbacea*, a parasite plant like *Loranthus;* and in the rivulet of Lullu, a frutescent *Oxalis*, ... *Antidisma*, and *Polypodium pteroides*, in small ravines.

The slate inclined this day more towards the west, and the hills were better covered with wood, and distin-guished by many new villages. The higher parts were covered with a red clay. From Mangoama Gomma a view opened over the upper part of the river, which is expanding itself over the surface of a country of less elevation. Near the upper Sangala is a narrow passage through the moun-tain, down which the river precipitates itself within a very contracted channel. For the first time I observed in the ravine here the transition to clay-slate, of which the hills are formed. Its inclination is uncertain, but generally towards the north. When on on the height of the banza Bomba Yanga, we were met by a party of slave-merchants. From thence we directed our way more towards the river, and proceeded over some smaller hills till we reached Condo Janga in the evening. The view above being un-interrupted over an open and flat country. The river had now the appearance of one of our northern lakes; its banks were covered with wood, and in some parts of it were rocks rising out of the surface in the form of a horse-shoe. Canoes again made their appearance, and renewed our hopes. We were at the beginning of a country evidently capable of an extensive cultivation, with a fine navigable river, with abundance of provisions for sale, and with an encreased population. The evening was spent in hunting after the hippopotami, which snorted close by us at the point of the bay; and they were so numerous, that we could not say it was their fault if the chase was unsuccessful.

Sept. 4. Our tour to-day was more agreeable than any we had made for a long time. After having crossed the bay in canoes, we walked along the level banks of the river. A delay however was occasioned by a quarrel with the carriers. In passing through two or three villages we were followed by a number of people attracted by curiosity. We proceeded over the bend of the river, which, higher up, again runs in a northerly direction. In the formation of the rocks clay-slate is still predominating. The river side was in some places covered with less slaty clay of a reddish colour, which higher up alternated with a compact blue, sometimes horny limestone. We pitched our tent at noon.

Our party begins again to be somewhat dispirited, and it is whispered that we shall return in about two days more, at a time when every thing seems to inspire fresh hopes. Many new plants were collected here, and two singular species of unknown fishes. We saw also a great many striped skins of a small species of antelope.

Sept. 5.   I made an excursion along the rugged banks of the river, which now form small sandy beaches between projecting rocks of clay-slate, with three or four alternating beds of the above-mentioned blue compact lime-stone.   I made my way over two ravines thickly covered with wood, and shot some pigeons, which flocked all around in great numbers.   I saw also some partridges, a species of the strandpiper, a vulture, which I fired at, and a corvus.   Of plants I saw a *Fern;* a *Frutex dioicus, stylo 3-partito plumoso;* a *Frutex spinosus debilis, syngenesia polygamia necessaria, flor. capitat;* a *Malamba* with young fruit, *seminibus in pulpa nidulantibus.*

We proceeded across the plain up to the foot of a fine hill, that limits it towards the west.   An excellent view of the course of the river here presented itself, comprehending its long course eastward from Condo Inga, which we had partly passed, with its . . . . small rocks, and the whole plain covered with scattered groups of palms. Higher up the river was seen turning northward round a point, above which the village Kabinda is situated at the foot of the eminences.   The northerly very expanded reaches of the river soon ceases, and the river turns again in a southerly direction, winding between small points, and disappearing behind the hills to the S. E., in which direction it seems to continue.

The country towards the north and north-east is somewhat precipitous, with hills of unequal height, but on the south side and south-east side it is more level.   No very considerable mountain has as yet been discovered in the back ground.   It is only along the river that trees are growing. The width of its expanded smooth surface is generally about one* English mile. It is here extremely beautiful, and if the hills were covered with wood, it would be another Ransfiord.† A number of spectators came down from the village. I descended quickly, but found no bargain made. The inhabitants had danced a Sanga. A gentleman promised us a canoe next morning. The dark long sky, which about a month ago at the horizon was shown to us as a sign of the approach of the rainy season, appeared in the evening at the north side of the river. The night was moonlight.

Sept. 6th. Having at last succeeded in hiring two canoes, the baggage was carried by them along the shore, while we proceeded by land through two villages over the plain and down to the river at the upper end of the rocks, which are disposed like a horse shoe. At this latter place the canoes were dragged over two rapids. Four hippopotami were seen here, at which I fired several shots, and hit one of them in the head, when he started up and disappeared. We proceeded round the point into the bay of Bobomga, and behind the first long rock, which was found to consist of crystallized lime-stone, of which perhaps the narrow tongue of land is also composed.

At the bottom of the bay is a small lake, into which the water flowed through a creek, which perhaps indicated the

---

* Dr. Smith is very loose and vague in all his descriptions and statements, that fall not within the sphere of Botany. Captain Tuckey makes the width of the River here from three to four English miles. ED.

† A firth in Norway, or a large bay.

rising of the river. Its shores were all around covered with *panicum*. An alligator was swimming in the lake, and another before the mouth. Traces of hippopotami were seen everywhere. Shoals of fish abounded in the small creeks. A *Hæmatopus* and several other birds were seen. All this gave to that spot a romantic appearance. Having crossed the projecting tongue of land situated under Kavinda, we pitched our tent under the hills in the sandy cove opposite the longest point; and here it was evident from the strand plants which projected into the water, that the river had risen from six to eight inches. The temperature was as low as $73\frac{1}{2}°$; observations were made morning and evening. An alligator was swimming about all night near the shore, and contrived to carry off the only goat we had left remaining. I found here again the beautiful *Tetrandria, corolla tubulosa, foliis multifidis*, which I had seen at Mampaya.

Sept. 7th. A third canoe was hired to day, and all embarked. Projecting into the bay are picturesque rocks, consisting of subverted alternating beds of clay and limestone.

Near some rocky islands, a number of small *Charadrii* were seen. Behind these rocks, a little higher up, the river is again somewhat contracted, and we were told that our people would not be able to pass in canoes. One of them was in fact upset, and broken by carelessness, which accident occasioned a long delay. At last we got over to the other side of the river, and landed in a beautiful sandy cove, at the opening of a creek, behind a long projecting point. It is called Sandi-Sundi. An immense number of hippopotami were seen here. In the evening a number of alligators were also seen.

Sept. 8. Our hunting excursions last night and this morning were equally unsuccessful. I fired at some hippopotami. A young *Mustela* was brought to us by the negroes. Some porters or coolies were again hired to carry the baggage over land. We proceeded up the hills till we reached the high land, which is here rather level, the soil consisting of clay. The clay-slate formation still continues. This side seems to be more populous than the other. After passing through three or four villages we again approached the river, which is here more than half a Danish mile in breadth. The shore is flat and sandy, with several varieties of lime-stone. We had now arrived at the end of the southern reach of the river, which again took a north-easterly direction, in consequence of a chain of undulating mountains being situated in that direction. The country on the north or east-side appears now flatly inclining. Towards the north-east are coves terminated by large sand-banks. We renewed our chase after the hippopotami.

Sept. 9th. We proceeded round the creek, into which at the upper end a large rivulet emptied itself, and over the first hills, when we found two villages. From the summit we had a view of the windings of the river, which, turning round the hills, takes a large sweep to the eastward, after which, according to the account given by the inhabitants, it runs to the southward. The ridge of hills consists

of clay-slate. The highest that were seen running east and
west - - - - - - - - - - - - - - - - - - -

[*At this place they turned their backs on the River, to the great
annoyance, as the Gardener states, of Dr. Smith, who had become
so much enraptured with the improved appearance of the Country, and the magnificence of the River, that it was with the utmost
difficulty he was prevailed on to return ; four days after this he was
attacked with fever.*]

A CONCISE VIEW *of the Country along the Line of the Zaire,—its Natural History and Inhabitants,—collected from the preceding Narratives, and from the Observations of the Naturalists and Officers employed on the Expedition.*

THE RIVER.—IF, from the lamentable and almost unaccountable mortality which brought to an untimely termination this ill-fated expedition, the grand problem respecting the identity of the Niger and the Zaire still remains to be solved ; we have at least, by means of it, acquired a more certain and distinct knowledge of the direction and magnitude of the latter river, in its passage through the kingdom of Congo, as well as a more extended and correct notion of the nature of the country, of its inhabitants and productions, than had hitherto been supplied in the accounts (and they are the only ones) of the early Catholic missionaries.

It now appears, that although this great river, which has been named promiscuously the Congo, the Zaire, and the Barbela (but which ought, as Captain Tuckey learned, to be called *Moienzi Enzaddi,* " the Great River," or " the river which absorbs all other rivers,") falls short, in some respects, of the magnificent character given to the lower part of its course ; yet in others, it has been much underrated. Its great velocity, for instance, its perpetual state of being flooded, and its effectual resistance of the tide, are exaggerations ; but in regard to its depth at the point

X x

of junction with the sea, it was found to exceed the highest estimate which had been given to it. In Maxwell's chart, which is the only one published deserving of notice, the soundings near to the mouth, and for a considerable distance upwards, are marked down at 100 fathoms; and the rate of the current at five, six, and even seven knots an hour. Captain Tuckey, when in the transport, says that they could get no bottom with 150 fathoms of line out; and Mr. Fitzmaurice was equally unsuccessful in the Congo sloop with a line of 160 fathoms. These attempts however are no proofs of the river being actually that depth, as the loose line floats away with the current; but Massey's sounding machine, which is so contrived by being thrown overboard, and unconnected entirely with the drift of the ship, as not to be influenced by it, indicated by its index, when hauled up, a descent to the depth of 113 fathoms; at which depth, the lead attached to it had not touched the ground; and it was observed, that although the current made a rippling noise, somewhat resembling that of a mill-sluice, yet, on trial, it was seldom found to exceed four and a half, or five knots an hour, and in many places not more than two and a half. It was however sufficiently strong in many parts of the channel to prevent the transport from entering the river for five days; and it was not until the sixth that, by taking the advantage of a strong sea breeze, which sets in regularly with more or less strength every afternoon, she was enabled, by creeping close to the shore, to stem the current, which is there less strong than in the middle where the water is deepest. The current however in the mid-stream must have been greater than it is generally stated ; as it is admitted by the surveyor, that, with every desire to complete the survey of the river in all its parts, he found it impossible, even with the aid of Massey's machine, to get the soundings in the mid-channel, though the river was, at that time, in its lowest state. Maxwell's chart was found to be incorrect in many respects, especially as to distances, which are generally too great. With regard to the flat islands formed by alluvial earth, and overgrown with the mangrove and the papyrus, constant changes are taking place, some gradually forming and encreasing in size, while others are wholly or partially swept by the current into the ocean.

The mistaken notion, which seems to have originated with the Portuguese, that the tide could make no impression on the current of the Zaire, is but partially true ; this mistake is now corrected by frequent observations of the tide forcing the reflux of the stream very perceptibly as high up as the commencement of the narrows at Sondie, where the rise and fall amounted from twelve to sixteen inches ; but though it caused the water to be dammed up, and a counter-current on one or both sides, yet, strictly speaking, the current in the middle of the river was never overcome by the tide.

The distance at which the narrows commence is about 140 English miles from the mouth of the river at Point Padron, and they continue as far as Inga, or forty miles nearly ; the width of the river being generally not more than from three to five hundred yards, throughout that extent, and in most parts bristled with rocks. The banks, between which the water is thus hemmed in, are, for the whole of this distance, every where precipitous, and composed entirely of masses of slate ; which, in several places, run in ledges across from one bank to the other, forming rapids or cataracts, which the natives distinguish by the name of Yellala. The lowest and the most formidable of these barriers was found to be a descending bed of mica slate, whose fall was about thirty feet perpendicular in a slope of 300 yards. Though in this low state of the river it was scarcely deserving the name of a cataract, it was stated by the natives to make a tremendous noise in the rainy season, and to throw into the air large volumes of white foam. Even now the foam and spray at the bottom are said to have mounted eighteen or twenty feet into the air.

On visiting this Yellala, Capt. Tuckey, Professor Smith, and Mr. Fitzmaurice were not a little surprised to observe, how small a quantity of water passed over this contracted part of the river, compared with the immense volume which rolled into the ocean through the deep funnel-shaped mouth ; the more so, as they had previously ascertained, in their progress upwards, that not a single tributary stream of water, sufficient to turn a mill, fell into the river on either side, between the mouth and the cataract ; and they concluded, that the only satisfactory explanation of this remarkable difference in the quantity, was the supposition that a very considerable mass of water must find its way through subterraneous passages, under the slate rocks; disappearing probably where the river first enters these schistose mountains, and forms the narrows, and rising again a little below their termination, at Point Sondie, where the channel begins to widen, and from whence to Lemboo Point, a succession of tornados and whirlpools were observed to disturb the regular current of the river. These whirlpools are described, both by Captain Tuckey and Mr. Fitzmaurice, to be so violent and dangerous, that no vessel could attempt to approach them. Even the eddies occasioned by them were so turbulent as frequently to resist both sails, oars, and towing, twisting the boats round in every direction; and it was with the utmost difficulty, that they were extricated without being swamped. The instances of rivers losing themselves for a time under ground are so common, in all countries, that there seems to be no particular objection to the hypothesis of the Zaire losing a great portion of its waters in its passage through the narrows, under its schistose bed. At the same time, the eye might be deceived in estimating the quantity of water forced into a narrow channel, and running with the rapidity of a mill-sluice till it falls over a cataract, by a comparison with that which flows in a deep and expanded bed, in one uniform and tranquil motion; having besides an eddy, or counter-current, on one or both sides, which carries a considerable portion of water in a retrograde direction.

Be this as it may, the Zaire, beyond the mountainous regions, was again found to expand to the width of two, three, and even more than four miles, and to flow with a current of two to three miles an hour; and near the place where Captain Tuckey was compelled to abandon the further prosecution of the journey, which was about 100 miles beyond Inga, or 280 miles from Cape Padron, it is stated that the river put on a majestic appearance, that the scenery was beautiful, and not inferior to any on the banks of the Thames; and the natives of this part all agreed in stating, that they knew of no impediment to the continued navigation of the river; that the only obstruction in the north-eastern branch, was a single ledge of rocks, forming a kind of rapid, over which however canoes were able to pass.

The opinion that the Zaire is in a constant state of flood, or, in other words, that it continues to be swelled more or less by freshes through the whole year, has been completely refuted by the present expedition. But the argument, which was grounded on this supposition, of its origin being in northern Africa, so far from being weakened, has acquired additional strength from the correction of the error. Like all other tropical rivers, the Zaire has its periodical floods; but the quantity of its rise and fall is less perhaps than that of any other river of equal magnitude. From the lowest ebb, at which the party saw it, to the highest marks of its rise on the rocks, the difference no where appeared to exceed eleven feet, and in many places was not more than eight or nine. The commencement of the rise was first observed above Yellala, on the 1st of September, to be three inches; and on the 17th of that month it had acquired, at the Tall Trees, near the mouth of the river, the height of seven feet; without the velocity having much, if at all encreased ; and without a single shower having fallen that deserves to be noticed. The little difference between the rise of seven feet, which then took place in the dry season, while the sun was still to the northward of the line, and that of eleven feet in the wet season, during which the sun is twice vertical, affords a solid argument for its northern origin ; and, when coupled with the particular moment at which it was first observed to rise, would seem to establish the fact, almost beyond a doubt, that one branch of the river, as was stated by the natives, must descend from some part of Africa to the northward of the Equator.

We find in Captain Tuckey's notes, after having observed the progressive rise of the river, the insertion of two words as a memorandum, " hypothesis confirmed." This hypothesis had previously been stated among the last notes of his Journal, which he did not live to reduce into a regular narrative, under these words; " extraordinary quiet rise of the river shows it to issue from some lake, which had received almost the whole of its water from the north of the line." But in a private letter written from Yellala, and brought home in the Congo, he dwells more particularly on this hypothesis; " combining" he says " my observations with the information I have been able to collect from the natives, vague and trifling as it is, I cannot help thinking that the Zaire will be found to issue from some large lake, or chain of lakes, considerably to the northward of the line ;" and he contends that, so far from the low state of the river in July and August militating against such an hypothesis, it has the contrary tendency of giving additional weight to it, " provided" he goes on to say, " the river should begin to swell in the early part of September, an event I am taught to expect, and for which I am anxiously looking out." The river did begin to swell at the precise period he had anticipated ; and that circumstance corroborating the previous conclusion he had drawn, induces him to note down in his journal, that " the hypothesis is confirmed."

It is evident that Captain Tuckey, on the latter part of his journey, could only put down a few brief notes to re-fresh his memory, which, from his exhausted state, on his return to the vessels, he was wholly unable to enlarge or explain ; and thus the reasoning on which he had built his hypothesis is lost to the world : he lamented, it seems, when on his death-bed, that he could not be permitted to live to put in order the remarks he had collected in tracing upwards this extraordinary river. Unfortunately none of the party has escaped to supply this deficiency ; the solidity, however, of Captain Tuckey's conclusion is not shaken, but rather corrobated, by what is known of physi-cal facts and the geographical probabilities, as connected with northern Africa. These may be briefly stated.

In the tropical regions, the rains generally follow the sun's course, and are not at their height till he approaches the tropics ; hence arises the exhausted state of the lakes of Wangara in the months of May, June, and July, and their overflowing in the middle and latter end of August, ac-cording to the observations of the Arabian geographers ;

and this late flooding of the lakes is obviously owing to the long *easterly* course of the Niger, collecting into its channel all the waters from the northward and the southward as it proceeds along. If, then, the ebb and flood of the Wangara lakes depend on the state of the Niger, it will follow, on the supposition of the identity of that river and the Zaire, that the flood and ebb of the latter, to the southward of the line, must correspond with the ebb and flood of the lakes of Wangara. The existence of those lakes has never been called in question, though their position has not been exactly ascertained; but supposing them to be situated somewhere between the twelfth and the fifteenth degrees of northern latitude, the position usually assigned to them in the charts, and that the southern outlet is under or near the 12th parallel, the direct distance between that and the spot where Captain Tuckey first observed the Zaire to rise, may be taken at about 1200 miles, which, by allowing for the windings of the river, and some little difference of meridians, cannot be calculated àt less than 1600 miles.

Admitting, then, that the lakes of Wangara should overflow in the first week of August, and the current in the channel of outlet move at the rate of $2\frac{1}{2}$ miles an hour, which is the average rate at which the Zaire was found to flow above the narrows, the flooded stream would reach that spot in the first week of September, and swell that river exactly in the way, and at the time and place, as observed by Captain Tuckey. No other supposition, in fact, than that of its northern origin, will explain the rise of the Zaire in the dry season; and if its identity with the Niger, or,

Y y

which amounts to the same thing, its communication with Wangara, should be disputed, Captain Tuckey's hypothesis of its issuing from some other great lake, to the northward of the line, will still retain its probability. The idea of a lake seems to have arisen from the " extraordinary *quiet* rise" of the river, which was from three to six inches in twenty four hours. If the rise of the Zaire had proceeded from rains to the southward of the line, swelling the tributary streams, and pouring, in mountain torrents, the waters into the main channel, the rise would have been sudden and impetuous; but coming on as it did in a quiet and regular manner, it could proceed only from the gradual overflowing of a lake.

There is, however, another circumstance in favour of a river issuing from Wangara, or the lakes and swamps designated under that name, and of that river being the Zaire. There is not a lake, perhaps, of any magnitude in the known world, *without an outlet*, whose waters are not saline—the Caspian, the Aral and the neghbouring lakes, the Asphaltites or Dead Sea, and all those of Asia, which have no outlet, are salt.* If therefore the lakes of Wangara had no outlet, but all the waters received into them spread themselves over an extended surface during the rains, and were evaporated in the dry season, there would necessarily be deposited on the earth, so left dry, an incrustation of salt, and the remaining water would be strongly impregnated with

---

* The *freshness* of the Zuré or Zurrah, the Aria Palus, in Seistan, rests on no authority—but if so, its waters are not evaporated, but pass off by filtration through the sand.

salt ; and both the one and the other would be encreased by every succeeding inundation. None of the African rivers are free from saline impregnations ; but the Niger, in its long easterly course, collecting the waters from the sandy and saline soil of the desert, where every plant almost is saturated with salt, must be particularly charged with it. No mention, however, is made by any of the Arabian writers of that indispensable article, salt, being procured in the mud or soil abandoned by the waters of Wangara ; on the contrary, it is well known that one great branch of the trade of Tombuctoo is that of obtaining salt from the northern desert, for the supply of the countries to the southward of the Niger. But if Wangara had no outlet, this could not be necessary, as both it and all the large inland lakes, so circumstanced, would afford more or less of salt ; and if so, the trade of the caravans proceeding with rock salt from Tegazza to Tombuctoo would not have existed ; as it is well known it has done, and still does, especially from the latter place to Melli and other countries south of the Niger, " to a great water," as Cadamosta says, " which the traders could not tell whether it was salt or fresh ; by reason of which (he says) I could not discover whether it was a river or the sea ; but," he continues, " I hold it to be a river, because if it was the sea, there would be no need of salt."

Edrisi, however, distinctly states them to be fresh water lakes, and says that the two cities of Ghana are situated on the two opposite shores of what the Arabs call a fresh water sea. This fresh water sea, therefore, must necessarily have an outlet ; or, like the Caspian, it would be no longer fresh ; and the conclusion is that, if the Niger runs into

these lakes of Ghana and Wangara, it does not there terminate, but that, in the season of the rains, it also flows out
of them.   In fact, Edrisi does not make the Niger to terminate in the swamps of Wangara or Vancara; he merely
describes them as being an island three hundred miles in
length, and one hundred and fifty in breadth, surrounded
by the Niger *all the year*, but that, in the month of August,
the greater part is covered with water as long as the inun-
dations of the Niger continue; and that when the river has
subsided into its proper channel, the negroes return to their
habitations, and dig the earth for gold, " every one finding
more or less, as it pleases God."   But not a word is mentioned of their finding salt, which indeed is the great inter-
changeable commodity for gold.

On the assumption, then, of Wangara discharging its
overflowing waters, the most probable direction of the
channel is to the southward; and as the evidence of the
northern origin of the Zaire amounts almost to the establishment of the fact, the approximation of the two streams is
in favour of their identity.   If the account of Sidi Hamet's
visit to Wassenah, as related by Riley, could be depended
on, a very few degrees only are wanting to bring the two
streams together; but with all the strong testimonies in
favour of Riley's *veracity*, every page of his book betrays
a looseness and inaccuracy, that very much diminish the
value of this Arab's narrative as it is given by him.   The
name of *Zadi*, given by this Arab merchant to the Niger at
Wassenah; that of *Zad*, which Horneman learned to be its
name to the eastward of Tombuctoo, " where it turned off to
the southward;" the *Enzaddi*, which Maxwell says is the
name given to the cataracts of the Zaire; and the *Moienzi*

enzaddi, which Captain Tuckey understood to be the name of the river at Embomma, are so many concurring circumstances which give a favourable though a faint colour to the hypothesis of the identity of the two rivers.

If any further exploration of the Zaire, upwards, should be undertaken, Captain Tuckey has sufficiently established the fact, that no naval equipment at home can avail in the prosecution of this object. All that appears to be necessary, is that of providing at the Cape de Verde islands a dozen or twenty asses and mules, and carrying them in a common transport up the river as far as Embomma ; from thence to make the best of the way over land direct for Condo Yanga, the place which has been assigned by Captain Tuckey, as possessing the greatest advantages for the necessary preparations for embarking on the river ; and these preparations would consist merely in purchasing or hiring half a dozen canoes, with the help of two or three ship carpenters, converting them into three double-boats, or twin-canoes, by a few planks, which would form a convenient platform for the accommodation of the party, the animals, and the baggage. In this way they would proceed where the river was navigable, and by land, with the assistance of the asses and mules, where interruptions occurred ; and thus they would avoid that degree of fatigue, which was unquestionably the principal cause of the death of those who fell on the late expedition. On the part of the natives, it is now pretty well ascertained, there would be no obstruction, unless they are of a very different disposition higher up in the interior, than what Captain Tuckey experienced them to be, which is not, as far as he could collect, very likely: the character of the negro having hitherto been every where stamped with mildness, simplicity, and benignity of disposition.

FACE OF THE COUNTRY—SOIL, CLIMATE, AND PRODUCTIONS. THE country named Congo, of which we find so much written in collections of Voyages and Travels, appears to be an undefined tract of territory, hemmed in between Loango on the north, and Angola on the south; but to what extent it stretches inland, it would be difficult to determine; and depends most probably on the state of war or peace with the contiguous tribes. All that seems to be known at present is, that the country is partitioned out into a multitude of petty states or Chenooships, held as a kind of fiefs under some real or imaginary personage living in the interior, nobody knows exactly where. Captain Tuckey could only learn that the paramount sovereign was named *Blindy N'Congo*, and resided at a banza named Congo, which was six days journey in the interior from the Tall Trees, where, by the account of the negroes, the Portuguese had an establishment, and where there were soldiers and white women. This place is no doubt the St. Salvador of the Portuguese. These chiefs have improperly been called kings: their territories, it would seem, are small in extent, the present expedition having passed at least six of them in the line of the river; the last is that of Inga, beyond which are what they call bush-men, or those dreadful cannibals whom Andrew Battel, Lopez, Merolla, and others, have denominated Jagas, or Giagas, " who consider human flesh as the most delicious food, and gob-

lets of warm blood as the most exquisite beverage ;" a ca-
lumny, which there is every reason to believe has not the
smallest foundation in fact. From the character and
disposition of the native African, it may fairly be doubted
whether, throughout the whole of this great continent, a
negro cannibal has any existence.

That portion of the Congo territory, through which the
Zaire flows into the southern Atlantic, is not very interesting, either in the general appearance of its surface, its
natural products, or the state of society, and the condition
of its native inhabitants. The first is unalterable; the
second and third are capable of great extension and
improvement, by artificial and moral cultivation ; but
with the exception of the river itself, there are probably
few points between the mouth of the Senegal and Cape
Negro, on that coast, which do not put on a more interesting appearance, in a physical point of view, than the
banks of the Zaire. The cluster of mountains, though in
general not high (the most elevated probably not exceeding two thousand feet), are denuded of all vegetation, with the exception of a few coarse rank grasses ; and
the lower ranges of hills, having no grand forests, as might
be expected in such a climate, but a few large trees only,
scattered along their sides and upon their summits, the
most numerous of which are, the Adansonia, Mimosa,
Bombax, Ficus, and palms of two or three species.

Between the feet of these hills, however, and the margins of the river, the level alluvial banks, which extend
from the mouth nearly to Embomma, are clothed with a

most exuberant vegetation, presenting to the eye one continued forest of tall and majestic trees, clothed with foliage
of never-fading verdure.   Numerous islands are also seen
to rise above the surface of the river, some mantled with the
thick mangrove, mingled with the tall and elegant palm,
and others covered with the Egyptian papyrus, resembling
at a distance extensive fields of waving corn.   Perhaps it
may be said, that the great characteristic feature of the
banks and islands of the lower part of the Zaire is the mangrove, the palm, the adansonia and the bombax, with intermediate patches of papyrus; and after the alluvial flats
have ceased, naked and precipitous mountains, resting on
micaceous slate, which, through an extent of at least fifty
miles, forms the two banks of the river; the only interruption to this extended shore of slate being a few narrow
ravines in which the villages of the natives are situated,
amidst clumps of the wine-palm, and small patches of cultivated ground.   On the summits of the hills, also, which
Captain Tuckey distinguishes by the name of plateaus,
there is a sufficiency of soil for the cultivation of the ordinary articles of food ; and here too numerous small villages
occur amidst the bombax, the mimosa, the adansonia and
the palm ; but the soil on the tops and sides is of a hard
clayey nature, incapable of being worked in the dry season,
but sufficiently productive when mellowed by the heavy
rains, and with the aid of a heated atmosphere.

The country however becomes greatly improved in every
respect, beyond the narrows of the river.   Hitherto the
general characteristic features of the geology of the coun-

try were mica-slate, quartz, and sienite ; but here the rock formation, though not entirely, was considerably changed ; the granite mountains and hills of pebbly quartz having given way to clay and ferruginous earth, and the mica-slate to lime-stone. The banks of the Zaire are now no longer lined with continued masses of mica-slate, but many rocky promontories of marble jut into the river, with fertile vales between them ; and the reaches of the river itself stretching out into broad expanded sheets of water, resembling so many mountain lakes. The greater part of the surface was now fit for cultivation, and towns or villages followed each other in constant succession, far beyond the limits of the Congo territory. Vegetation was more generally diffused, as well as more varied ; and rills of clear water trickled down the sides of the hills, and joined the great river. It was just at the commencement of this improved appearance of the country, where, from the sickly state of the party, and the loss of their baggage, Captain Tuckey was reluctantly compelled to abandon the further prosecution of the objects of the expedition ; and in some respects it was fortunate he did, as had he proceeded two or three days longer, the whole party must unquestionably have perished in the interior of Africa, and might perhaps never more have been heard of.

The account which the missionaries have given of the climate, corresponds exactly with that which was experienced by Captain Tuckey. " The winter," says Carli, " of the kingdom of Congo, is the mild spring or autumn

Z z

of Italy; it is not subject to rains, but every morning there falls a dew which fertilizes the earth." None of the party make any complaint of the climate; they speak, on the contrary, in their notes and memoranda, of the cool, dry, and refreshing atmosphere, especially after the western breezes set in, which they usually do an hour or two after the sun has passed the meridian, and continue till mindight; and when calm in the early part of the day, the sun is said so seldom to shine out, that for four or five days together, they were unable to get a correct altitude to ascertain the latitude. So much, however, depends on locality, that at the place where the Congo was moored, the range of the thermometer differed very materially from that on board the transport lower down, and also from that observed in the upper parts of the river. The former vessel was moored in a reach surrounded by hills, and what little of the sea breeze reached her, had to pass over a low swampy island. Here, Mr. M'Kerrow noticed the range of the thermometer to be from 70° to 90° in the shade; sometimes, though but seldom, as low as 67° in the night, and as high as 98° at noon; and one day on shore it rose to 103° under the shade of trees; at the same time, above Inga, the temperature seldom exceeded 76° in the day, and was sometimes down to 60° at night. He seems to think, that partly owing to a better position of the transport, which remained at anchor lower down the river, nearly opposite to the Tall Trees, where she had the benefit of the sea breeze without interruption, and partly by preventing her crew from going on shore, this vessel continued healthy until she received on board the unfortunate people belonging to the Congo.

The following table exhibits the state of Fahrenheit's thermometer, at three periods of the day, for one month, from 20th July to 20th August, in different parts of the river, from the entrance to the Cataract, and of the water of the river at noon.

Date.	Air. 8 A. M.	Air. Noon.	Air. 8 P. M.	Water. Noon.	Date.	Air. 8 A. M.	Air. Noon.	Air. 8 P. M.	Water Noon.
	°	°	°	°		°	°	°	°
July 20	72	74	73	75	Aug. 5	71	78	78	77
21	71	76	74	76	6	72	80	78	76
22	72	75	74	76	7	72	77	77	78
23	72	75	75	76	8	71	77	76	76
24	72	74	73	77	9	69	78	78	78
25	71	76	76	77	10	69	76	76	78
26	72	78	80	76	11	70	76	75	77
27	73	78	77	77	12	68	77	78	76
28	69	80	76	76	13	70	76	77	76
29	70	78	74	75	14	73	78	76	77
30	70	76	76	76	15	72	78	76	77
31	71	76	74	76	16	72	77	75	76
Aug. 1	69	73	75	76	17	70	76	76	76
2	71	73	76	76	18	71	76	77	78
3	71	74	76	76	19	71	78	77	77
4	69	76	76	76	20	69	78	75	76

It is remarked in a meteorological journal, imperfectly kept by Captain Tuckey, in proceeding up the river, that from the mouth to Embomma, the temperature of the river was almost invariably at 76°.

The alimentary plants are very various, and for the most valuable of them, the natives are indebted to the Portuguese. The staple products of the vegetable world consist of manioc or cassava, yams, and maize or Indian corn; to which may be added sweet potatoes, pumpkins, millet of two or three species, and calavanses: they have besides cabbages, spinach, pepper, capsicum, the sugar-cane, and tobacco. Of fruits they have the plaintain or banana, papaw, oranges, limes, and pine-apples. The latter fruit was met with by Captain Tuckey growing on the open plains near the extreme point of his journey, and far beyond where any Europeans had advanced. This fruit, therefore, as well as the bananas, the one being from the West, the other from the East Indies, (or both perhaps from the *West*), must have been carried up into the interior by the natives. The only beverage used by the inhabitants, except when they can get European spirits, is the juice of the palm tree, of which there are three distinct species. It is usually known by the name of palm wine, and was considered by the whole party as a very pleasant and wholesome liquor, having a taste, when fresh from the tree, not unlike that of sweetish cyder; is very excellent for quenching the thirst, and for keeping the body gently open. When tapped near the top, the juice runs copiously out during the night, but very little is said to exude in the day time. One of the species yields a juice sweeter than the rest, and this being suffered to ferment, is said to produce a liquor of a very intoxicating quality. The trees are remarkably tall, and are ascended by means of a flexible hoop which encloses, at the same time, the body of the person intending to mount and the stem of the tree, against the latter of which the feet are pressed, while the back rests against the hoop. At each step the hoop is moved upward with the hand, and in this way they ascend and descend the highest trees with great expedition : should the hoop give way, the consequence must be fatal.

They have no want of domestic animals to serve them as food, though very little care appears to be bestowed on them. They consist chiefly of goats, hogs, fowls, the common and Muscovy duck, and pigeons; a few sheep, generally black and white, with hair instead of wool. The Chenoo of Embomma had obtained from the Portuguese a few horned cattle, but no pains whatever were taken to increase the breed. They have no beasts of burden of any description. Of wild animals the country produces great variety, but the natives are too indolent and inexpert to convert them to any useful purpose. They have elephant's, leopards, lions, buffaloes, large monkeys with black faces, and numerous species of antelopes, with which Africa every where abounds ; wild hogs, porcupines, hares, and a great variety of other quadrupeds, from which an active people would derive important ádvantages. Guinea fowl and red legged partridges are also abundant, large, and fine ; and wild pigeons, of three or four species, very plentiful.

The country appears to be remarkably free from teazing and noxious insects, excepting bugs and fleas in the huts, and the black ants, which erect those singular mushroom-shaped habitations, some of which have two or three domes, and sometimes occur in whole villages. The party suffered no annoyance from scorpions, scolopendras, musquitoes, which are almost universally swarming in warm climates. From the abundance of bees, and the hills being well clothed with grass, Congo might be made a land " flowing with milk and honey."

The lower part of the river abounds with excellent fish, which would appear to be an important article of subsistence to those who inhabit the woody banks occupied by the mangrove. Bream, mullet and cat-fish are the most abundant. A species of *Sparus*, of excellent flavour, was caught by the party in large quantities, each of them weighing generally from thirty to forty pounds, and some of them even sixty. Mr. Fitzmaurice observes that, near Draper's islands, he fell in with three or four hundred canoes, in which the people were busily employed in dragging up a species of shell-fish, which he compares to what is usually in England called the clam, and which is stated by Captain Tuckey to be a species of Mya. Most of these fishermen, it was thought, had no other abode than the shelter which the woods afforded them ; that they form a kind of hut by bending and entwining the living branches, in the same manner as is sometimes practised by the roving Caffres bordering on the colony of the Cape of Good Hope ; others make the caverns in the rocks the abodes of themselves and families during the fishing season ; for it would seem that these huts and retreats were but temporary, as the shells of these fish were opened, the animal taken out, and dried in the sun. In the upper parts of the river, women were frequently seen fishing with scoop nets, made from the fibres of some creeping plants ; and in one village, a woman was observed spinning cotton for nets; the herbaceous cotton plant growing every where wild. In some places the fish were caught in pots; in others they took them by means of a poisonous plant.

A fish resembling the Silurus electricus was brought on board the Congo from Embomma, which, by the account of the natives, when alive and touched, communicates a severe shock to the hand and arm, or to use their own expression " it shoot through all the arm." It is thus described by Mr. M'Kerrow: length three-feet six inches; head large, broad and compressed; mouth furnished with six long cirrhi, four on the under and two on the upper jaw; mandibles dentated; tongue short, and eyes small; body without scales; pectoral fins near the branchial openings, the ventrals near the anus; dorsal fin soft, and placed near the tail; upper parts of the body thickly spotted black, and the under of a yellowish white; skin exceedingly thick.

The Zaire swarms with those huge monsters the hippopotamus and the alligator, or rather crocodile, (for it appears to be of the same species as the animal of the Nile,) and particularly above the narrows. Both these animals seem to be gregarious, the former being generally met with in groups of ten or twelve together; the latter in two or three, sometimes five or six. The flesh of the hippopotamus is excellent food, not unlike pork; but it does not appear that the negroes are particularly fond of it, as the only one killed by the present party was suffered to putrify on the margin of the river; though it is stated that the flesh is sometimes sold in the market. One crocodile only was killed, whose length was nine feet three inches, and girt across the shoulders three feet seven inches.

Food, Lodging, Utensils and Clothing.—The staple articles of subsistence, at least in the dry season, appear to be manioc, ground-nuts, and palm wine; to which may probably be added Indian corn and yams, the latter of which are stated to be remarkably fine; and of Indian corn they have regularly two crops in the year. Animal food is not in general use, though sold in the daily market held at Embomma, which is at a village distinct from the banza, or residence of the Chenoo, and at which from a hundred to three hundred persons are said to assemble; in this market, the party observed a supply of goats, fowls, eggs, besides vegetables, fish, and salt. It must be recollected, however, that this place is the grand mart for conducting the slave trade of the Zaire, and these supplies may be chiefly intended for the crews of the European ships.

The negroes of Congo are exceedingly foul feeders, and particularly filthy in their preparation and their eating of animal food; they broil fowls with the feathers on, and pieces of goat without being at the trouble of removing the skin, or even the hair; and they devour them when scarcely warmed, tearing the flesh in pieces with their teeth in the most disgusting manner. Mr. Fitzmaurice relates that one day, as their butcher had taken off the skin of a sheep, the Mandingo slave purchased by Captain Tuckey, had slily conveyed away the skin, which, with the wool (or rather the hair) he had thrown over a smokey fire, and when discovered, he had nearly eaten the whole skin in a state scarcely warm. There do not appear, however, to be the slightest grounds for supposing that they ever eat human flesh, not even that of their enemies, but that all the accusations of this nature are totally false.

None of the banzas or villages seen by the party were of great extent; the largest probably not exceeding one hundred huts. Embomma, Cooloo, and Inga, are each the residence of a Chenoo; the first was supposed to consist of about sixty huts, exclusive of the Chenoo's inclosure, and about five hundred inhabitants; the second, one hundred huts, and from five to six hundred inhabitants; and the third, being the last in the line of the river within the kingdom of Congo, of seventy huts, and three hundred inhabitants. The party stationed at this banza understood, that the Chenoo could command about two hundred fighting men, one hundred of whom he can arm with musquets; and with this force he conceives himself to be the dread and terror of his enemies. These banzas are usually placed amidst groves of palms and adansonias.

The huts in general consist of six pieces, closely woven or matted together, from a reedy grass, or the fibres of some plant; the two sides exactly corresponding, the two ends the same, excepting that in one is the door way, an opening just large enough to creep in at, and the two sloping sides of the roof also correspond. The sides and ends are made fast to upright posts stuck in the ground; and the two pieces of the roof are bound to the sides, and also

3 A

to each other; and as each piece is very light, a house can, at any time, be removed from one situation to another with great ease; sometimes the roofs are semi-circular. The value of one of these moveable houses is stated to be not more than the price of five or six fowls, and in five minutes may be put together. Permanent houses, however, such as those of the Chenoo, are made of the palm leaves with considerable skill, having several posts along the sides and ends, and covered externally with the blades or back rib of the palm leaf, bound together with a creeping plant in regular zig-zag figures. They are also generally inclosed within a fence of reeds matted together.

Their household utensils are very few, and as simple as the houses themselves. Baskets made of the fibres of the palm tree; bowls and bottles of gourds or calabashes, or of the shell of the monkey bread-fruit (Adansonia) to hold their provisions and water, earthen vessels of their own making to boil their victuals, and wooden spoons to eat them; a mat of grass thrown on a raised platform of palm-leaves, their only bedding. The articles of dress are equally sparing and simple, the common people being satisfied with a small apron tied round their loins, of a piece of baft, if they can get it, or of native grass-matting, made by the men; of the same grass they make caps, whose surface is raised and figured in a very beautiful manner, and the texture so close that they will hold water. Rings of brass or iron are welded on the arms and ankles, and sometimes bracelets of lion's teeth; and the women generally contrive to have strings of beads round their necks and arms and legs, and in default of these, strings of the cowrie shell, or of the round seeds of various plants.

Their canoes are generally hollowed out of the trunk of the bombax or cotton tree, or of a species of ficus, the common size being about twenty-four feet in length, and from eighteen to twenty inches in width; and they are all pushed forwards with long paddles, the men standing upright; they use neither sails, nor any substitute for them.

A rude hoe of iron, stuck into a wooden handle, is the implement used for agricultural purposes; but the climate is so fine, that, by merely scratching the surface of the ground, they succeed in raising good crops. The great scarcity of provisions, experienced by the party who proceeded up the river, was occasioned entirely by the long drought, and that want of precaution in laying up a stock against such a contingency, which, it would seem, is here rather the effect of indolence and thoughtlessness, than any distrust in the right and security of property; which indeed is so well understood, that almost all the disputes among the natives arise from their tenacity in the division of property, whether in land or stock. This participation is frequently so minute, that, as Captain Tuckey observes, a fowl or a pig may sometimes have three or four proprietors.

POPULATION AND CONDITION OF THE PEOPLE.— Though the population evidently increased, the farther the party proceeded into the interior, the banks of the river were but thinly inhabited in the very best and most productive parts ; and nothing appeared that could give the least colour to those exaggerated statements of the Catholic missionaries, who speak of such masses of men collected together as are not to be met with in the most populous parts of Europe. Carli, for instance, states the " Grand Duke's" army to amount to 160,000 men ; and he accounts for the vast population of Congo from the indulgence of every man being allowed to take as many wives as he pleases, and the absence of all those religious institutions and societies which, in Europe, consign their members to a state of celibacy. Nay, we are told, that the king, Don Antonio, could muster an army of 900,000 men, and that he actually brought 80,000 against the Portuguese, who with 400 Europeans and 2000 negroes, with the help of the Virgin Mary, easily put to route this great force, dethroned the king, and set up a new one of their own. Whether such a population ever existed, or if so what became of it; whether wars, pestilence, or famine, swept those vast multitudes away, or whether their progeny were sent off to other lands, the Portuguesé, who best could tell, have been silent on the subject; it is quite clear, however, that no such population exists at the present day.

Leaving out the paramount sovereign of Congo, whose existence seems to be rather doubtful, the component parts of a tribe or society, would appear to consist of—1. the Chenoo ; 2. the members of his family ; 3. the Mafooks ; 4. Foomos ; 5. fishermen, coolies and labouring people ; 6. domestic slaves.

The title and authority of the Chenoo are hereditary, through the female line, as a precaution to make certain of the blood royal in the succession; for although the number of the Chenoo's wives is unlimited, none but the offspring of her who is descended from royal blood, can inherit; and in default of issue from any such, the offspring of any other princess married to a private person, lays claim to the chiefship, and the consequences are such as might be expected; feuds and civil broils arise, which terminate only in the destruction of the weaker party. A Chenoo's daughter has the privilege of chusing her own husband, and the person she fixes upon is not at liberty to refuse; but it is a perilous distinction which is thus conferred upon him, as she has also the privilege of disposing of him into slavery, in the event of his not answering her expectations. Aware of his ticklish situation, he is sometimes induced to get the start of her, and by the help of some poisonous mixture, with the efficacy of which the people of Congo are well acquainted, rids himself of his wife and his fears at the same time.

When a Chenoo appears abroad, one of his great officers carries before him his scepter or staff of authority, which is a small baton of black wood about a foot in length, inlaid with lead or copper, like the worm of a screw, and crossed with a second screw, so as to form the figures of rhomboids. What their native dresses may be beyond the sphere of communication with European slave-dealers, is not exactly known, but little more probably than an apron of some skin-cloth, or grass-matting; the lion's skin to sit upon, was said to be sacred to the Chenoo, the touching of which by the foot of a common person is death or slavery. From the cataract downwards, the ridiculous cast-off dresses of French and Portuguese generals, form no part of the native costume of Congo, which, with the exception of an apron, anklets, bracelets, and necklaces, may be presumed to be neither more nor less than sheer nakedness.

The members of the Chenoo's family are his councillors, by whose advice he acts in all matters of importance ; and it is remarkable, that their consultations are generally held under the boughs of the *ficus religiosa.* In case of war, the elders remain behind to take care of the village, while the brothers, sons, or nearest relations of the Chenoo are usually selected to conduct, under him, their warlike expeditions.

The Mafooks are the collectors of the revenues, which are chiefly derived from trade; towards the lower part of the river, they begin by acting as linguists or interpreters between the slave dealers of the interior, and the European purchasers; but having made a fortune, which was frequently the case in this once lucrative employment, they purchase the rank of Mafook, and from that moment are said to be dumb, and utterly unable any longer to interpret.

The *Foomos* are composed of that class of the society who have houses and lands of their own, two or three wives, and perhaps a slave or two to work for them ; they are in fact the yeomanry of the country.

The fishermen, coolies and labouring people appear to

consist of those who have no fixed property of their own, but act as the labourers and peasantry of the country, and are very much at the disposal of the Chenoo or chief, though not slaves.

Domestic slaves do not appear to be numerous, and are not considered as common transferable property, and only sold for some great offence, and by order of the council, when proved guilty. Saleable slaves are those unhappy victims who have been taken prisoners in war, or kidnapped in the interior by the slave catchers, for the sake of making a profit of them; or such as have had a sentence of death commuted into that of foreign slavery.

THE SLAVE TRADE.—The banks of the Zaire are not the part of Africa where the slave trade, at present, is carried on with the greatest activity, though there were three Portuguese schooners and four pinnaces at Embomma, on the arrival of the expedition. The two great vents are the Gulf of Guinea to the northward, and Loango and Benguela to the southward of this river. The chiefs and their Mafooks were, however, all prepared to trade on the appearance of the ships, and much disappointed on learning that the object of the expedition was of a very different nature. They had heard at Embomma, overland from the coast, some vague rumours concerning the nature of the expedition, which they did not well comprehend; and when the Mafook of the Chenoo first came on board, he was very inquisitive to know, whether the ships came to make trade, or make war; and when he was distinctly told that the object was neither the one nor the other, he asked, " what then come for; only to take walk and make book ?"

As it would appear, that the state of slavery is a condition inherent in the principles on which the society of every negro tribe is founded, the gradation from domestic to foreign slavery is so easy, that as long as a single door remains open for disposing of human beings, it is to be feared, that very little progress has actually been made towards the abolition of this disgraceful and inhuman traffic. It is of little use to dam up the mouths of the Senegal and the Gambia, and turn the current into the channels of Lagos, Formosa, Calabar and Camaroons ; or to stop up these vents, while the Zaire, the Coanza, and the Guberoro remain open. The prolonged march of the kafilas over land may somewhat increase the prices to the purchaser, and prolong the misery of the slave, but the trade itself will not be much diminished on that account; while there is but too much reason to fear, that the passage across the Atlantic will be attended with circumstances of aggravated cruelty and inhumanity. Indeed nothing short of a total and unqualified prohibition of the traffic by every power in Europe and America, can afford the least hope for a total abolition of the foreign trade; and even then, there is but too much reason to believe, that the Mahomedan powers of Egypt and northern Africa will extend their traffic to the central regions of Soudan, which in fact, since the nominal abolition, has very considerably encreased in those quarters.

STATE OF SOCIETY.—The state of society among the negro nations seems to be pretty nearly the same, and their moral character not very different; the people of Congo would appear, however, to be among the lowest of the negro tribes. The African black is by nature of a kindly, cheerful, and humane disposition, entirely free from that quick, vengeful and ferocious temper which distinguishes the savages of the Pacific and South Sea islands, particularly those of New Guinea, which most resemble the negroes in external appearance. Contented with very humble fare, his happiness seems to consist in a total relaxation from all bodily exertion; excepting when animated by the sound of his rude native music calling him to the dance, in which he is always ready to join with the greatest alacrity. But indolence is the negroe's bane; and until some strong motive for shaking it off shall take possession of his mind, and convince him of the utility of industrious pursuits, by bettering his condition, little hope can be entertained of the civilization of Africa, even should a total and radical abolition of the slave trade be effected. The vast shoals of Catholic missionaries poured into Congo and the neighbouring parts of Southern Africa, from Italy, Spain, and Portugal, in the sixteenth and seventeenth centuries, appear not to have advanced the natives one single step in civilization; and the rude mixture of Catholic with Pagan superstitions, which were found among the Sognio people on the left bank of the Zaire, close to the sea coast, was all that could be discovered of any trace of Christianity, after the labours of these pious men for three hundred years. Some of these people came

3 B

off to the vessels, and they are represented as being the very worst in every respect of all the tribes that were met with on the banks of the river, being dirty, filthy, and over-run with vermin. One of them was a priest, who had been ordained by the Capuchin monks of Loando, and carried with him his diploma, or letters of ordination; he could just write his name, and that of St. Antonio, and read the Romish litany; but so little was he of a Catholic, that his rosary, his relics, and his crosses were mixed with his domestic fetiches; and so indifferent a Christian, that this " bare-footed black apostle," as Dr. Smith calls him, boasted of his having no fewer than five wives.

Captain Tuckey seems to think that the plan of sending a few negroes to be educated in Europe, for the purpose of returning to instruct their countrymen, is as little likely to succeed, as that of sending missionaries among them; and that colonization holds out the only prospect of meliorating their civil and moral condition. How far this might succeed with the negroes, remains to be tried; in all other countries, inhabited by a savage or half-civilized people, extirpation has followed close on the heels of colonization. The unconquerable avidity for spirituous liquors on the part of the savages, and the same propensity for their possessions on that of the colonists, have produced contentions, encroachments and spoliation, which terminate invariably to the detriment of the natives, and too frequently to their utter extermination. It might at the same time be well worth the experiment, of prevailing on a few of the

Moravian missionaries to settle themselves in a negro village, to instruct the natives in the useful arts of agriculture, manufactures and trade; to make them feel the comforts and advantages of acquiring a surplus property; to instil into their minds sound moral precepts; and to divert their attention from their gross and senseless superstitions to the mild and rational principles and precepts of the Christian religion.

The worst feature in the negro character, which is a very common one among all savage tribes, is the little estimation in which the female sex is held; or, rather their esteeming them in no other way than as contributing to their pleasures, and to their sloth. Yet, if this was the extent to which female degradation was subject, some palliation might perhaps be found in the peculiar circumstances of the state of the society; but the open and barefaced manner in which both wives and daughters were offered for hire, from the Chenoo or chief, to the private *gentleman*, to any and all of the persons belonging to the expedition, was too disgusting to admit of any excuse. Some of the Chenoos had no less than fifty wives or women, and the Mafooks from ten to twenty, any of which they seemed ready to dispose of, for the time, to their white visitors; and the women most commonly, as may well be supposed, were equally ready to offer themselves, and greatly offended when their offer was not accepted. It would seem, however, that whether they are lent out by their tyrants, or on their own accord, the object is solely that of obtaining the wages of prostitution; the heart and the passions had no share in the transaction. It is just possible, that this facility in transferring women to the embraces of strangers, is confined to those parts of the country where they have had communication with Europeans, who have encouraged such connections; though it must be admitted that, on the present occasion, very little difference, in this respect, appears to have been observed on the part of the women, in places beyond where slave dealers are in the habit of visiting. Captain Tuckey, however, says, that in no one instance, beyond Embomma, did they find the men *allant en avant* in their offer of the women; but the Embomma men said, falsely it is to be hoped, that it was only their ignorance, and the little intercourse they had with white men, that prevented it; and that any of them would think themselves honoured by giving up his wife or daughter to a white man.

No such licentious conduct it would seem is sanctioned among themselves; where natives are the only parties concerned, an intrigue with another man's wife entails slavery on both the offenders; and if the wife of a Chenoo should go astray, he inflicts what punishment he may think fit on the lady, but the paramour must suffer death. Mr. Fitzmaurice states, that an instance of this kind occurred while he was stationed at Embomma. The man was first carried to Sherwood, the mate of a slave ship then trading in the river, and offered to him for sale; but on being rejected, those who had charge of him bound his hands and feet, and, without further ceremony, threw him into the river.

MORAL AND PHYSICAL CHARACTER.—It is a strange inconsistency of human feeling that, in all uncultivated societies, the weaker sex should be doomed to perform the most laborious drudgery. In Congo, the cultivation of the land, and the search after food in the woods and on the plains, frequently the catching of fish, devolve wholly on the women ; while the men either saunter about, or idle away the time in laying at full length on the ground, or in stringing beads, or sleeping in their huts: if employed at all, it is in weaving their little mats or caps, a kind of light work more appropriate to the other sex, or in strumming on some musical instrument.

Their indolent disposition, however, does not prevent them from indulging an immoderate fondness for dancing, more especially on moon-light nights. No feats of activity are displayed in this species of amusement, which consists chiefly in various motions of the arms and gesticulations of the body, not altogether the most decent. The pleasure it affords is announced by hearty peals of laughter. They are also fond of singing, but it is only a monotonous drawling of the voice, not very well calculated to delight the ears of the auditors. Their musical instruments are, a sort of guitar or lyre of the rudest kind, horns, shells and drums ; and sometimes calabashes filled with small stones to make a rattling noise. They have songs on love, war, hunting, palm wine, and a variety of subjects, some of which have been attempted to be written down and translated by Captain Tuckey, but in so imperfect a manner and so much defaced, as not to admit of being made out.

In all the memoranda of the gentlemen employed on the expedition, the natives of Congo are represented as a lively and good-humoured race of men, extremely hospitable to strangers, and always ready to share their pittance, sometimes scanty enough, with the passing visitor. In one of the notes only, they are characterized as shrewd, cunning, and thievish. Men living in a state of society like theirs, have occasion for all their shrewdness and cunning ; but with respect to their thievish propensity, though common to almost all savage and half-civilized tribes, the testimony of Captain Tuckey is rather in favour of their honesty. It is true, that when returning down the river in a sickly and helpless condition, and in great haste and anxiety to reach the vessels, some trifling advantage was taken to pilfer part of their baggage ; but it is in favour of these people that, considering all the circumstances of the distressed situation of the party, they were able to bring away with them any part of their scattered property.

The stature of the men of Congo is that of the middle size, and their features, though nearest to those of the negro tribe, are neither so strongly marked, nor so black as the Africans are in general. They are not only represented as being more pleasing, but also as wearing the appearance of great simplicity and innocence. Captain Tuckey could not discover among the people any national physiognomy ; but few mulattoes ; and many had the features of southern Europeans. The discovery, by the party, of burnt bones, and of human sculls hanging from trees, might have led to the injurious idea of their being addicted to the eating of human flesh, had no further enquiries been made concerning them : accounts of cannibalism have been inferred by travellers on appearances no better founded than these : and it is probable, that the many idle stories repeated by the Capuchin and other missionaries to Congo, of the Giagas and Anzicas, their immediate neighbours, delighting in human flesh, may have had no other foundation than their fears worked upon by the stories of the neighbouring tribes, who always take care to represent one another in a bad light, and usually fix upon cannibalism as the worst.

SUPERSTITIONS.—Ignorance has always been accounted the prolific mother of superstition. Those of the negroes of Congo would be mere subjects of ridicule, if they were harmless to society ; which however is not the case. Every man has his *fetiche,* and some at least a dozen, being so many tutelary deities, against every imaginable evil that may befal them. The word is Portuguese, *feitiço,* and signifies a charm, witchcraft, magic, &c.; and what is remarkable enough, it is in universal use among all the negro tribes of the Western Coast.

There is nothing so vile in nature, that does not serve for a negro's fetiche ; the horn, the hoof, the hair, the teeth, and the bones of all manner of quadrupeds ; the feathers, beaks, claws, skulls and bones of birds ; the heads and skins of snakes ; the shells and fins of fishes ; pieces of old iron, copper, wood, seeds of plants, and sometimes a mixture of all, or most of them, strung together. In the choice of a fetiche, they consult certain persons whom they call fetiche-men, who may be considered to form a kind of priesthood, the members of which preside at the altar of superstition. As a specimen of these senseless appendages

ges to the dress, and the dwelling of every negro, the following represents one which the wearer considered as an infallible charm against poison ; the materials are, an European padlock, in the iron of which they have contrived to bury a cowrie shell and various other matters, the bill of a bird, and the head of a snake ; these are suspended from a rosary consisting of the beans of a species of *dolichos*, strung alternately with the seeds of some other plant.

Others, with some little variation, are considered as protections against the effects of thunder and lightning, against the attacks of the alligator, the hippopotamus, snakes, lions, tigers, &c. &c. And if it should so happen, as it sometimes does, that in spite of his guardian genius, the wearer should perish by the very means against which he had adopted it as a precaution, no blame is ascribed to any negligence or want of virtue on the part of the fetiche, but to some offence given to it, by the possessor, for which it has permitted the punishment. On this account, when a man is about to commit a crime, or do that which his conscience tells him he ought not to do, he lays aside his fetiche, and covers up his deity, that he may not be privy to the deed. Some of the persons of the expedition shewed to one of the chief men a magnet, which he said was very bad fetiche for black man ; he was too lively and had too much *savey*.

This would be all well enough, if an opinion of their virtues in warding off evil affected only themselves ; and they might even be useful when considered as a guard upon their actions ; but their influence does not stop here ; they are considered in one sense as a kind of deity, to whom prayers are addressed for their assistance, and if afforded, thanksgivings are returned ; for the honour of the fetiche also, abstinence is performed, and penalties inflicted ; but if unsuccessful in any enterprize on which the fetiche has been consulted, the owner immediately parts with him, and purchases another from the priest. These cunning men have gone a step further, and have succeeded in per- suading the silly people, that by their means, any part of a man's property may be fetiched or made sacred, in the same manner, or nearly so, as the *tabboo*, which is so uni- versally practised in all the Pacific and South Sea islands ; and their mode of detecting a thief, bears a very remark- able resemblance to that which Campbell describes to be used among the people of the Sandwich islands.

But the evil does not end here. Mr. Fitzmaurice, while he stopped at Banza Cooloo, was witness to a trans-

3 C

action, which will best explain the ill effects of these sense-less superstitions. A woman had been robbed of some manioc and ground nuts; she applied to a gangam or priest for a fetiche, which would compel the robber to restore the property; and the manner of doing it is as follows. The fetiche being exposed in some public place, the people of the village dance round it, and with the most hideous howlings invoke it to produce the thief, or to direct that within a certain time, and at a certain place, he shall deposit the stolen goods, in failure of which, that this newly created divinity will be pleased to destroy both him and his relations. If at the expiration of the time, which is usually two days, the property is not restored, the fetiche is removed, and the first person of the village who dies, is considered to be the thief. It usually happens, that the goods are restored, but this was not the case in the present instance. The morning after the removal of the fetiche, the most dismal howlings were heard in the village, and, on sending the interpreter to enquire into the cause, he returned and reported, that the fetiche had killed the thief, and that the noise proceeded from the relations mourning over the body. " The deceased," says Mr. Fitzmaurice, " had been one of my coolies, and was a fine strong young man, apparently about twenty-four years of age. I had seen him the preceding evening walking about in good health, which, together with the circumstance of his having died in convulsions, leads me to suspect that, rather than suffer the efficacy of the fetiche to be questioned, the priest had selected this poor fellow as

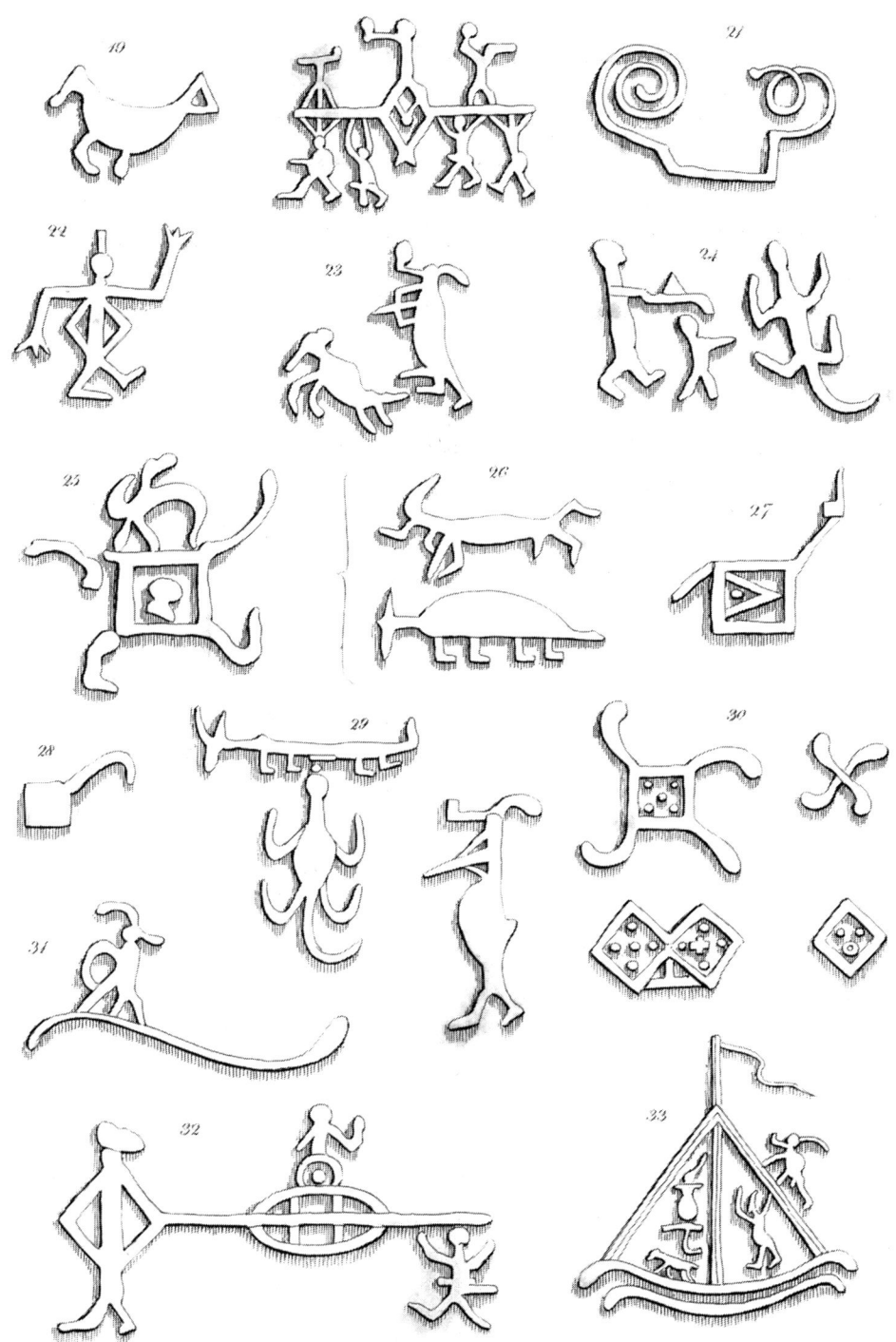

From a sketch by Lieu.ᵗ Hawkey.

I Clarke sc

Copies of Figures in low-relief on the face of the Fetiche Rock.

CAPTAIN TUCKEY'S VOYAGE IN AFRICA.

Page 380

Published by John Murray, Albemarle Street London Nov.ᵗ 1.ˢᵗ 1817.

the victim to his imposture, and had contrived to send him out of the world by poison; an opinion in which I am the more confirmed, from the relations of the deceased having found it necessary to present the priest with a larger quantity of manioc and nuts than what had been stolen, a necessary precaution, as my interpreter assures me, to preserve their own lives."

The following circumstance, which passed between Mr. Fitzmaurice and his friend the Chenoo of the village, is a curious trait of simplicity or cunning in the manners of these people. This Chenoo had boasted of a war fetiche, which if any one attempted to shoot at, the flint would fall out, and the person so attempting would fall down dead. On Mr. Fitzmaurice and Mr. Hodder expressing a wish to have a shot at this redoubtable deity, he observed, that he loved them too much to let them try; on telling him however that if, on firing, they missed it, or if they sustained any harm, they would give him a whole piece of baft and two bottles of brandy, his fears for their safety immediately vanished before the prospect of gain, and he consented; six yards was the distance measured off. The fetiche was the figure of a man rudely carved in wood and covered with rags, about two feet high, and one foot broad, and the time appointed was the following morning. In the course of the evening, the interpreter, who had a great regard for the strangers, appeared extremely sad and pensive, and being asked the cause, replied, that he very much feared his good masters were going to die, and intreated in the most urgent manner, that they would give the baft and brandy, and let the fetiche alone. Being absent for some time, he said, on his return, that he had been at the village ; that the King and his nobles were holding a palaver, whether they should venture the fetiche or not, and that they had asked him, whether he thought white men would dare fire at it, and on his answering in the affirmative, they exclaimed, " mindeele zaambie m'poonga," white men are gods." The Chenoo made his appearance the following morning, but without the fetiche, and was very desirous to see the fowling piece fired, in which he was gratified, and on perceiving the ball strike the mark fired at, he seemed very much astonished, and went away without saying a word. In the evening he returned, with nearly the whole of the inhabitants; begged they would not think of firing at his fetiche, for if they should hit it, and this was known to the neighbouring Chenoos, they would all make war upon him immediately ; an intreaty which was uttered with so much real axiety in his countenance as to leave no doubt of his being in earnest.

Besides the individual fetiches which are selected by a priest, or by the caprice of the wearer, various striking objects of nature are held in general estimation. The Taddi Enzazzi, or lightning stone, and the fetisch rock, are objects of this kind. The latter is considered as the peculiar residence of Seembi, the spirit which presides over the river. On the side of some rocks inhabited by fishermen, round the point of Soonda, are a number of raised figures, formed apparently with sand and ashes and laid on wet, which, when indurated, appear like stone sculptured in low relief. The annexed plates are fac-similes of those figures copied by Lieutenant Hawkey, respecting which, he observes, that he could not learn, from any inquiries he was able to make, whether they had any connection with the religious notions of the people, though they went by the name of fetiches. They were said to be the work of a learned priest of Nokki, who taught the art to all those who chose to pay him. The names of the objects, corresponding with the numbers on the plates, are mentioned by Lieutenant Hawkey, as under.

1. A gentleman in his hammock and guard,
2. A gentleman borne by his slave.
3. 4. 5. } Unknown.
6. A lizard.
7. An alligator.
8. 9. } Unknown.
10. A hippopotamus.
11. Unknown.
12. A buffalo.
13. A chasseur.
14. A buffalo.
15. A bird.
16. Unknown.
17. An alligator.
18. A hunter killing a deer.
19. A bird.

20. A gentleman in his hammock.
21. A snake.
22. Unknown.
23. A man shooting a bird.
24. An old man and a young one killing an alligator.
25. Unknown.
26. A hunter and hippopotamus.
27. An elephant.
28. Unknown.
29. A hunter, a deer, and an alligator.
30. Tattooing figures.
31. A man and snake.
32. A gentleman in his hammock,
33. A ship.

In several other places, figures of a similar kind were met with, cut into the face of the slaty rock, or into wood, or on the surface of the gourds or pumpkins, most of which had something of the fetiche or sacred character attached to them. They have some vague notion of a future paradise, in which they shall all be happy; they also entertain some idea of a good and an evil principle; the former is distinguished by the name of Zamba M'Poonga; the latter by that of Caddee M'Peemba; but they seem to pay more veneration to, and to feel a greater dread of, their substantial fetiches, than these imaginary personages.

The most inoffensive part of their superstitions is the respect which they show to the dead; and absurd as it may appear, a veneration for deceased friends and relations is always a favourable trait in the character of a people. Those who can afford, and they omit no endeavours to obtain it, cover the dead bodies of their relations with many folds of clothing, and keep them above ground, till, from the quantity of wrappers added from time to time, they have arrived at an immense bulk; in this state they are then deposited in a hut; they mourn their loss at stated times of the day with howlings and lamentations; and at length they bury them in graves of vast depth, with the view probably of preventing the possibility of their being scratched up by beasts of prey; they plant trees and shrubs round the graves, and like the Welsh and the Chinese, decorate them with flowers or place fetiches upon them. An elephant's tusk placed at the head and another at the foot, mark the grave as belonging to a person of some distinction.

Copies of Figures in low-relief on the face of the Fetiche Rock.

CAPTAIN TUCKEY'S VOYAGE IN AFRICA.

Published by John Murray, Albemarle Street London Nov.r 1 1817

CRIMES AND PUNISHMENTS.—The only capital crimes are stated to be those of poisoning and adultery, the latter of which is singular enough, considering in what little estimation women are held. Murder and theft are punished by retaliation and restitution, or selling the criminal into slavery. The Gangam and his Kissey are the grand jury who find the bill, but the accused undergoes a trial by ordeal before the elders of the community. He is made to chew a certain poisonous bark; if guilty, he keeps it in his stomach and it occasions his death; if innocent, he throws it up again and he is acquitted of the charge; and thus the guilt or innocence of a man is made to depend on the strength of his stomach. The practice of poisoning is so common, that the master of a slave always makes him taste his cooked victuals before he ventures to eat of them himself.

DISEASES AND REMEDIES.—The natives in general appeared to be healthy; the diseases under which they mostly laboured, were of the cutaneous kind, few being free from the itch, and scrofula; leprosy, and elephantiasis were observed, and some few cases of fever and fluxes occurred. They appeared to be subject also to indolent tumors, and most of them were observed to have large navels. Among the people of the neighbouring towns who came down to Inga to see the white men that were stationed there, a Mafook brought with him his daughter, a girl of about twelve years of age, whose skin was perfectly white, but of a pale sickly colour, though the father said she was quite stout and healthy; she had curly hair and negro features.

The only medicines used by them, and those but sparingly, are infusions and decoctions of native plants; and among others the root of a species of dioscorea, of a very strong bitter taste, is very much chewed by them as a preventive of fluxes; but the Gangam Kissey and various fetiches are mostly resorted to for the cure of diseases; and when the Gangam, who acts in the threefold capacity of priest, public accuser, and physician, sees the case to be desperate, he gives the patient over to Zamba M'Poonga.

It is not easy to conceive for what purpose the shoals of missionaries were sent among the Congo negroes, nor in what manner they passed their time in the country. Their accounts are filled with the multitudes they baptized, and they baptized all who offered themselves; but it is a very extraordinary fact that they should not have instructed some of them to read and write. No trace of any such instruction appeared along the banks of the Zaire, except in the instance before mentioned; nor did it appear that they had any mode of registering time or events, except by the moon, and in this way only for a very few years.

LANGUAGE.—The language of the Congo and the neighbouring states, differs very materially from all the known languages of the negroes of northern Africa; but from the copious vocabularies obtained by Captain Tuckey, there would seem to be a radical affinity between all the languages on the western coast of Southern Africa, and that these languages have pervaded the greater part of that portion of the Continent, and extended even to the eastern coast.

The letter of Mr. Marsden, referred to in Captain Tuckey's instructions, contains some curious information on this subject; as well as some remarks on the language in general, which may be useful to future travellers; the following is an extract from it.

" Knowing so little, as we do, of the countries on the " banks of the Zaire (which I observe is also called by " D'Anville, the Barbela river), few particular instructions " can be given, regarding the language spoken in that " quarter; and it will depend upon Captain Tuckey to avail " himself of the information that circumstances may place " within his reach. In most cases the opportunity will be " little more than that of collecting a few of the most com- " mon words, which may, however, be sufficient to shew " whether the people speaking them, have or have not an " original connection with others geographically and poli- " tically separated from them; and comparisons of this " kind will be much facilitated by having uniform lists " which not only suggest the proper words at the momen " of enquiry, but place them mechanically beside each " other. Where a longer residence admits of freer inter- " course, and the means of acquiring a more perfect know- " ledge of the language, it will be desirable, besides at- " tempting to fill up the larger vocabulary,* that pains " should be taken to examine its grammatical structure,

* A printed selection of English words.

3 D

" and to ascertain, for instance, how the nominative and
" subjunctive words in a sentence are placed with respect
" to the verb ; how the adjective with regard to the sub-
" stantive ; how plurals and degrees of comparison are
" formed ; whether there is any kind of inflexion or varia-
" tion of syllables of the same word according to its posi-
" tion in the sentence and connection with other words ;
" whether the pronouns personal vary according to the
" rank or sex of the person addressing or person addressed ;
" and whether they are incorporated with the verb ; and to
" observe any other peculiarities of idiom, that the lan-
" guage may present ; noting the degree of softness, harsh-
" ness, indistinctness, intonation, guttural sounds, and the
" prevalence or deficiency of any particular letters of the
" alphabet, as we should term them, such as R and F.
" The extent of country, over which a language is under-
" stood to prevail, should also be a subject of investigation ;
" and, by what others it is bounded at every side. Also,
" whether there may not be a correct language of com-
" munication between nations, whose proper languages are
" distinct.

" I observe that the name of *Congo* belongs to the coun-
" try on the southern side of the Zaire; and that Loango,
" Kokongo, N'Goio, Tomba, and N'Teka, are the names
" of kingdoms or districts on the northern side. The spe-
" cimens I have of the language of Loango (apparently the
" most considerable of these) shews it to be radically the
" same with that of Congo, although, as dialects, they vary
" a great deal. It will probably be found, that this is the

" case with regard to the others also ; and I am the more
" inclined to believe the language very general in that
" part of Africa from the following circumstance : I had
" formerly a negro servant from Mosambique, who came
" by the way of Bombay to Bencoolen, and having taken
" down from his mouth the words of his native tongue, I
" was afterwards much surprized to find them correspond,
" in many instances, not only with the language of the
" Caffers, as given by Sparrman, but more especially with
" that of Congo, as will be seen on comparing a few of the
" words of the latter, as given by Benjamin (the Congo
" black) with those taken from my servant.

English.	Congo.	Mosambique.	Kaffer.
Three.	Tatoo.	Atatoo.	
Ten.	Coomy.	Kumir.	
Four.	Me-sana.		Sanu.
Man.	Momtoo.	Muntu.	
Woman.	Makaintu.	Muke.	
Foot.	Cooloo.	Mo-guru.	
Day.	Booboo.	Riubu.	
Dead.	Cufoy.	Kufoa.	
Water.	Maza.	Madje.	Maazi.

" But it was not my intention to have gone into this de-
" tail; the fact, however, is very curious, the distance being
" so considerable."

It is sufficiently remarkable, however, that while this agreement is found between the languages of tribes so very distant
from each other, so great a difference should prevail in different parts of the same district, and at so short a distance, as

appears by the Vocabulary (Appendix, No. I.) collected and filled up by Captain Tuckey; the first column of which are the words of the Malemba language, on the coast and near the mouth of the river, the second those of Embomma ; and it is stated that the language beyond Inga differed very considerably from that of Embomma.

Mr. Marsden, who obligingly furnished the list of English words in a printed form, and whose extensive knowledge of languages, stamps a value on any opinion he may give on that subject, has communicated the following observations on Captain Tuckey's vocabulary.

" The very copious and apparently accurate vocabulary " of the *Congo* language, collected by Captain Tuckey, " has furnished the means of comparison with the other " languages and dialects prevailing in the southern por- " tion of Africa, and has thereby served to establish the " fact of an intimate connexion between the races of peo- " ple inhabiting the western and the eastern coasts of the " peninsula ; although in that parallel, its breadth is little " less than thirty degrees of longitude.

" Upon selecting some of the most familiar terms, and " comparing them with the specimens we possess, it will " be seen, in the first place, that the words as written down " by Captain Tuckey, from the mouths of the natives " of *Congo*, agree generally with those given by Brusciotto, " Oldendorp, and Hervas; allowance being made for the " differences of European orthography. They also cor- " respond with those of the neighbouring countries of " *Loango* and *Angola*, with some variety of labial pro-

" nunciation ; and less perfectly with the languages of the
" *Mandongo* (not to be confounded with the *Mandinga* of
" Northern Africa) and the *Camba* people; both of the
" same western coast. It is highly probable, that all
" these mutually understand each other in conversation.
" Between the Congo language and that of the tribes on
" the eastern side, the affinity, although radical is much
" less striking, and the people themselves must consider
" them as quite distinct; but the following instances of
" resemblance, in words expressing the simplest ideas,
" may be thought sufficient to warrant the belief, that the
" nations by whom they are employed, must, at a remote
" period, have been more intimately connected."

	Congo. Tuckey.	Congo. Brusciotto.	Congo. Oldendorp.	Mosambique. Native.	De Lagoa. White.	Kaffer. Sparrman.
Three	Tatoo	Tattu	Si-tattu	Ba-tatu	—	A-tatu
Four	M'na	Ya	Sija	Me-sana	Moonaw	Sanu
Five	Tanoo	Tanu	Sit-tan	—	Thanou	—
Ten	Coomy	Icumi	Si-kumi	Kumi	Koumaw	Sumi
Eye	Mieso	—	—	Meso	Teesho	—
Tooth	Meno	—	—	Meno	Menho	—
Dead	Foi	—	Affua	Ku-foa	—	Ufile
Water	Maza	Mase	Masa	Madji	Matee	Maesi
Hog	Gorolooboo	—	Engulo	Guruay	Gulloway	—
Sun	Tangua	N'Tazi	Tangu	—	—	Langga
Moon	Mooezy	—	—	Moysé	Moomo	—
Salt	Moon-qua	—	—	—	Mun-you	—

There does not seem to be the least truth in the com-
plicated mechanism of the Congo language, which some
fanciful author thought he had discovered, and which has
been repeated by succeeding writers; none of " those idioms

of which the syntax and grammatical forms, ingeniously combined with art, indicate, in the opinion of Malte-Brun, " a meditative genius, foreign to the habitual condition of these people."

These few observations contain a summary of the knowledge of the moral circumstances and condition of the people, and their means of subistence, as obtained by the expedition to the point of the river where its researches terminated. The physical information acquired is, on account of its scientific form, kept separate, and follows in the Appendix

APPENDIX, No. I.

A Vocabulary of the MALEMBA and EMBOMMA Languages.

English.	Malemba.	Embomma.
...ve,	Tanda	Teleema,
...sent,	Ieli	Ieli-kouka
...use,	N'Doke	
...mire,	Equaila	
...vice,	Wenapee	
...iltery,	Wavuca, Ng-Cazganie	Sougam casan-gana
...aid,	Wonga	Boema
...ont, v.	N'Sone	nganzey
...er,	Quonema	
...ernoon,	Masseca	maseaka
...ain,	Quandee	
...ree,	Ioca Chivueede	
...,	E'Zoola	zeelo
...ke,	Deddy Deddy	
...ve,	N'Chema	monio
...,	ionsou. M'Venu-ionsou, I give all	Yo
...ne,	Caca ; Meno caca, I am alone	
...vays,	Loumbau E'on-sou, all times	Tangibana
...use,	Queembela	
...chor,	Boam-poutou	boam-poutou
...l,	Isha	
...gry	N'Zalla, Lengula, sulky	
...other	Lequa, Lequa chanca, another thing.	
...swer, v.	Tamboudede leoua, chouso-le-qua, any thing	boulem-beembo
...y,		
...roach, v.	Queamena or Wesadea	Isa
...u,	Coco	
...ny,		Cacomunta
...ive,	N'Chemosouca	
...amed,	N'Sone	sony-zakaleka
...ore,	Vanase, vanasse	vananze
...,	Couvaula	uvroola
...eep,	Leca	lélé
...ist,	Cousadesa	
	M'pou	

English.	Malemba.	Embomma.
Aunt,	Cacandee	Menkaze
Avoid,	Souama	
Awake,	Catoumauca	cotouka
Axe,	M'Peebe	tawly
Back,	Nema	booza
Back again,	Vantauca	
Bad,	Mabee, Moontoa N'zambico, bad man	mambee
Bag,	Ecouba	Kouba
Bake,	Zampaimbe, Bolo Zampaimbe, bake bread	
Bald,	Vandou	vandou
Bargain,	Saomba	soomba
Bark (rind)	Taunda	babosy
Barren	Seeta	
Barter, v.	Taubeengana-quetau	Vinja
Basket,	M'Bangou	
Bathe,	N'Younga	Sookoola
Battle,	Nouana	N'ousna
Bawl, v.	Beconoa	
Bay,	Londo	N'zeela
Beads,	M'Sanga, or Sanga	p'sanga
Beard,	N'Deva	Devoe
Beat, v.	Yaita, beco-yaita, don't strike	Bolo
Bed,	M'Foulou	Cheea
Before,	Ovetide quande	
Beg,	M'panou	M'cootoo
Begin,	Davove	Tona
Behind,	Oquinema	quanima
Believe,	Eande	
Belly,	Voumou	Voomoo
Below,	Quonsee	quoonsee
Bend,	Voumbama	beenza
Betray,	Moueve	
Between,	Fouloumose	
Beyond,	Valla	
Big,	Ounene	tolo
Bind,	Cangama	
Bird,	Noone	Noonee

English.	Malemba.	Embomma.	English.	Malemba.	Embomma.
Bite, v.	Lavata	tibila	Cold,	Chazee	cheosey
Bitter,	Cazau	nooly	Comb, v.	Sanoo	Sanoo
Black,	M'fiote	M'fiot	Come,	Weesa	ouise
Blind,	Mesoumafoa	mafoi	Conquer,	cheena waate	boogazy
Blood,	Menga	Menga	Cook, v.	Lambe	lamba
Blue,	Chandombe	Chinomba	Copper,	Sango	Songo
Boat	N'Zaza	N'zaza	Corner,	Fouma	
Body,	Solango	Avia	Country,	Seame	N'zee
Boil, v.	Lamba (meno lamba, I boil)	laamb	Cow,	Gombe	pacheza
			Cry, v.	Dela	leela
Bold,	Oumolo, or qu-angolo		Cure, v.	Sambouca	belola
			Curse, v.	Lakelaca	lokala
Bone,	Vissee, or Vese	vezze	Cut,	Veengoana	Yango
Bottom,	Coonansee	Coonansee			
Box, n.	Lookata		Dance, v.	Keena	keena
Boy,	N'Taoude	leeze, toadi	Dark,	Tombe ; Night time	M'boi
Brave,	Quangolo or Ou-molo	pandé	Daughter,	Chincoomba	Coomba
Bread,	Bolo	Bolo	Day,	Laumbau	Moinee
Break, v.	Baudede	bourica	Dead,	Fauede	foi
Breasts,	N'Toulou	Maemi maboi	Deaf,	Matoo, Mafou (ear blind)	(the same)
Bridge,	Saaoka	Subooka			
Bring,	Twala		Deceive,	Maueve	M'poonizea
Broad,	Tamamase		Deep,	M'peenda	Vinda
Brother,	Pangame		Depart,	Wenda quakoo	Yenda quako
Build,	Taunganza		Devil,	Cadde M'Pemba	Coolam pam'
Burn, v.	E. Veede	Monovia	Dew,	Desa or Deza	Lizee
Busy,	Salansalanga		Die, v.	M'foa	foca foi
Buy,	Soumbaquacou	Soomba	Dig,	Sema. Sema au-loo, dig grave	sicum
Cable,	Seenga	Seenga	Diligent,	kebba-bene	
Call, v.	Bokela	Lundoo	Discourse,	yako. Palaver	zoco
Calm,	Bacanam-pemba-quano, no wind	bauano pemo	Dispute,	N'ganzy	ganzy
			Divide,	chakeky	chakeky
Careful,	Kbea	bongo leeko	Dog,	M'Boa	M'boi
Carry,	Nata	Nata	Door,	Kaveloo, door-place, E'Vitoo.	Vitoo
Cat,	Boude	boodé			
Catch, v.	Bacca	Seemba	Down, ad.	Coonasse. Wenda conasse, go down	
Change,	Veenga. veenqua lequa (change something)	aviengeza	Dream, v.	N'dazee, dosentou	lota
			Drink, v.	Noa	noi
Cheeks,	Matamma	Matamma	Drop, v.	Bauede	soonoquezy
Chew,	Dade	Casu	Drown,	Seendede	fomo
Chief, n. s.	Menta		Drunk,	Calelau, or Coloa Malavou, drunk from wine	Coloi
Child,	Mauana	moana			
Chin,	Bevau	bevo			
Choose,	Zona		Ear,	Cooto-Matoo	Matoo
Circle,	Zounga(Zoonga)	zegoomaneena	Earth (soil)	N'tato	toto
City,	Banza	M'banza	Earth (globe)	sionso	sionso
Clean,	Soucoula, (means also wash)	Neaveze	East,		akoo
			Eat,	Dea	lia
Cloth,	chindele	blele	Edge,	Maino	tova
Cloud,	E'Sanche	tooty	Egg,	Makee	makee maso
Coarse,	Catyauwataco	voonga			

English.	Malemba.	Embomma.
ht,	E'Nana	N'ana
hty,	Nana longcamma	lonvois
pty,	champabala	bacana leevco
, n.	Seena	Seena
my,		Giahelady
ugh,	Fouaing	Foiny
er,		Cota
ape, v.	cheena	teena
use, n.	cabely	liezo
,	Mesau	
brow,	N'daou	davu
e,	loosi	loozie
, v.	Bouede	booide
e,	Voona or Ovoo-nene	boisey
ily,		cunda
,	N'Seke	tanzy
t, v.	cadedeco, stomach empty	
	Tola	tolezy
ner,	tata	taata
r, v.	Wonga (cheenico do not run away.)	cheonico
st,	dela quoomosee	mocu beeza
ther,	N'Salla	caia muza
l, v.	Seembede	seembelé
iale,	Kentoo	chemta
ch, v.	Twala	voola
er,	Yaila, sick	
een,	Ecaume tanoo	macooma tanoo
y,	Macaumatanoo	Sambanoo
ht, v.	Nooana	noana
	Zonga. Zonga Maza, fill water	Ouazia
l,	Tomba	tomba
ger,	M. blembo	loozala
e,	Bazao	bazoo
a,	Bishe or Bizhy	M'foo
e,	Tanoo	toanoo
,	Vavawoote	bassa
h,	Gombai	M'psoonia
it, v.	E'folo. E'folo de Maza, float on the water	
	Mazely. Maza mazely water flood.	
od,	Foundee	foondia
ver,	Lavooka	catooka
v.	Seeda-quonema, I follow you.	londa
ow,		

English.	Malemba.	Embomma.
Fond,	N'Zona. N'Zona Kentou, fond of woman	n'zoolozy
Food,		belia
Fool,	Laoo	booba
Foot,	Tambee	tambee
Forbid,	Zoueneco	seembeedi
Forget,	Zeembakeene	zimbancoonie
Forgive,	Mangene	vanica
Fork,	Soma	soma
Forty,	Macoomaya	macoomana
Four,	Yaea Quea or kea	m'na
Fourteen,	Ecoameaeya	coom m'na
Fowl,	Soosoo	soo soo
Free,	Foomoo, free man	foomongana
Fresh,	Enasoodeca (no stink)	
Friend,	Dequame	Coondiamy
Fruit,		Cooia
Full,	Ezaily (Glossa Ezaily Maza, glass full of water)	zala
Fur,	Meca	meeka
Girl,		caintoo
Give,	vana	vama
Glad,	Tondele	tondiza
Go, v.	Wenda	ouenda
Goat,		combo
God,	Zambe M' Poungoo	yambee
Gold,	Ola	voola
Good,	Maboote	tibooty mavooté
Great,	Foonioo-a-Moote	keenani
Green,	Chambeo	kankoososo
Grow,	E'Menene	coola
Guard, v.		lunglula
Hair,	N'Sooke	M'sootchy
Half,	Cachanseea	teeny
Hand,	Candase	coco
Handsome,	Mamoote	quenevezey
Hang,	Keteca,	zungalaquoi
Happy,		oobooeli
Hard,	Golozeenge	bala
Head,	N'Too	M'too
Hear,	Weloo(Oweloowe do you hear that)	oneloo
Heart,	N'Cheema	monio
Heaven,	Ezooloo	coozolo
Heavy,	Zeeta	zeeta
Hen,	Soosoo N'kentoo woman fowl	

English.	Malemba.	Embomma.	English.	Malemba	Embomm
Herb,	Foundee	teel,	Lake,	Eanga	Cooly
Here,	Wesa-ba (come here)		Lame,	Tolooca	toloca
			Land,	Zela	n'se
Hide, v.	Soo-aimy	souka	Last,	Quenema	lequampe de
High,	M'Saiky	nankoo	Laugh, v.	Saiba	seva
Hill,	M'Zanza	vemongo	Law,	Yoco, and Palaver	m'cusa
Hire, v.	Salla Ecofeeta, work and I'll pay you, Ooeza poota, come and I'll pay you		Lead, n.	choomboo	choomboo
			Learn, v.	Longua or Cou-camba	longua
			Leg,	Veende,	maloo
			Lie (down) v.	Daile,	bleka
Hit, v.	Oungetele		Lie (falsity)	Vouna	m'voonoo
Hog,	N'Gooloo	gooloobo	Lift,	Nata	naogoova
Hold, v.	Seemba	seemba	Light (not heavy)	Bacana Zitaco	zelaco
Hole,	Nooa	cooloo	Light (not dark)	Mouene	mooini
Hollow, n. a.	Lequa champa-bala, empty	voovooloo	Lightning,	N'Zaza	lusiemo
			Lips,	Bevau	bleelee
Home,	Coompootoo		Little,	Chakai	chepehow
Honest,	Moontoo N'Zam-be	moonta	Live, v.	Ena-Wa-wautee, I live	
Horn,	M'Poca and M'Poonge	m'poka	Long,	Chella	chicolezy
			Look, v.	Talla	
Horse,	Cavalo	cavallo (Portug.)	Love, v.	Laou	zolozy
Hot,	Mooene	bazoo	Low,	Vousee	toola
House,	M'Zo	m'zo			
Hundred,	changcamma	m'cama	Mad,	E'Laou or Ai-leooa	laooka
Hungry,	N'Zalla	zala			
Hurt, v.	N'Gansey	coontanty	Make, v.	Saneca	vanga
Husband,	Etoco	nooniani	Male,	Moontou,	
			Man (homo)	Moontau	boocala
I,	Meno	meenoo	Many,	Enjecaca	benga
Idle,	Casasalaco		Market,	E'Zandou,	zando
Jealous,	Fontavouke cas-same	chimpala	Mat,	Teba	teva
			Meet, v.	Baulasenna	boolanjana
If,	Onso-Onso Zo-nene, if you like	vo	Melt, v.	Manze	quabooka
			Mend,	Londa	londo
Industrious,	Salla woete		Middle,	Counzee	cawty
Innocent,		mandico	Milk,	Chimvooma	chialy
Interpret,	N'Camba	sencamba	Mine, pr.	Wamee or E'Chame	chamy
Join,	Ecca or Yeca				
Journey,	Diata	tanzey	Money,	Bango	bongo
Iron,	Loocaneba	saangua	Month,	N'Gondai	} gondé
Island,	Zoonga	zoonga	Moon,	N'Gondai	
Jump, v.	Zotooka		More,	Lequa	boola
Justice,		coticounda	Morning,	Kensouca	menamena
			Mother,	Mamma	mama
Keep,	Loonda	saonou	Mountain,	M'Zanza	m'zanza
Kick, v.	Waita	tockensy	Mouth,	Noua	m'no i
Kill,	M'foa, or M'foua	bonda	Much,	Panega	yenzy
King,	Nemboma or fooma, cheeno	m'cheeno	Mud,	Folo	m'teachy
			Musick,	Sambe	yeoola
Knife,	M'Baily	belée			
Knot,	Acolo or E'colo	colo	Nails, (ongles)	Sonso	n yula
Know,	Ounzoi	N'zabizy	Name,	Zena	zena

English.	Malemba.	Embomma.
rrow,	Voucoufe	cheechow
ar, nigh,	Calavou	chevolagaya
ck,	E'Laca	m'singoo
ighbour,	Voumosetweena	boleamba
st,	Mounguanza	jula
t,	N'Zalo	condy
ver,	Bacana Vanaco	ooenoquako
w,	Chacheva	chamona
ght,	Masseca	fookoo
ne,	E'Vaua	nana
nety,	Louvoua Long-camma	lunana
',	Nana	bucanaco
ise,	Bouba	yoko-beke yoko, don't make noise
ne,	Gonguame	chunuvalaututo
rth,		velo
se,	Mazaumau	yoono
w,	Waau	booboo
,	Manze	mazey
d,		seemba
e,	Basé	mosey
ly,	(Lequa chemasi, no more than one)	
en, v.	Zibaula	zaboquely
t,	Lava	bussykissey
ve, v.		boncooa
in,	Malau tanta	yela
int, v.	Cousandeemba	vela
per,	Papalla	papela (Port.)
y, v.	Feta	fitezy
ace,	Ele-Nauana	noinidecau
ople,	Peendouame	bantoo
erce,	(Zeka Aulou make Hole)	
nch,	Acoutanta	jongana
ain, n. s.	Voulelamene or Vouyanzala	
ty, v.	Choboubo	sacoona
asure,	Tondela	tooendacoit
enty,	Ingee	yengabeeni
ison,	Daukee	goolioongo
or,	Machanzambe	beezycunda
ssess,	Doundedeca	
t,	Sea	cheenzo
egnant,	Acuemeta	mavoomocavo
esent, (gift)	Ta	vana
etend,	N'Sallaco	
event,	Ounseembede	

English.	Malemba.	Embomma.
Price,	E'faunda	bongoqua
Priest,	Wecheche	gonga
Private,	Sauama	
Profit,	Keta	
Promise, v.	Sompeea	getu coovana
Proud,	Venda wawoote	
Pull,	(anoar) Vouela. voula bene, pull well voulla Eoumose, pull together	
Quarrel,	N'Dokee	zouza
Quarter part,	N'Dambouka	
Queen,	Camma-Foumou	foo moonchainto
Question,	Balounge dede	
Quick,	Sampouea	yong'nana
Quiet,	Molo Molo	
Rain,	Voula	vola
Rat,	N'Coumbe	pooloo
Raw,	Yangtounzau	yancoonzo
Read,	Soneca	chimboiky
Ready,	Panga	
Rebel, n.	N'Couta	
Receive,	Tamboude	
Red,	Yampaimbe	
Rejoice,	Veca Monacoo	
Return, v. n.	Avotoquede	voo taloo
Reward, n.	Ounzetou	
Rich,	Sena	vovama
Ride,	Sambela	cundama
Ring, n.	Loangana	longa
Ripe,		soowondoo
Rise, v.	Katomaca	talama
River,	Moela	moela
Road,	Mozeila	enzala
Roast,	Coka or Coca	roka
Rob,	Mowee	lovenda yeba
Rope,	Seenga	singa
Rotten,	Chawola	kabowle
Rough,	Meca	
Round,	Chenzaongolo	soolama
Row, v.	Zongoloca	vooila
Run,	Zoucooloca	zuola
Safe	Chinavona	
Sail, n.	Voola	voola
Salt,	Mongua	moongua
Sand,	Yengasee	neengy
Savage	Ganze	
Say,	Vova	
Sea,	Embou or M'Bou	boo
Seat,	Voanda	chansoo
Secure,	Yeco-baca	

English.	Malemba.	Embomma.	English.	Malemba.	Embomma.
See,	Tala	mona	Strong,	Golo	golo
Seek,	Tomba	tomba	Sun,	Mouene	tangua
Sell,	Zeca	loombeeca	Swear,	Gozee	deffy
Send,	Toma	tooma	Sweet,	Chinzsilla	
Servant,	Toudeamme	moonaleze	Swim,	Yonga	coivela
Seven,	Sambouady	Sambody	Sword,	Tanzee	soma
Seventy,	Lousambouady	loosambody			
Shade,	Pozee		Tail,	M'Kela	keela
Shake,	Necona	nicocka	Take.	Bonga	
Shame,	Somee	sonee	Tear, v.	Masanga	baka
Share, n.	Auncoeya	cayana	Tell,	Camba	
Sharp,	Looca	etooide	Ten,	Ecau-me	coomy
Shell,	Chimpenga	encaissoi	There,	Chinna	ouvana
Short,	Cofee	cooffey	They,	Ana	doo
Shut,	Zeca	zeca	Thief,	Moevee	moivy
Sick,	Yela ; yela yela, very sick		Thigh,	Ebooboo	boodou d'acou
Side,	Louvate	mona	Thin,	Enka	lovilo
Silent,	Beca yoca, be silent	canganikoota	Thirsty,	Pouilla	pooina
			Thirteen,	Coomee é tatou	macoomatatoc
Silver,	Plata	parata (Port.)	Thirty,	Macoum a tatou	macoomasamb noo
Sing,	Wimbela	quimbela	This,		eki
Sink.	Cheseendede	seendissa	Throat,	Gongolo	elaka
Sister,	Panga M'Kentau	pangankainto	Throw,	Looza	lasa
Sit,	Voenda	ovanda	Thunder,	Mandazee	moindozy
Six,	Sambanou	sambanoo	Three,	Tatau	tatoo
Sixty,	Macouma Sambanou		Tie,	Kanga	cavaga
			To	Oula	
Sky,	E'Zooloo	zooloo	To-morrow,	Baze mene	Bazimeney
Sleep, v.	Laika or Laica	leeka	Tongue,	Loodeemee	loodimee
Slow,	Conka	neké	Tooth,	Manoo	menoo
Smell, n.	Noucouna	soody	Touch, v.	Vepatacanna	touta
Smoke,	Moiscee	moisy	Town,	Banza	Banza
Smooth,	Lelamma	vendoomona	Tree,	N'Chee	chee
Soft,	Labella	bootaboota	True,	Chillica	kelica
Son,	Moene	moonayakala	Turn,	Votola	viloka
Soon,	Kainga	oo	Twelve,	Ecoume Eole	coomy emioly
Sore,	Tanta	bezy	Twenty,	Macoumolee	macoomoly
Sorry,	Cardee	keady	Two,	Cole	meoly
Sour,	Gangomona	gongoomona	Village,	Deemba	voota
Sow, (grain)	Coona	zeka moongé	Virgin,	Toubola	
Speak,	Ovova (short)	vova			
Spit, v. n.	Chaca	taoulamete	Under,	Cama	cooianda
Split,	Tongona	babo	Understand,	Ocuvanga	outooway
Square,	Shanana	conzoia	Unhappy,	Eango	Mimamby obc edico
Stab,	Chonda	coonzoka			
Stand, v.	Talama	telema			
Star,	Bota	botelé	Unjust,	Zemba Canee	
Steal,	Queya	moocey			
Stink, v. n.	Soode	soody	Voice,	Deenga	
Stone,	Tadee	etudy			
Storm,	Voolazambe	teemboi	Up,	Twoinda	
Stranger,	Zenzee	zainza	Upon,	Tandoo	
Strike,	Boola	yundi weeta			

English.	Malemba.	Embomma.	English.	Malemba.	Embomma.
e, v.	Catomoko	Catoomoka	Wind, n.	M'Paibe or M'Paima	pemo
k,	Diata	deuta			
,	Veta	zingoo	Wing,	Evee	evevê
m,	Quamona	cafoota	Wise,	Lookee	quandooka
h, v.	Soucoula	Yonga	Woman,	Kentou	quinto
ch, v.	Wingala	lanjedilla	Wood, (lignum)	Bala	coony
er,	Maza	maza	Wood, (sylva)	Lebala	chencootoo
k,	Goloco		Wool,	Meza	maka
ry,	Deembalou	necton coongely	Word,	Dinga	diambo
p,	Yenza	beela	Work, v.	Salla	salo
l, n. a.	Sambocadee	quamy	Wrong,	Zimbacainna	diambo deady
en,	Chalombo	oongatoo	Year,		m oo
te,	Pamba	pembei	Yes,	Enga	eenga
,	Nanèe	ounanie	Yesterday,	E. Zono	zono biokelly
ked,	Mabe (bad)	untoonga	You,	Gaia (plural, Yeno)	gaiyay
e,	Cazammee	casamy			
l,	Sittau	booloo			

This Vocabulary I do not consider to be free from mistakes, which I cannot now find time to discover: all the objects of the senses are, however, correct.—J. TUCKEY."

APPENDIX, No. II.

Observations on the Genus OCYTHOE *of Rafinesque, with a Description of a new Species.* By WILLIAM ELFORD LEACH, M. D. F. R. S.

From the Philosophical Transactions.

PLINY, ALDROVANDUS, LISTER, RUMPHIUS, D'ARGENVILLE, BRUGUIERE, BOSC, CUVIER, and SHAW, have described a species of this genus, that is often found in the *Argonauta argo* (common paper-nautilus), and which they have regarded as its animal, since no other inhabitant has been observed in it.

Sir JOSEPH BANKS, and some other naturalists, have always entertained a contrary opinion, believing it to be no more than a parasitical inhabitant of the argonaut's shell, and RAFINESQUE, (whose situation on the shores of the Mediterranean, has afforded him ample opportunities of studying this animal, and of observing its habits) has regarded it as a peculiar genus, allied to the *Polypus** of Aristotle, residing parasitically in the above mentioned shell.

Dr. BLAINVILLE, ten months since, when speaking of the *Argonauta*, said, " animal unknown," and he has lately informed me, that he has written a long dissertation to prove, that the *Ocythoë* of RAFINESQUE, does not belong to the shell in which it is found.

The observations made by the late Mr. JOHN CRANCH, zoologist to the unfortunate Congo expedition, have cleared from my mind any doubts on the subject. In the gulf of Guinea,

* Sepia octopodia LINNE'.

and afterwards on the voyage, he took by means of a small net, (which was always suspended over the side of the vessel) several specimens of a new species of *Ocythoë*, which were swimming in a small *argonauta*, on the surface of the sea.

On the 13th of June he placed two living specimens in a vessel of sea water; the animals very soon protruded their arms and swam on and below the surface, having all the actions of the common *polypus* of our seas; by means of their suckers, they adhered firmly to any substance with which they came in contact, and when sticking to the sides of the basin, the shell might easily be withdrawn from the animals. They had the power of completely withdrawing within the shell, and of leaving it entirely. One individual quitted its shell, and lived several hours, swimming about, and showing no inclination to return into it; and others left the shells, as he was taking them up in the net. They changed colour, like other animals of the class cephalopoda: when at rest the colour was pale flesh-couloured, more or less speckled with purplish; the under parts of the arms were bluish grey; the suckers whitish.

The *Ocythoë* differs generically from the *polypus*, in having shorter arms, with pedunculated instead of simple suckers; the superior arms too are dilated into, or furnished with, a wing-like process on their interior extremities.

All the internal organs are essentially the same as in the *polypus*, although they are somewhat modified in their proportion; but as these differences may be the result of the contraction caused by the spirits, in which they are preserved, it may be more prudent not to dwell on them. Two characters, however, which I could not discover in the *polypus*, may be mentioned, namely, four oblong spots on the inside of the tube, resembling surfaces for the secretion of mucus; two inferior and lateral, and two superior, larger, and meeting anteriorly. On the rim of the sac, immediately above the branchiæ, on each side, is a small, short, fleshy tubercle, which fits into an excavation on the opposite side of the sac. This character, which, with slight modifications, is common to this genus, to *loligo* and *sepia*, does not exist in the *polypus*.*

Although the superior arms are stated to perform such different functions from those of the *polypus*, yet they are supplied in the same manner, and from the same source with nerves. The muscles of these parts were in too contracted a state, to enable me to ascertain if they were in any degree different from those in the same parts of its kindred genus.

The general form of the body of this species of *ocythoë* is the same as that of the common *polypus*, and it is covered by the same integuments, without any surface adapted either to adhere to, or to secrete, the shell in which it is found. The sexes differ as in the *polypus*.

Ocythoë Cranchii.

O. corpore purpureo-punctato, brachiis subtus cerulescente-griseis ; superioribus membranâ spongeosâ pallidâ maculatâ.

The superior arms are generally attached to the side of the membranes (fig. 5. Pl. XII.); but in one specimen the membranes adhere only by their base, below the apex of the arm, fig. 6. The membrane is subject to great variation in size and form, and is often different on the arms of the same individual.

One male only was sent home, all the others were females, which had placed their eggs in the spiral part of the shell.

One female, that had deposited all her eggs, withdrew completely within the shell, as in fig. 3; her body on one side had

* The rudiment of the bone, which occurs in the *polypus*, (as has been observed by Cuvier) is not to be found in the *Ocythoë*.

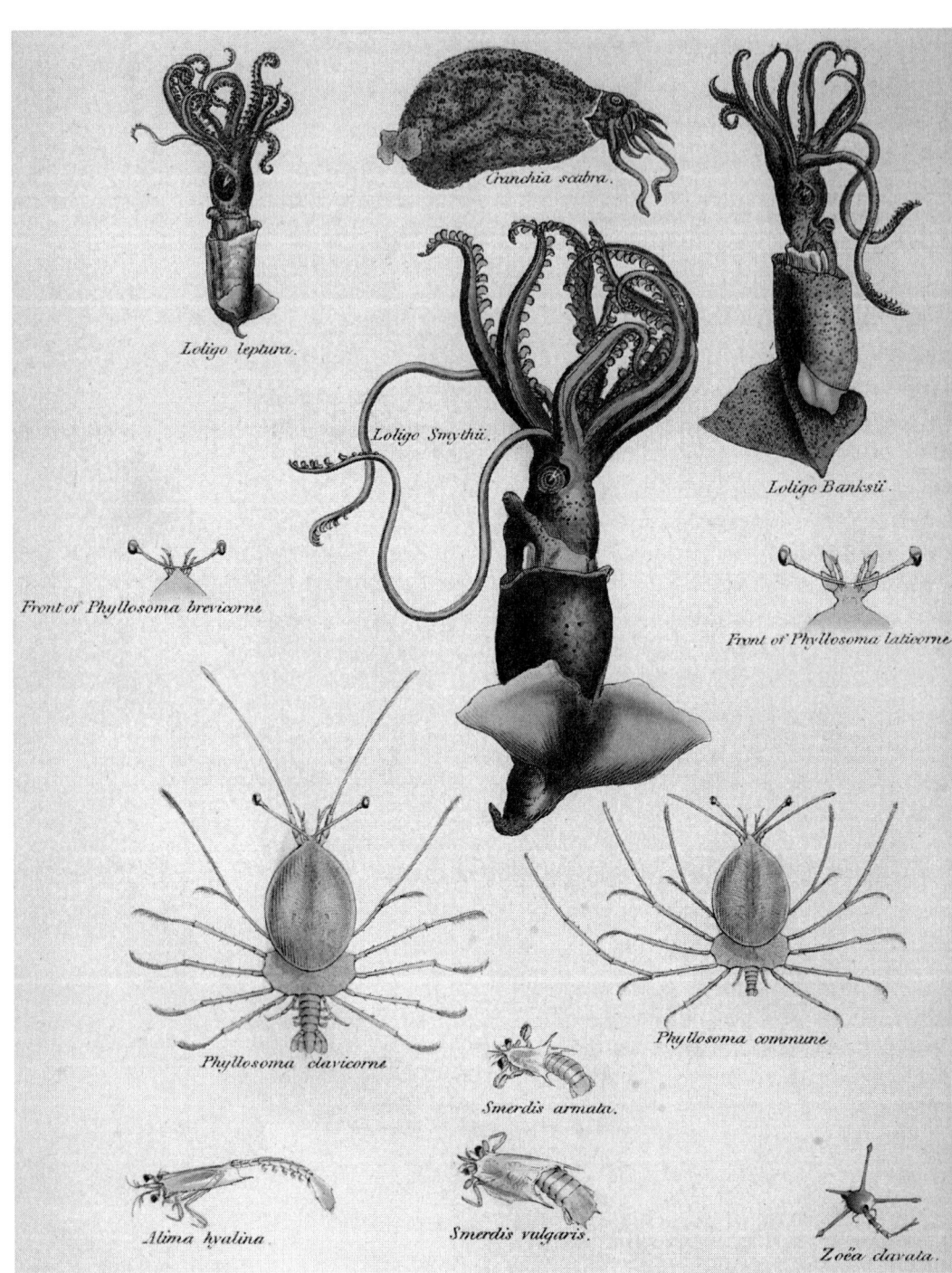

Loligo leptura.

Cranchia scabra.

Loligo Smythii.

Loligo Banksii.

Front of Phyllosoma brevicorne

Front of Phyllosoma laticorne

Phyllosoma clavicorne

Phyllosoma commune

Smerdis armata.

Alima hyalina.

Smerdis vulgaris.

Zoëa clavata.

Published Nov. 1. 1817. by John Murray, London.

all the impressions of the shell, and the suckers on all the arms were diminished in size, as if from pressure.

EXPLANATION OF PLATE XII.

Fig. 1. Ocythoë CRANCHII sitting within the shell.

Fig. 2. The animal without the shell.

Fig. 3. One completely retracted within the shell.

Fig. 4. Ditto taken out of the shell showing the impressions of the shell on the body.

Fig. 5. Left superior arm (common appearance) magnified.

Fig. 6. Right superior arm (variety) magnified.

APPENDIX. No. III.

The distinguishing characters between the OVA *of the* SEPIA, *and those of the* VERMES TESTACEA, *that live in water explained.* By *Sir* EVERARD HOME, *Bart. V. P. R. S.*

[*From the Philosophical Transactions.*]

LINNÆUS was led into an error respecting the animal that forms the shell argonauta, by the circumstance of a species of sepia having been often found in this shell. This erroneous opinion has been adopted by many naturalists upon the Continent, even those conversant in comparative anatomy.

Whether the argonauta is really an internal shell, which I have asserted it to be, may possibly never be determined by direct proofs, as the animal belonging to it has not been met with. The present observations are confined to the question of the probability of its being formed by the species of sepia frequently found in it; and the materials of the present Paper, which are furnished from the specimens of natural history collected in the late expedition to the Congo, enable me to prove, in contradiction to such an opinion, that the ova of this particular species of sepia are not those of an animal of the order vermes testacea, that live in water.

The young of all oviparous animals, while contained in the ovum, must have their blood aerated through its coats, but in the vermes testacea, if the shell were formed in the ovum, the process of aerating the blood must be very materially interfered with; for this reason, the covering or shell of the egg first drops off, and the young is hatched before the shell of the animal is formed; this I have seen taken place in the eggs of the

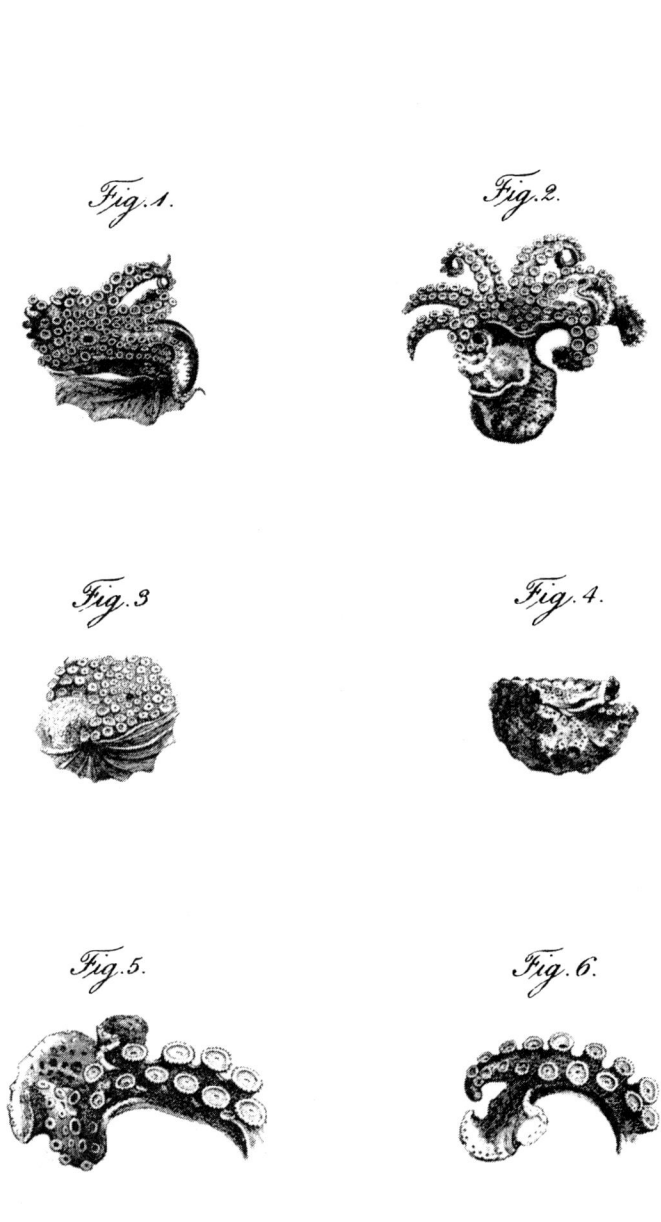

Fig.1.

Fig.2.

Fig.3

Fig.4.

Fig.5.

Fig.6.

J.ᵗ Basire sc.

garden snail, but in the testacea that live in water, the young requires some defence in the period, between the egg being hatched, and the young acquiring its shell, which is not necessary in those that live on land; for this purpose, the ova are enclosed in chambers of a particular kind.

This camerated nidus in the larger animals of this tribe, must be familiar to all naturalists, since specimens in a dried state, containing the young shells completely formed, are to be met with in collections of natural history; but I am not aware that all the purposes for which such a nidus is supplied by nature, have ever been explained.

I have been informed by a friend, who while in the East Indies saw the chank (a shell belonging to the same genus with the *voluta pyrum* of Linnæus,) shed its eggs, that the animal discharged a mass of mucus, adapted to the form of the lip of the shell, and several inches in length; this rope of eggs, enclosed in mucus, at the end which is last disengaged, was of so adhesive a nature, that it became attached to the rock, or stone on which the animal deposited it. As soon as the mucus came in contact with the salt water, it coagulated into a firm membranous structure, so that the eggs became enclosed in membranous chambers, and the nidus having one end fixed and the the other loose, was moved by the waves, and the young in the eggs, had their blood aerated ; when the young were hatched, they remained defended from the violence of the waves, till their shells had acquired strength.

What passes under the sea, few naturalists can be so fortunate as to have an opportunity of observing, and although what I have stated was communicated to me by an eye witness, it required confirmation, as well as an opportunity of examining the nidus, before I could give it my assent. Since that time, I have procured from my friend Mr. LEE, the Botanist, of Hammersmith, a portion of a camerated nidus brought from South Carolina, containing shells of an univalve not very different from the chanks of the East Indies. This nidus is represented in the annexed drawing. (Plate XIII. fig. 7.)

I have also, which is still more satisfactory, seen the camerated nidus of the helix janthina. This animal not living at the bottom of the sea, like the vermes testacea in general, deposits its ova upon its own shell, if nothing else comes in its way; one of the specimens of the shell of the janthina caught in the voyage to the Congo, fortunately has the ova so deposited, as will be seen in the annexed drawings, made by Mr. BAUER, who was so pleased with the appearance the parts put on in the field of the microscope, that he was desirous of making a representation of them. (Pl. XIII. fig. 1, 2, 3, 4, 5, 6.)

In this instance, the ova are single, but in other tribes, several ova are contained in one chamber. In the land snail, the eggs have no such nidus. The following observations respecting them were made in the year 1773, the first year that I was initiated in comparative anatomy, under Mr. HUNTER. He kept snails to ascertain their mode of breeding, and the notes that were made at the time in my own hand writing, I now copy.

August 5, 1773. A snail laid its eggs, and covered them over with earth; Mr. HUNTER took one out and examined it; the egg was round, its covering strong, and of a white colour, with a degree of transparency; it had no yelk; a small speck was observable with a magnifying glass in the transparent contents.

On the 9th no apparent change had taken place. On the 11th the speck had enlarged, but was too transparent to admit

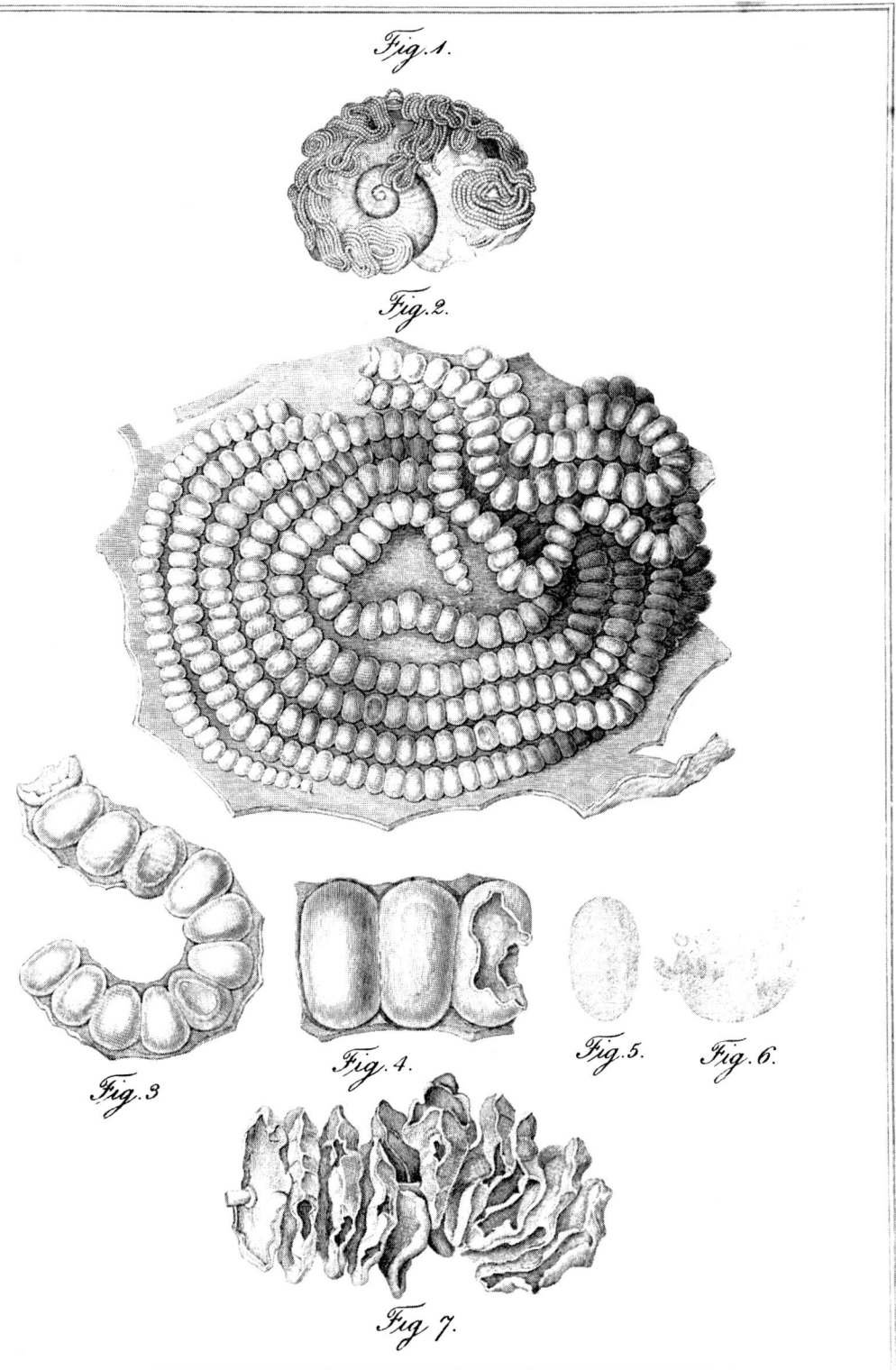

Fig. 1.

Fig. 2.

Fig. 3.

Fig. 4.

Fig. 5.

Fig. 6.

Fig. 7.

of its form being distinguished; upon moving the speck it fell out of its place.

On the 12th the embryo was indistinctly seen.

On the 15th the embryo filled $\frac{1}{4}$ part of the egg, but the different parts were still indistinct.

On the 18th the body of the embryo had become larger, and the covering thicker.

On the 19th, the coverings or shells of all the eggs were more or less dissolved, so much so that Mr. Hunter thought all the eggs were rotting, and the whole brood of young would be lost.

On the 20th, the young were hatched, and the shells completely formed.

On the 23d, when the young snails were put in water, their bodies came out of the shell as in full grown snails.

On the 24th, they all deserted their nests.

The specimens of the sepia found in the argonaut shell, which, was caught by Mr. Cranch, in this expedition to the Congo, had deposited some of its eggs in the involuted part of the shell, and the animal being fortunately caught in the shell identified the eggs to belong to it; (Pl. XIV.) they are united together by pedicles, like the eggs of the sepia octopus, and in all other respects resemble them; they differ from those of the helix janthina and the other vermes testacea, that live in water, in having no camerated nidus, and in having a very large yelk to supply the young with nourishment, after they are hatched.

Upon these grounds, this animal must be resolved into a species of sepia, an animal which has no external shell, and only uses the shell of the argonaut, when it occasionally gets possession of one.

Some naturalists, unacquainted with comparative anatomy, have asserted that in these eggs they saw the argonaut shell partly formed ; they must have mistaken the yelk, which will be seen in the drawing to be unusually large, for the new shell.

EXPLANATION OF THE PLATES.

PLATE XIII.

Fig. 1. The shell of the helix janthina, with the ova in its camerated nidus, attached to it ; magnified twice in diameter.

Fig. 2. A portion of the nidus magnified 12 times in diameter.

Fig. A string of the same nidus magnified 25 times in diameter.

Fig. 4. Two of the same ova and one empty chamber, magnified 50 times in diameter.

Fig. 5. One of the same ova, and

Fig. 6. The same slightly bruised, both magnified 50 times in diameter.

Fig. 7. A portion of the camerated nidus, in a dried state, belonging to the ova of a univalve from south Carolina, of the natural size.

PLATE XIV.

Fig. 1. The shell of the argonauta, with the ova of the octopus deposited in it, magnified twice in diameter.

Fig. 2. A cluster of the same ova, as they are seen when immersed in water, magnified 12 times in diameter.

Fig. 3. One of the same ova with its pellicle, magnified 25 times in diameter.

Fig. 4. The yelk of the egg.

Fig. 5. A transversal section of the same.

Fig. 6. A longitudinal section of the same. The three preceding figures are magnified 50 times in diameter.

Fig. 7. A collapsed egg, as seen when taken out of the water, magnified 25 times in diameter.

Fig.1.

Fig.2.

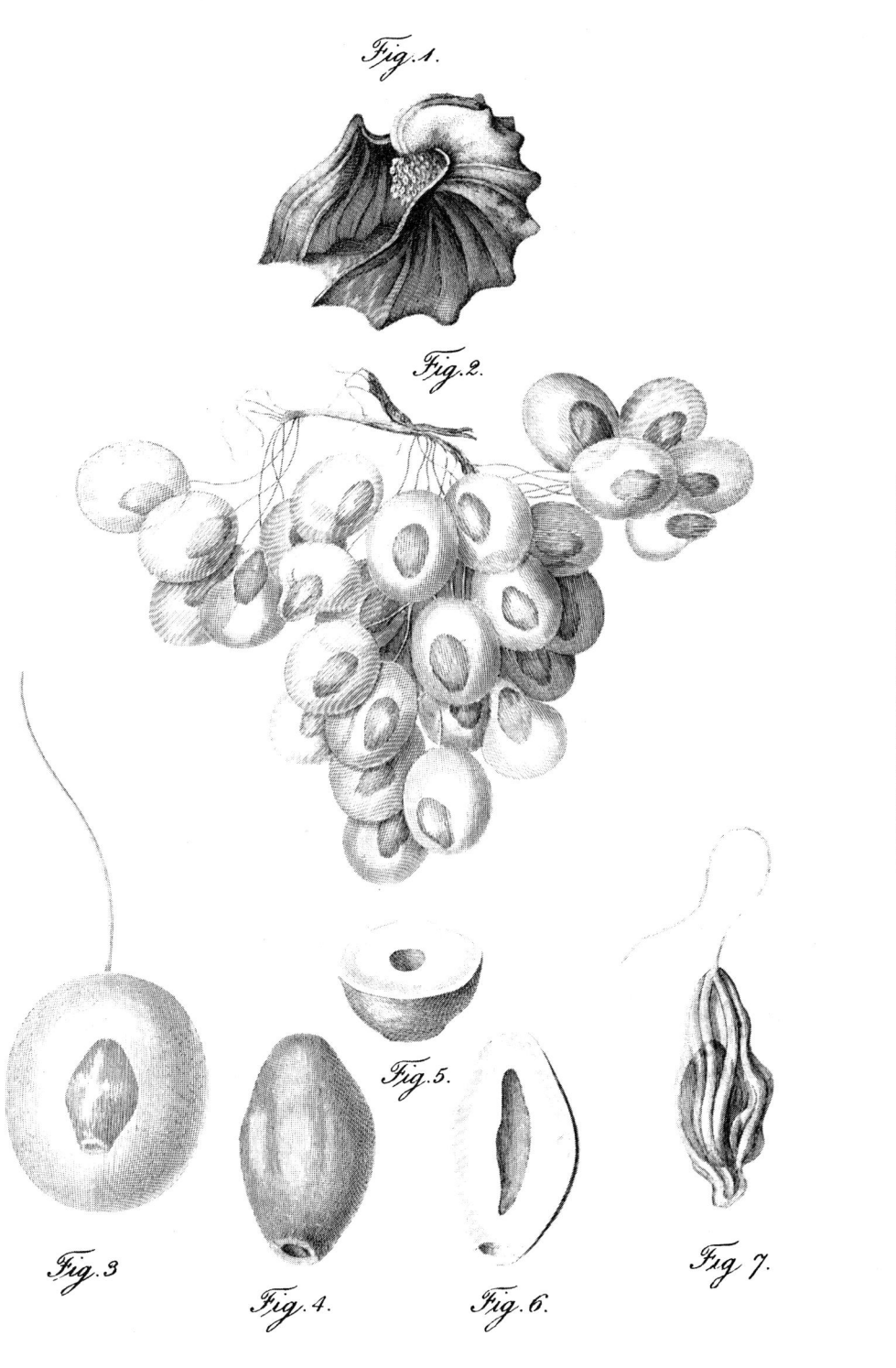

Fig.5.

Fig.3

Fig.4.

Fig.6.

Fig 7.

APPENDIX. No. IV.

A general Notice of the Animals taken by Mr. John Cranch, *during the Expedition to explore the Source of the River Zaire.*

MAMMALIA.

Calitrix sabœa (Audebert, iv. ii. f. 4.) one of the various species of *monkies* that are generally denominated *green*, was found in great plenty at Tall Trees.

Three very young *lions* (probably of the Senegal sort) were brought to Mr. Cranch by the natives, who called them *boulaces.* They were kept alive three days and fed on soaked bread, which doubtless caused their death.

BIRDS.

Aquila melanœtos,* (Savignys Oiseaux d'Egypt. pl. ii. f. 2.)
Ierax musicus, Singing hawk, *(Le Vail. Ois. d'Afr. i. pl. 27.)*
Circus ——, (L'Acoli, *Le Vail.*)
Elanus melanopterus, (Sav. Ois. d'Eg. pl. ii. f. 2.) In great plenty.
Milvus œtotius, (Sav. Ois. d'Eg. pl. iv. f. 1.)
Polophilus ——, (Sav. Ois. d'Eg. pl. iv. f. 1.)
Corvus scapularis, (Le Vail. Ois. d'Af. ii. pl. 53.)
Coracias afra, African roller.
Passer, (Savig. Ois. d'Eg. pl. v. f. 7.)
Hirundo Savignii, (Sav. Ois. d'Eg. pl. iv. f. 4.)
Hirundo Smithii. (New species.) Black colour glossed with steel-blue, whitish beneath the tail, and wing-quills black; the former with a white band, the upper part of the head chestnut, the outermost tail feathers very long. A single specimen was killed off Chisalla island.
Alauda, (Le Vail. Ois. d'Afr. pl. 196.)
Sylvia, (Savig. Ois. d'Eg. pl. v. f. 3.)
Sylvia, (Le Vail. Ois. d'Afr. 121.)
Certhia cincta, (Ois. Dor. ii. pl. 10.)

* *Plin. lib. x. cap. 3. sec. 3. et seq.*

Certhia chalybea, (*Ois. Dor. ii. pl.* 13 *et* 14.)

Merops erythropterus, (*Pl. énl.* 318.)

Upupa Epops, Common Hoopoe, not varying in the slightest degree from that of Europe.

Alcedo maxima var.? With the breast ferruginous, the belly varied with black and white, the throat white. In other respects it agrees exactly with the common varieties from Senegal.

Alcedo Senegalensis, (*Pl. énl.* 594.)

Alcedo ——, (*Pl. énl.* 556) probably a variety of *Senegalensis,* or the other sex.

Alcedo rudis, (*Pl. énl.* 62.)

Buceros ——, (*Le Vail. Ois. d'Afr. pl.* 233.)

Perdix Cranchii, (new species.) Cinereous-brown beneath, whitish, freckled with dark-brown; the spots on the belly elongate and inclining to ferruginous; throat naked.

Columba ——, (*Savig. Ois. d'Eg. pl.* 5. *f.* 9) common.

Vanellus ——, (*Savig. Ois d'Eg. pl.* 6. *f.* 3.)

Scopus umbretta, Tufted Umber; not uncommon.

Ardea ——, (*Savig. Ois. d'Eg. pl.* 8, *f.* 1.)

Ardea Senegalensis, (*Pl. énl.* 315.)

Parra Africana, (*Lath. Syn. tab.* 87.)

Recurvirostra —— Very much destroyed, but from the parts remaining, not to be distinguished from our European species, *R. Avosetta,* the common Avoset.

Phalacrocorax ——, (*Savig. Ois. d'Eg. pl.* 8. *f.* 2.)

Plotus Congensis, (new species.) Black; head and neck brownish chestnut; back and wing coverts streaked with white. One only was killed.

Anas ——, (*Savig. Ois. d'Eg. pl.* 10. *f.* 1.)

Sterna senex, (new species.) Cinereous-black, top of the head gray, belly with a very faint and obsolete teint of chestnut.

Rhynchops niger, (*Pl. énl.* 357.)

REPTILES.

Trionyx Egyptiacus, (*Geoff. St. Hill. Rept. d'Eg. pl.* 1.) The head only of this extraordinary animal was sent home, in spirits.

Coluber Palmarum, (new species.) Reddish; beneath whitish, the scales of the sides and back very long-ovate and carinated. Found in palm trees at Embomma.

Coluber Smythii, (new species.) Brown-gray beneath whitish, the sides, especially anteriorly, with triangular whitish spots, bordered with sooty-black; the scales of the sides and back hexagonal, rather narrower at their extremities. This species was found in great plenty near Embomma on the ground. The back is very faintly marked with some transverse narrow whitish bands, spotted with black.

FISHES.

About eighty species of this class were taken during the voyage; but as I have not yet studied the marine fishes, I can say but little about them. Two species of a genus (which appears to be new) allied to *Leptocephalus* were taken off the African coast. Their head is smaller and more pointed than that of *Leptocephalus;* their bodies are even more compressed, but are marked in the same manner by transverse zigzag lines, and their teeth are similar. Rudiments only of the dorsal and anal fins exist towards the extremity of their bodies, and no pectoral fins can be discovered.

In the river itself three new species were discovered, namely:

Sp. 1. *Silurus Congensis.* With the upper nostrils the angles of the mouth and each side of the chin furnished with a filament, the first ray of the dorsal and pectoral fins serrated towards the point, which is unconnected with the second ray; the second ray very much elongated and attenuated, the daciniæ of the tale acute.

> OBS. The first ray of the dorsal fin is only serrated towards its point, the unconnected apex itself being destitute of teeth. The first ray of the pectoral fins, is serrated above the unattached part, and the teeth are continued downwards to near its middle. It is akin to *Silurus mystus* (*Geoff. Poiss. de Nile.*) but may very easily be distinguished from it by the characters of the pectoral fins, and by the presence of the filaments on the chin. The filaments of the chin and nostrils are nearly of equal length; those of the angles of the mouth are very long.

Sp. 2. *Pimelodus Cranchii.* Chin on each side nostrils and angles of the mouth furnished with a filament, pectoral fins with the first ray shorter than the second, very strong and sulcated; behind very strongly serrated, anterior

3 G

dorsal fin, with the first ray thick striated without teeth, caudal fin with lanceolate laciniæ.

> Obs. The front of the head is obtuse and rounded; the upper part is irregularly sulcated, and the vertex is striated: the striæ being disposed in rays; the mouth is large; the filaments of the nostrils are very short, and those of the angles of the mouth are a third longer than those of the chin. The hinder dorsal fin is short and not very fleshy.

Sp. 3. *Oxyrhynchus deliciosus.* The scales concentrically sculptured, the dorsal ones rounded; those of the sides and belly very broad, the teeth linear acuminated behind and before.

> Obs. This animal is doubtless referable to the genus *Oxyrhynchus* of Athenæus.* It differs from its congener *Momyrus anguilloides* (*Geoff. Poiss. de Nil.* pl. vii.) in the form of its scales, (which in that species are of the same size and form on all parts of the body) and in the shape of the dorsal fin, which in *O. deliciosus* is more acute in its hinder upper edge. This fish is very common in the river, and its flesh is of a most exquisite flavour.

Cephalopoda.†

Of this class one new genus, and six new species were discovered; four of which are figured in the annexed plate.

Genus I. Ocythoë. *Sp.* 1. *Ocythoë Cranchii,* of which a description is given in Appendix, No. III.

Genus II. Cranchia.‡ Body oval, sack-shaped: fins approximating, their extremities free: neck with a frenum behind, connecting it with the sack, and with two other frena connecting it with the sack before.

Sp. 1. *Cranchia scabra.* Sack rough, with hard rough tubercles.

Sp. 2. *Cranchia maculata.* Sack smooth, beautifully mottled with distant ovate spots.

---

* *Deipnos,* lib. iii. 116; viii. 356; vii. 312.

† For a synopsis of the genera of this class see Zoological Miscellany, vol. iii.

‡ The localities of the two species sent home were unfortunately lost.

Genus III. LOLIGO. The characters presented by three new species, are very different from those observed in the *Loligines* of the European seas. The distal suckers of the larger or supplementary arms, are produced into hooked processes, and in two of them all the suckers of the shorter arms are formed in the same manner.*

Sp. 1. *Loligo Banksii.* Shorter arms with globose simple suckers, the fins forming, by their union, a rhomboidal figure.

OBS. The colour of this, when alive, is pale flesh. The body is yellowish behind, sprinkled irregularly with blackish spots teinted with purple. The external aspect of the arms is freckled with purplish. The under parts of the fins without spots. One specimen was taken in the Gulph of Guinea.

Sp. 2. *Loligo leptura.* Shorter arms with hooks on their suckers, longer arms with free hooks on the distal suckers, tail abrupt and slender.

OBS. The body and external aspects of the arms are smooth, with a few tubercles arranged into longitudinal lines. Two were taken in 1, 8, 0 N. lat. 7, 26, 30 E. long.

Sp. 3. *Loligo Smithii.* Shorter arms with hooks on their suckers, larger arms with the hooks of the distal suckers furnished anteriorly with a membrane, tail gradually attenuated.

OBS. Body and arms externally tuberculated; the tubercles purple with white lips, and arranged into longitudinal lines.

PTEROPODA.†

Of this division of the molluscous tribe of animals, two species of Peron's genus *Cleodora* were taken in south lat. 2, 14, 0 E. long. 9, 55, 15, and S.

---

* In the museum of the College of Surgeons is preserved part of the arm of a large and unknown animal of this class, in which the suckers are all furnished with distinct strong and free hooks.

† Of the genus FIROLA (whose situation has not yet been satisfactorily ascertained, but which, with Cuvier, I am disposed to consider as more nearly allied to the GASTEROPODA, than to any other class) a new species was found in S. lat. 3, 15, 0, E. long. 9, 38, 0, viz. *Firola arcuata.* Dorsal fin simple, vermiform appendage none, tail arched above, without any vermiform appendage. Two other species were sketched by Lieut. Hawkey, but were not received.

lat. 2, 41, 0, E. long. 9, 16, 0, both having a spinous process on each side of their shell, near its opening. One species is beautifully sulcated transversely, and the other but slightly so.

Hyalœa tridentata (vulgarly called the chariot Anomia) was also taken in abundance in the Gulph of Guinea.

GASTEROPODA.

Janthina fragilis was the only species of this class that was brought home; all the rest, as well as the collection of the species of the following class,

ACEPHALA,
were lost.

CIRRIPEDES.

Nine new species of *Barnacles* were discovered, all of which are very interesting; since they augment especially the genus *Cineras*, of which but two species only were known, and also two divisions of Hill's genus *Pentalasmis*, of which likewise very few have been described.

Sp. 1. *Cineras Chelonophilus.* Body lanceolate, peduncle abrupt, upper scales small and acuminated behind, the hinder scale straight and linear.

OBS. The purplish stripes of this species are very faint, and the scales beneath the legs are covered by a thin membrane, which renders them very opaque. The space between the superior and posterior scales is very great. A large quantity occurred adhering to the legs, neck, and shell of some turtles that were taken in 36, 15, 0 N. lat. 16, 32,0 W. long. See page 9.

Sp. 2. *Cineras Cranchii.* Body obliquely truncated above; the peduncle rather abrupt, upper scales linear with obtuse extremities, hinder scale with a subgibbose apex.

OBS. The vittæ are three on each side, very strong; the two anterior ones are often interrupted.

Sp. 3. *Cineras Olfersii.* Body above acuminated, upper scales with both extremities (especially the hinder one) acuminated, hinder scale at its middle subgeniculated. Found on Fucus natans (*Linn.*)

PENTALASMIS.*

Divisions of the genus.

* *Hinder scale simply arcuated. Lateral scales smooth.*

** *Hinder scale simply arcuated. Lateral scales costated.*

*** *Hinder scale abruptly bent below the middle.*

Sp. 1. *Pentalasmis* (*) *Cheloniæ.* Superior scales broad, rounded at their points, hinder scale convex. Found on turtles in N. lat. 36, 15, 0 W. long. 16, 32, 0. Page 9.

Sp. 2. *Pentalasmis* (*) *Hillii.* Superior scales narrow, anteriorly obliquely-truncated ; (hence as if produced behind), hinder scale carinated below.

Sp. 3. *Pentalasmis* (**) *Spirulæ.* Rather convex, upper scales with their points anteriorly produced.

β With the ribs spined. Found in great abundance adhering to the floating shells of SPIRULÆ. (to which in several specimens part of the animal still adhered) 22, 0, 0 N. lat. 19, 17, 0 W. long.

Sp. 4. *Pentalasmis* (**) *dilatata.* Larger scales anteriorly dilated, hinder scale with granulated striæ (often behind with two or four teeth.) 0, 14, 0, N. lat. 6, 18, 52, E. long. adhering to *Janthina fragilis.*

Sp. 5. *Pentalasmis* (***) *Donovani.* Hinder scale, with a longitudinal elevated little line ; angle rectangular ; bend obtuse with a transverse elevated little line. Taken in 0, 38, 0 S. lat. 7, 50, 0 E. long.

Sp. 6. *Pentalasmis* (***) *Spirulicola.* Hinder scale narrow carinated from the apex to the angle ; angle rectangular, geniculated, prominent. Found on shells of Spirula, 22, 0, 0 N. lat. 19, 17, 0 W. long.

CRUSTACEA.

Portunus, (a new species,) without spines on the front aspect of its arms, was taken in the Gulph of Guinea.

Lupa ; of this genus three new species were discovered, all of which belong to that section in which the hinder lateral spine of the shell is very much elongated.

Machœrus ; a new genus allied to *Gonoplax,* but differing in having short

* The peduncle of those of the first division is very long, or moderately so ; of the two other divisions extremely short.

peduncles to its eyes, which are inserted into the same part of the shell as in that genus.

Pilumnus, (a new species.) Gulph of Guinea.

Grapsus minutus, and a new species. Gulph of Guinea.

Dorippe. Species not determined.

Sp. 1. *Megalopa* * *Cranchii* (new species) with a broad, entire, porrected rostrum, having its point terminating in one spine, and each side armed with a tooth, hinder coxæ armed with a straight spine. Gulph of Guinea. This species belongs to the same division of the genus with those of our seas.

Sp. 2. *Megalopa maculata.* (new species) Shell smooth, and spotted with black, rostrum narrow and abrubtly deflexed.

Sp. 3. *Megalopa sculpta.* (new species) Shell sculptured (like that of *Cancer floridus Herbst*) and very hairy, rostrum narrow and abruptly deflexed.

OBS. These two species were likewise taken in the Gulph of Guinea; they form a new division of the genus characterised by the deflexed rostra.

Scyllarus. Of this genus, a common species was taken during the voyage, and having been preserved in spirits, allowed me to ascertain by dissection, that its nervous system is in all respects similar to that of the other macrourous crustacea. Its lamelliform broad antennæ send their nerves to the same ganglion. The optic nerves are more curved in their course.

Of the large group of macrura comprehending the shrimps and prawns there are eleven new species, and seven new genera.

A new genus allied to *Nebalia.*

Zoëa. The type of this genus was discovered in the Atlantic by Bosc, who believed it to hold an intermediate situation between the crustacea with pedunculated, and those with sessile eyes. By Latreille it was referred to the *Entomastraca.* In N. lat. 1, 36, 0 E. long. 8, 46, 37, Mr. Cranch took a new species of this interesting genus, by which I have been enabled to verify the opinion published in the Supplement to the Encyclopædia Britannica, (vol. i. p. 423) where I have referred it to the crustacea with pedunculated eyes.

---

* The last segment of the abdomen on each side is furnished with two moveable plates, which I formerly overlooked.

Zoëa clavata. The eyes of this species, like that of its congener, are large, with very short peduncles. The shell is somewhat triangular; the front being terminated by a long spiniform rostrum.* The middle of the back and the sides are armed with a long clavate spine.

    Obs. It differs from Bosc's *Zoëa pelagica* in having clubbed instead of acute spines. Its situation is certainly in the same group with *Nebalia*.

Two new genera of the same natural family with *Squilla*, have established the situation of that genus. They have in common with it sixteen locomotive legs: the anterior pair is elongate and slender; the second pair much elongated and raptorious; the three following pairs are short, with their last joint compressed, and terminated by a moveable claw; the three hinder pairs are short, and remote from the rest, the last joint but one being furnished with a moveable appendice at its base.† Mouth with two mandibles and four maxillæ. Upper antennæ with three articulated setæ. Under antennæ with an elongate lamella at their base. Abdomen with two moveable foliaceous appendages arising from a common peduncle, attached to each side of the belly: the peduncle of those of the last joint is produced into a spine; the exterior lamella composed of two joints. The second pair of legs of the following new genera, *Smerdis* and *Alima*, have none of those denticulations which afford so striking a character in those of *Squilla*.

Gen. I. Smerdis. Sides of the shell approximate beneath. Mouth anterior.

Sp. 1. *Smerdis vulgaris.* Shell with a very short spine on the hinder part of its back.

    This animal was found in great plenty every day from the latter end of April to the beginning of June.

Sp. 2. *Smerdis armata.* Shell with a very long spine on the hinder part of its back. A few specimens of this species were taken between the latter end of

---

* Which is broken in the only tolerable specimen that was sent home.

† All the legs of these genera, as well as of *Squilla,* have each a foliaceous appendage at their base, which are certainly the organs of respiration. In Squilla, the outer foliaceous appendages beneath the abdomen, have filamentous processes, which the French naturalists have considered to be the respiratory organs. The two new genera want these filaments, but have those appendages (common to all the malacostraca with pedunculated eyes) at the bases of their legs.

April and twentieth of May, and were not seen afterwards, although the preceeding sort were still abundant.

Gen. II. ALIMA. Thorax elongate with the sides not approximating. Mouth placed towards the hinder part of the thorax.

Sp. 1. *Alima hyalina.* Occurred abundantly at Porto Praya, and in 7, 37, 0 N. lat. 17, 34, 15 W. long.

PHYLLOSOMA,* the most curious genus of crustacea that has yet been discovered, and of which there are at least four very distinct species, occurred in the greatest profusion from the 10th April to the 30th May. The shell of this genus is membranaceous and as thin as a leaf; the part containing the mouth, and from whence the legs arise, is drawn backwards and projects beyond the hinder part of the shell. The front of the shell bears the eyes and antennæ: the eyes have the first joint of their peduncles very much elongated; the second joint is short, and the eyes themselves are abruptly larger than their peduncles. The superior antennæ are (as in all the other malacostraca with pedunculated eyes) bifid. The inferior antennæ are variable in their projection, and form, affording characters which, for the present, I shall only venture to use for the purpose of specific distinctions. The abdomen has the usual appendages beneath, and those of the last joint are converted into swimming or rather steering lamellæ. The mouth, when first viewed, appears to be trilobate; this arises from a clypeus similar to that covering the mouth of *Squilla*, and the prominence of the exterior sides of the mandibles, which are much curved and dilated towards their middle; their points are bifid, and one lacinia is unidentate within. Two pairs of maxillæ are very distinctly to be seen; the outer ones are terminated by three spines. I have not had time to ascertain the modification of the interior ones, nor to ascertain the existence and insertion of the palpi. The front pair of legs is extremely short and dilated at its base, with all the joints (the first excepted?) confluent. The second pair is short; the third joint at its base has a flagrum which is articulated towards its point; the last joint is terminated by long spines and a claw with unequal spines. The five following pairs of legs are very long, and the hinder ones gradually

---

* The third, fourth, fifth, sixth, and seventh pairs of legs, in the numerous specimens sent home, were for the most part broken off at their third joint, the flagrum only remaining. See the plate.

encrease in length : at the base of the second joint, each is furnished with a bipartite flagrum, the second division of which is articulated and ciliated : the third pair is terminated by a simple ciliated joint; the three next pair by claws, which in some of the species are ciliated with spines, and meet little spines on the interior side of the apex of the joints to which they are attached : the last pair is abruptly shorter than the preceding legs, and varies in the number of its joints, from two to five. The organs, termed ears by the French naturalists, are very large and prominent. I have not examined the nervous system.

1. *Phyllosoma brevicorne.* Inferior antennæ shorter than the superior ones, with the second division slightly dilated externally; the two last divisions setaceous; hinder pair of legs two-jointed; the second joint simple.

2. *Phyllosoma laticorne.* Inferior antennæ a little longer than the superior ones, the second joint very much dilated externally, and produced at its external apex, the last division lanceolate, hinder pair of legs five-jointed, the last joint with a simple slightly curved claw.

Two specimens only were taken, and their locality was not set down.

3. *Phyllosoma commune.* Inferior antennæ filiform rather more than twice the length of the superior ones, hinder pair of legs four-jointed; the last joint terminated by a straight simple claw.

Taken at Porto Praya and during the voyage until 2, 58, 0 S. lat. 9, 21, 22 E. long. in the greatest profusion.

4. *Phyllosoma clavicorne.* Inferior antennæ filiform, half as long again as the thorax, with the extremity of the last division clavate, hinder pair of legs four-jointed; the last joint terminated by a simple and very slightly curved claw. Occasionally taken with the preceding species.

From the above very general observations, it will be very evident to entomologists, that *Phyllosoma* constitutes a family of crustacea macroura, to which no other discovered genus can be referred.

Amongst the sessile-eyed crustacea, with compressed bodies, there are four new species which constitute the types of as many genera, and of those with depressed bodies, there are; of

SPHÆROMA, a new species.

CYMOTHOA, a new species.

And a new species of an unnamed genus, intermediate betwixt the genera *Æga* and *Eurydice.*

3 H

ENTOMOSTRACA.

Two new species of the genus *Caligus* of Müller were found on fishes.

MYRIAPODA.

IULUS, two species, one of which is new.

SCOLOPENDRA, one nondescript species.

INSECTA.

Thirty-six species only reached England in a tolerable state, the rest were entirely destroyed by insects and damp. Amongst them is a new genus of the family *Scarabæidea*, and probably there are five or six new species, which I have not yet found time to examine.

ANNELEIDES.

A new species of *Nereis* was taken in a bit of floating wood 0, 21, 0 N. lat. 5, 49, 37 E. long, together with a genus not known to me.

ENTOZOA.

One species of this parasitical class, was taken out of the intestinal canal of an albicore.

ACALEPHÆ.

PORPITA. Disc cartilaginous, round, composed of rays. Stomach central and round. Mouth slightly prominent and capable of very great distention. Whole underside covered by tentacula, those of the middle terminated by suckers, those next the margin larger (and simple or at least terminated by indistinct suckers.*)

Of this genus, to which the above characters are now given, a new species, was found in 8, 12, 0 N. lat. 18, 13, 7 W. long. viz,

Porpita granulata. The rays of the upper part of the disc granulated by pairs. The stomach of one specimen was filled with the debris of a fish.

---

* In specimens of very soft animals preserved in spirits of wine, where the organs are much contracted, it is generally impossible to ascertain all the characters of each part. Naturalists should therefore take every opportunity to describe them whilst alive, since all are not gifted with the extraordinary powers of a Savigny.

VELELLA. To this genus likewise some additional characters may be added. The disc is oval and cartilaginous, having an oblique crest on its upper side. The disc itself is composed of two thin oval plates joined together by several concentric septa. The whole of the cartilaginous part is covered by a dense membrane, and its inferior surface is covered by tentacula, which surround its stomach ; those towards the centre are terminated by suckers, those towards the margin are longest, and appear to be simple. The stomach is oblong, and the mouth very prominent. The membrane in passing from the upper part of the disc, to the lower, is produced beyond its margin, and the produced part is consequently composed of two membranes, which are united towards the margin of the disc.*

1. *Velella scaphidea.* Crest set on the disc from left to right ; its apex abruptly produced.

       Velella scaphidea   *Peron et Le Seuer. Atlas, pl. xxx.*

2. *Velella pyramidalis.* Crest set on the disc from right to left ; its apex gradually produced, pyramidal.

       Taken in plenty in 26, 34, 0 N. lat. 18, 28, 0 W. long.

From the MS. observations made by Mr. Cranch, it is evident that a box containing specimens of marine animals, preserved in spirits, and a very large portion of the birds, have been lost. I have before remarked, that of the birds received, those enumerated, were the only specimens in a state fit for examination ; the greater part being totally destroyed by insects.

---

* This produced membrane is to be observed in all the species, and therefore cannot be taken as a specific character, as has been done by Lamarck, for our European species.

APPENDIX.   No. V.

Observations, Systematical and Geographical, on Professor CHRISTIAN SMITH's *Collection of Plants from the Vicinity of the River Congo, by* ROBERT BROWN, *F. R. S.*

THE Herbarium formed by the late Professor Smith and his assistant Mr. David Lockhart, on the banks of the Congo, was, on its arrival in England, placed at the disposal of Sir Joseph Banks; under whose inspection it has been arranged; the more remarkable species have been determined; and the whole collection has been so far examined as the very limited time which could be devoted to this object allowed.

In the following pages will be found the more general results only of this examination; descriptions of the new genera and species being reserved for a future publication.

In communicating these results I shall follow nearly the same plan as that adopted in the Botanical Appendix to Captain Flinders's Voyage to Terra Australis:

1st. Stating what relates to the three Primary Divisions of Plants.

2dly. Proceeding to notice whatever appears most remarkable in the several Natural Orders of which the collection consists; and,

3dly. Concluding with a general comparison of the vegetation on the line of the river Congo, with that of other equinoctial countries.

I. The number of species in the herbarium somewhat exceeds 600; the specimens of several of which are, indeed, imperfect; but they are all referable with certainty to the primary divisions, and, with very few exceptions, to the natural orders to which they belong.

Of the Primary Divisions, the Dicotyledonous plants amount to 460

The Monocotyledonous to   -   -   -   -   113

And of the Acotyledonous, in which Ferns are included, there are

only   -   -   -   -   -   33 species.

It is a necessary preliminary, with reference especially to the first part of my subject, to determine whether this herbarium, which was collected in a period not exceeding two months, and in a season somewhat unfavourable, can warrant any conclusions concerning the proportional numbers of the three primary divisions, or of the principal natural orders in the country in which it was formed.

Its value in this respect must depend on the relation it may be supposed to have to the whole vegetation of the tract examined, and on the probability of the circumstances under which it was formed, not materially affecting the proportions in question.

Its probable relation to the complete Flora of the country examined, can at present be judged of only by comparing it with collections from different parts of the same coast of equinoctial Africa.

The first considerable herbarium from this coast, of which we have any account, is that formed by Adanson, on the banks of the Senegal, during a residence of nearly four years. Adanson himself has not given the extent of his collection, but as he has stated the new species contained in it to be 300,* it may I think, be inferred, that altogether it did not exceed 600, which is hardly equal to that from Congo. Limited as this supposed extent of Adanson's herbarium may appear, it is estimated on the most moderate calculation of the proportion that new species were likely to bear to the whole vegetation of that part of equinoctial Africa, which he was the first botanist to examine; allowance being at the same time made for the disposition, manifested in the account of his travels, to reduce the plants which he observed to the nearly related species of other countries.

From the herbarium, and manuscripts in the library of Sir Joseph Banks, it appears, that the species of plants collected by Mr. Smeathman at Sierra Leone, during a residence of more than two years, amounted to 450.

On the same authority I find that the herbarium formed in the neighbourhood of Cape Coast by Mr. William Brass, an intelligent collector, consisted of only 250 species.

And I have some reason to believe, that the most extensive and valuable collection ever brought from the west coast of equinoctial Africa, namely, that formed by Professor Afzelius, during his residence of several years at Sierra Leone, does not exceed 1200 species: although that eminent naturalist, in the course of his researches, must have examined a much greater extent of country than was seen in the expedition to Congo.

From these, which are the only facts I have been able to meet with respecting

* *Fam. des Plant.* 1. *p. cxvi.*

the number of species collected on different parts of this line of coast, I am inclined to regard the herbarium from Congo as containing so considerable a part of the whole vegetation, that it may be employed, though certainly not with complete confidence, in determining the proportional numbers both of the primary divisions and principal natural orders of the tract examined; especially as I find a remarkable coincidence between these proportions in this herbarium and in that of Smeathman from Sierra Leone.

I may remark here, that from the very limited extent of the collections of plants above enumerated, as well as from what we know of the north coast of New Holland, and I believe I may add of the Flora of India, it would seem that the comparative number of species in equal areas within the tropics and in the lower latitudes beyond them, has not been correctly estimated : and that the great superiority of the intratropical ratio given by Baron Humboldt, deduced probably from his own observations in America, can hardly be extended to other equinoctial countries. In Africa and New Holland, at least, the greatest number of species in a given extent of surface does not appear to exist within the tropics, but nearly in the parallel of the Cape of Good Hope.

In the sketch which I have given of the botany of New Holland, I first suggested the enquiry respecting the proportions of the primary divisions of plants, as connected with climate; and I then ventured to state that " from the equator to 30° lat. in the northern hemisphere at least, the species of Dicotyledonous plants are to the Monocotyledonous as about 5 to 1, in some cases considerably exceeding and in a very few, falling somewhat short of this proportion, and that in the higher latitudes a gradual diminution of dicotyledones takes place until in about 60° N., and 55° S. lat. they scarcely equal half their intratropical proportion." *

Since the publication of the Essay from which this quotation is taken, the illustrious traveller Baron Humboldt, to whom every part of botany, and especially botanical geography, is so greatly indebted, has prosecuted this subject further, by extending the enquiry to the natural orders of plants: and in the valuable dissertation prefixed to his great botanical work,† has adopted the same equinoctial proportion of Monocotyledones to Dicotyledones as that given

---

* *Flinders' Voyage to Terra Australis*, 2. *p.* 538

† Nova Genera et Species Plantarum, quas in perigrinatione orbis novi collegerunt, &c. *Amat. Bonpland* et *Alex. de Humboldt.* ex sched. autogr. in ord. dig. *C. S. Kunth.* 1815, *Parisiis.*

in the Paper above quoted; a ratio which seems to be confirmed by his own extensive herbarium.

I had remarked, however, in the Essay referred to, that the relative numbers of these two primary divisions in the equinoctial parts of New Holland appeared to differ considerably from those which I had regarded as general within the tropics; dicotyledones being to monocotyledones only as 4 to 1. But this proportion of New Holland very nearly agrees with that of the Congo and Sierra Leone collections. And from an examination of the materials composing Dr. Roxburgh's unpublished Flora Indica, which I had formerly judged of merely by the index of genera and species, I am inclined to think that nearly the same proportion exists on the shores of India.

Though this may be the general proportion of the coasts, and in tracts of but little varied surface within the tropics, it seems at the same time probable from Baron Humboldt's extensive collections, and from what we know of the vegetation of the West India islands, that in equinoctial America, in tracts including a considerable portion of high land, the ratio of dicotyledones to monocotyledones is at least that of 11 to 2, or perhaps nearly 6 to 1. Whether this or a somewhat diminished proportion of dicotyledones exists also in similar regions of other equinoctial countries, we have not yet sufficient materials for determining.

Upon the whole, however, it would seem from the facts of which we are already in possession, that the proportions of the two primary divisions of phænogamous plants, vary considerably even within the tropics, from circumstances connected certainly in some degree with temperature. But there are facts also which render it probable, that these proportions are not solely dependent on climate. Thus the proportion of the Congo collection, which is also that of the equinoctial part of New Holland, is found to exist both in North and South Africa, as well as in Van Diemen's Island, and in the south of Europe.

It is true indeed that from about 45° as far as to 60°, or perhaps even to 65° N. lat. there appears to be a gradual diminution in the relative number of dicotyledones; but it by no means follows, that in still higher latitudes a further reduction of this primary division takes place. On the contrary, it seems probable from Chevalier Giesecke's list of the plants of the west coast of Greenland,* on different parts of which, from lat. 60° to 72°, he resided several

* Article Greenland, in Brewster's Edinburgh Encyclopædia.

years, that the relative numbers of the two primary divisions of phænogamous plants are inverted on the more northern parts of that coast;*
dicotyledones being to monocotyledones, in the list referred to, as about
4 to 1, or nearly as on the shores of equinoctial countries. And analogous
to this inversion it appears, that at corresponding Alpine heights, both in
the temperate and frigid zones, the proportion of dicotyledones is still further
increased.

The ACOTYLEDONOUS or cryptogamous plants of the herbarium from Congo,
are to the phænogamous as about 1 to 18. Some allowance is here to be
made for the season, peculiarly unfavourable, no doubt, for the investigation of this class of plants. But it is not likely that Professor Smith, who had
particularly studied most of the cryptogamous tribes, should have neglected
them in this expedition; and the circumstance of the very few imperfect specimens of Mosses in the collection being carefully preserved and separately
enveloped in paper, seems to prove the attention paid to, and consequently the
great rarity of, this order at least; which, however, is not more striking than
what I have formerly noticed with respect to some parts of the north coast of
New Holland.†

I have in the same place considered the Acotyledones of equinoctial New
Holland, as probably forming but one-thirteenth of the whole number of
plants, while the general equinoctial proportion was conjectured to be one-
sixth. This general ratio, however, is certainly over-rated, though it is probably an approximation to that of countries containing a considerable portion
of high land. Within the tropics therefore, it would seem that the ratio of
acotyledonous to phænogamous plants, varies from that of 1:15 to 1:5; the
former being considered as an approximation to the proportion of the
shores, the latter to that of mountainous countries.

* That some change of this kind takes place on that coast might perhaps have been
conjectured from a passage in Hans Egede's Description of Greenland, where it is
stated, that although from lat. 60° to 65° there is a considerable proportion of good
meadow land, yet in the more northern parts, " the inhabitants cannot gather grass
" enough to put in their shoes, to keep their feet warm, but are obliged to buy it from
" the southern parts." (English Translation, p. 44, and 47.)

† *Flinders' Voyage*, 2. *p.* 539.

II. The NATURAL ORDERS of which the herbarium from Congo consists, are 87 in number; besides a very few genera not referable to any families yet established. More than half the species, however, belong to nine orders, namely, to *Filices*, *Gramineæ*, *Cyperaceæ*, *Convolvulaceæ*, *Rubiaceæ*, *Compositæ*, *Malvaceæ*, *Leguminosæ*, and *Euphorbiaceæ*; all of which have their greatest number of species in the lower latitudes, and several within the tropics.

I now proceed to make some observations on the orders above enumerated, and on such of the other families, included in the collection, as present any thing remarkable, either in their geographical distribution, or in their structure; more especially where the latter establishes or suggests new affinities; and I shall take them nearly in the same order, as that followed in the botanical appendix to Captain Flinders s Voyage.

ANONACEÆ. Only three species of this family are contained in the collection. One of these is *Anona senegalensis*, of which the genus has been considered doubtful even by M Dunal in his late valuable Monograph of the order. That it really belongs to Anona, however, appears from the specimen with ripe fruit preserved in the collection. It is remarkable therefore as the only species of this genus yet known which is not a native of equinoctial America: for Anona asiatica, of which Linnæus had no specimen in his herbarium, when he first proposed it under this name, according to the original synonyms, is nothing more than Anona muricata: and A. obtusiflora, supposed by M. Tussac† to have been introduced into the American Islands from Asia, does not appear to differ from A. mucosa of Jacquin, which is known to be a native of Martinica.

The second plant of this order in the collection is very nearly related to *Piper Æthiopicum* of the shops, the Unona æthiopica, and perhaps also Unona aromatica of Dunal :‡ these with several other plants already published, form a genus, which, like Anona, is common to America and Africa, but of which no species has yet been observed in Asia.

Of MALPIGHIACEÆ, an order chiefly belonging to equinoctial America, there are also three species from Congo.

* *Monogr. de la famille des Anonacées, p. 76.*

† *Flore des Antilles, 1. p. 193.* ‡ *Anonac. p. 113 et 112.*

3 I

One of these is *Banisteria Leona,* first described, from Smeathman's speci-mens, by Cavanilles,* who has added the fruit of a very different plant to his figure, and quotes the herbarium of M. de Jussieu as authority for this species being likewise a native of America, which is, I believe, equally a mistake.

The two remaining plants of Malpighiaceæ, in the collection, with some additional species from different parts of the coast, form a new genus, having the fruit of Banisteria, but with sufficient distinguishing characters in the parts of the flower, and remarkable in having alternate leaves.   From this disposition of leaves, in which the genus here noticed differs from all others decidedly belonging to the order, an additional argument is afforded, for referring *Vitmannia* to Malpighiaceæ, as proposed by M. du Petit Thouars ;† and the approximation, though perhaps not the absolute union of Erythroxylon to the same family is confirmed.

It may not be improper here to notice a very remarkable deviation from the usual structure of leaves in Malpighiaceæ, which is supposed to occur in a plant of equinoctial Africa, namely *Flabellaria pinnata* of Cavanilles (the *Hiræa pinnata* of Willdenow.)   It is certain, however, that the figure given by Cavanilles of this species is made up from two very different genera ; the pin-nated leaf belonging to an unpublished *Pterocarpus ;* the fructification to a species of *Hiræa,* having simple opposite leaves.   The evidence respecting this blunder, which was detected by Mr. Dryander, is to be found in the herbarium of Sir Joseph Banks.

In Malpighiaceæ the insertion of the ovulum is towards its apex, or consi-derably above its middle ; and the radicle of the embryo is uniformly superior. In these points Banisteria presents no exception to the general structure, though Gærtner has described its radicle as inferior, and M. de Jussieu does not appear to have satisfied himself respecting the fact.‡   It appears, how-ever, that M. Richard is aware of the constancy in the direction of the embryo in this order.§

HIPPOCRATICEÆ. M. de Jussieu has lately proposed this as a distinct family,‖ of which there are two plants in the collection.   The first is a species of Hippocratea ; the second is referable to Salacia.

* *Dissert.* 424, *t.* 247.   † *In Nov. gen. Madagasc.* n. 46, (Biporeia.)
‡ *Annal. du Mus. d'Hist. Nat.* 18, *p.* 480.   § *Mem. du Mus. d'Hist. Nat.* 2, *p.* 400.
‖ *Annal. du Mus. d'Hist. Nat* 18, *p.* 183.

In Hippocraticeæ, the insertion of the ovula is either towards the base, or is central; the direction of the radicle is always inferior. In these points of structure, which are left undetermined by M. de Jussieu, they differ from Malpighiaceæ, but agree with Celastrinæ, to which, notwithstanding the difference in insertion and number of stamina, and in the want of albumen, they appear to me to have a considerable degree of affinity; especially to Elæodendrum, where the albumen is hardly visible, and to Ptelidium, as suggested by M. du Petit Thouars, in which it is reduced to a thin membrane.

SAPINDACEÆ. Only four plants of this natural family, which is almost entirely equinoctial, occur in the herbarium. Two of these are new species of Sapindus. The third is probably not specifically different from *Cardiospermum grandiflorum* of the West India Islands. And the fourth is so nearly related to *Paullinia pinnata*, of the opposite coast of America, as to be with difficulty distinguished from it. M. de Jussieu,† who probably intends the same plant, when he states P. pinnata to be a native of equinoctial Africa, has also described a second species from Senegal.‡ No other species of this genus has hitherto been found, except in equinoctial America; for Paullinia japonica of Thunberg, probably belongs even to a different natural order. The species from Congo, however, seems to be a very general plant on this line of coast; having been found by Brass near Cape Coast, and by Park on the banks of the Gambia.

In Sapindaceæ there is not the same constancy in the insertion of the ovulum and consequent direction of embryo, as in the two preceding orders. For although, in the far greater part of this family, the ovulum is erect and the radicle of the embryo inferior, yet it includes more than one genus in which both the seeds and the embryo are inverted. With this fact it would seem M. de Jussieu is unacquainted; § and he is surely not aware that in his late Memoir on *Melicocca** he has referred plants to that genus differing from each other in this important point of structure.

TILIACEÆ. It is remarkable that of only nine species belonging to this

---

* *Hist. des Véget. des Isles de l'Afrique, p. 34.*

† *In Annal. du Mus. d'Hist Nat. 4, p. 347.*   ‡ *loc. cit. p. 348.*

§ *Annal. du Mus. d'Hist. Nat. 18, p. 476.*   ‖ *Mém. du Mus. d'Hist. Nat. 3, p. 179.*

family in Professor Smith's herbarium, three should form genera hitherto unnoticed.

The *first* of these new genera is a shrub, in several of its characters related to Sparmannia, like which, it has the greater part of its outer stamina destitute of antheræ: in the structure of its fruit, however, it approaches more nearly to Corchorus.

The *second* genus also agrees with Corchorus in its fruit; but differs from it sufficiently in the form and dehiscence of the antheræ; as well as in the short pedicellus, like that of Grewia, elevating its stamina and pistillum.

The *third*, of which the specimens are in fruit only, fortunately, however, accompanied by the persistent flower, is remarkable in having a calyx of three lobes, while its corolla consists of five petals; the stamina are in indefinite number; and the fruit is composed of five single-seeded capsules, connected only at the base. In the want of symmetry or proportion between the divisions of its calyx and corolla, it resembles the *Chlenaceæ* of M. du Petit Thouars,[*] as well as *Oncoba* of Forskael and *Ventenatia* of M. de Beauvois.[†] The existence of this new genus decidedly belonging to Tiliaceæ, and having a considerable resemblance to Ventenatia, whose place in the system is, indeed, not yet determined, but of which the habit is nearly that of Rhodolæna, seems in some degree to confirm M. du Petit Thouars's opinion of the near relation of Chlenaceæ to Tiliaceæ; though M. de Jussieu, in placing it between Ebenaceæ and Rhodoraceæ,[‡] appears to take a very different view of its affinities.

MALVACEÆ. Of this family 18 species were observed on the banks of the Congo. It forms, therefore, about one thirty-fourth part of the Phænogamous plants of the collection; which is somewhat greater than the equinoctial proportion of the order, as stated in Baron Humboldt's dissertation,[§] but nearly agrees with that of India, according to Dr. Roxburgh's unpublished Flora Indica.

The greater part of the Malvaceæ of the collection, belong to *Sida* and *Hibiscus*; and certain species of both these genera are common to India and America. *Urena americana* and *Malachra radiata*, hitherto supposed to be

---

[*] *Hist des Végét. des Isles de l'Afrique, p.* 46.   [†] *Flore d'Oware,* 1, *p.* 29, *t.* 17.

[‡] *Mirbel Elem. de Physiol. Veg. et de Bot.* 2, *p.* 855.

[§] *Prolegomena, p. xviii. De Distrib. Geogr. Plant. p.* 43.

natives of America only, are also contained in the collection; and the loftiest tree seen on the banks of the Congo, is a species of *Bombax*, which, as far as can be determined from the very imperfect specimens preserved in the herbarium, does not differ from *Bombax pentandrum* of America and India. I have formerly remarked* that Malvaceæ, Tiliaceæ, Hermanniaceæ, Butneriaceæ, and Sterculiaceæ, constitute one natural class; of which the orders appear to me as nearly related as the different sections of Rosaceæ are to each other. In both these, as well as in several other cases that might be mentioned, there seems to be a necessity for the establishment of natural classes, to which proper names, derived from the orders best known, and differing perhaps in termination, might be given.

It is remarkable that the most general character connecting the different orders of the class now proposed, and which may be named from its principal order Malvaceæ, should be that of the valvular æstivation of the Calyx: for several, at least, of the genera at present referred to Tiliaceæ, in which this character is not found, ought probably, for other reasons likewise, to be excluded from that order: and hence perhaps also the Chlenaceæ, though nearly related, are not strictly referable to the class Malvaceæ, from all of whose orders, it must be admitted, they differ considerably in habit.

LEGUMINOSÆ. According to Baron Humbolt,† this family, or class, as I am rather disposed to consider it, constitutes one-twelfth of the Phænogamous plants within the tropics. Its proportion, however, is much greater in Professor Smith's herbarium, in which there are 96 species belonging to it, or nearly one-sixth of the whole collection. And, ample allowance being made for the lateness of the season when the collection was formed, which might be supposed to reduce the number of this family less than many of the others, Leguminosæ may be stated as forming one-eighth of the Phænogamous plants on the banks of the Congo. In India, it probably forms about one-ninth, which is also nearly the proportion it bears to Phænogamous plants in the equinoctial part of New Holland.

I have formerly proposed to subdivide Leguminosæ into three orders.‡

Of the first of these orders, MIMOSEÆ, there are only eight species from

---

* *Flinders's Voy.* 2, *p.* 540.   † *op. citat.*   ‡ *Flinders's Voy.* 2, *p.* 551.

Congo, seven of which belong to Acacia, as it is at present constituted: the eighth is a sensitive aculeated *Mimosa* very nearly allied to M. aspera of the West Indies, as well as to M. canescens of Willdenow, found by Isert in Guinea ; and perhaps is not different from the species mentioned by Adanson as being common on the banks of the Senegal.

Of the second order, CÆSALPINEÆ, the collection contains 19 species, among which there are four unpublished genera.   One of these is *Erythrophleum* of Afzelius, the Red Water Tree of Sierra Leone ; another species of which genus is the ordeal plant, or *Cassa* of the natives of Congo. *Guilandina Bonduc* and *Cassia occidentalis*, are also in the herbarium ; the former, I believe, is unquestionably common to India and America ; whether Cassia occidentalis be really a native of India and equinoctial Africa, in both of which it is now at least naturalized, is perhaps doubtful.

Among PAPILIONACEÆ, which constitute the principal part of Leguminosæ in the collection, there is only one plant with stamina entirely distinct. This decandrous plant forms a genus very different from any yet established, but to which *Podalyria bracteata* of Roxburgh * belongs.

The genera composing Papilionaceæ on the banks of the Congo have, upon the whole, a much nearer relation to those of India than of equinoctial America.   To this, however, there is one remarkable exception.   For of the only two species of *Pterocarpus* in the collection, one is hardly to be distinguished from P. *Ecastophyllum*, unless by the want of the short acumen existing in the plant of Jamaica.   The second agrees entirely with Linneus's original specimen of P. *lunatus* from Surinam, and seems to be not uncommon on the west coast of equinoctial Africa ; having been observed by Professor Afzelius at Sierra Leone, and probably by Isert in Guinea :† while no species of Pterocarpus related to either of these has hitherto been observed in India. On the other hand *Abrus precatorius* and *Hedysarum triflorum*, both of which occur in the collection, are common to equinoctial Asia and America.

TEREBINTACEÆ, as given by M. de Jussieu, appears to be made up of several orders nearly related to each other, and of certain genera having but little affinity to any of them.   Of this, indeed, the illustrious author of the

* *Coromand. Plants, 3 tab.*   † *Reise naeh Guinea, p.* 116.

Genera plantarum seems to have been aware. He probably, however, had not the means of ascertaining all their distinguishing characters, and therefore preferred leaving the order nearly as it was originally proposed by Bernard de Jussieu in 1759.

One of the orders included in Terebintaceæ, and which is proposed by M. de Jussieu himself, under the name of CASSUVIÆ, consists of Anacardium, Semecarpus, Mangifera, Rhus and Buchanania, with some other unpublished genera.

The perigynous insertion of stamina in *Cassuviæ* (or *Anacardeæ*) may be admitted in doubtful cases from analogy, there being an unpublished genus belonging to it even with ovarium inferum. And the ovarium, though in all cases of one cell, with a single ovulum, may, at least in those genera in which the style is divided, be supposed to unite in its substance the imperfect ovaria indicated by the branches of the style, and which in Buchanania are actually distinct from the complete organ. The only plant belonging to this order in the herbarium, is a species of *Rhus*, with simple verticillate leaves, and very nearly approaching in habit to two unpublished species of the genus from the Cape of Good Hope.

AMYRIDEÆ, another family included in Terebintaceæ, and to which the greater part of Jussieu's second section belongs, may, like the former order, be considered as having in all cases perigynous insertion of stamina; this structure being manifest in some of its genera. Of Amyrideæ, there are two plants in the collection. The first of these is a male plant, probably of a species of Sorindeia:* the second, which is the *Safu* of the natives, by whom it is cultivated on account of its fruit, cannot be determined from the imperfect state of the specimens; it is, however, probably related to Poupartia or Bursera.

CONNARACEÆ, is a third family which I propose to separate from Terebintaceæ: it consists of Connarus *Linn.* Cnestis *Juss.* and Rourea of Aublet or Robergia of Schreber. The insertion of stamina, in this family, is ambiguous; but as in a species of Cnestis from Congo, they originate from, or at least firmly cohere with, the pedicellus of the ovaria; they may be con-

* *Aubert du Petit Thouars nov. gen. madagas. n.* 80.

sidered perhaps in all the genera rather as hypogynous than perigynous. The most important distinguishing characters of Connaraceæ, consist in the insertion of the two collateral ovula of each of its pistilla being near the base; while the radicle of the embryo is situated at the upper or opposite extremity of the seed, which is always solitary.   In *Connarus* there is but one ovarium, and the seed (figured by Gærtner under the name of Omphalobium,) is destitute of albumen. *Rourea* or Robergia, has always five ovaria, though in general one only comes to maturity.   Its seed, like that of Connarus, is without albumen, and the æstivation of the calyx is imbricate.

Of *Cnestis,* there are several new species in Professor Smith's herbarium. This genus has also five ovaria, all of which frequently ripen; the albumen forms a considerable part of the mass of the seed; and the æstivation of the calyx is valvular.   The genera of this group therefore differ from each other, in having one or more ovaria, in the existence or absence of albumen; and in the imbricate or valvular æstivation of calyx.   Any one of these characters singly, is frequently of more than generic importance, though here even when all are taken together, they appear insufficient to separate Cnestis from Connarus.

In considering the place of the Connaraceæ in the system, they appear evidently connected on the one hand with Leguminosæ, from which Connarus can only be distinguished by the relation the parts of its embryo have to the umbilicus of the seed.   On the other hand, *Cnestis* seems to me to approach to *Averrhoa,* which agrees with it in habit, and in many respects in the structure of its flower and seed; differing from it, however, in its five ovaria being united, in the greater number of ovula in each cell, in the very different texture of its fruit, and in some degree in the situation of the umbilicus of the seed.

But *Averrhoa* agrees with *Oxalis* in every important point of structure of its flower, and in most respects in that of its seed.

Oxalis indeed differs from Averrhoa in the texture of its fruits, in some respects in the structure of its seed; and very widely in habit, in the greater part of its species.   The difference in habit, however, is not so great in some species of Oxalis; as for example, in those with pinnated and even ternate leaves from equinoctial America; and in that natural division of the genus including *O. sensitiva,* of which there are two species in the Congo herbarium.

This latter section of Oxalis* agrees also with *Averrhoa Carambola*† in the foliola when irritated, being reflected or dependent, which is likewise their position in the state of collapsion or sleep, in all the species of both genera.

To the natural order formed by Oxalis and Averrhoa, the name of OxaLIDEÆ may be given, in preference to that of *Sensitivæ*, under which, however, Batsch‡ was the first to propose the association of these two genera, and to point out their agreement in sensible qualities and irritability of leaves.

M. de Jussieu, in a memoir recently published,§ has proposed to remove Oxalis from Geraniaceæ, to which he had formerly annexed it, and to unite it with Diosmeæ.

It appears to me to have a much nearer affinity to *Zygophylleæ*,‖ though it is surely less intimately connected with that order than with Averrhoa.

I am aware that M. Correa de Serra, one of the most profound and philosophical botanists of the present age, has considered Averrhoa as nearly related to Rhamneæ ¶ or rather to Celastrinæ; from which, however, it differs in the number and insertion of stamina and especially in the direction of the embryo, with respect to the pericarpium.

In all these characters Averrhoa agrees with Oxalis; its relation to which is further confirmed on considering the appendage of the seed or arillus, whose modifications in these two genera seem to correspond with those of their pericarpia.

CHRYSOBALANEÆ. The genera forming this order are Chrysobalanus, Moquilea, Grangeria, Coupea, Acioa, Licania, Hirtella, Thelira, and Parinarium, all of which are at present referred by M. de Jussieu to Rosaceæ, and the greater part to his seventh section of that family, namely, Amygdaleæ. If Rosaceæ be considered as an order merely, these genera will form a separate section, connecting it with Leguminosæ. But if, as I have formerly proposed, both these extensive families are to be regarded as natural classes, then they will form an order sufficiently distinct from Amygdaleæ, both in fructification and habit, as well as in geographical distribution.

The principal distinguishing characters in the fructification of *Chrysobalaneæ*, are the style proceeding from the base of the ovarium; and the ovula (which,

---

* Herba sentiens, *Rumph. Amboin.* 5, *p.* 301.   † *Bruce in Philos. Transact.* 75, *p.* 356.

‡ *Tab. affin. p.* 23.   § *Mém. du Mus. d'Hist. Nat.* 3, *p.* 448.

‖ *Flinders's Voy.* 2, *p.* 545.   ¶ *Annal. du Mus. d'Hist. Nat.* 8, *p.* 72.

3 K

as in Amygdaleæ, are two in number,) as well as the embryo being erect. The greater part of Chrysobalaneæ have their flowers more or less irregular; the irregularity consisting in the cohesion of the foot-stalk of the ovarium with one side of the tube of the calyx, and a greater number, or greater perfection of stamina on the same side of the flower.

Professor Smith's herbarium contains only two genera of this order, namely, *Chrysobalanus* and *Parinarium*.* One species of the former is hardly distinguishable from *Chrysobalanus Icaco* of America, and is probably a very common plant on the west coast of Africa : *Icaco* being mentioned by Isert † as a native of Guinea, and by Adanson‡ in his account of Senegal.

Of *Parinarium,* there is only one species from Congo, which agrees, in the number and disposition of stamina, with the character given of the genus. In these respects M. de Jussieu § has observed a difference in the two species found by Adanson at Senegal, and has moreover remarked that their ovarium coheres with the tube of the calyx. In that species most common at Sierra Leone, and which is probably one of those examined by M. de Jussieu, the ovarium itself is certainly free, its pedicellus, however, as in the greater part of the genera of this order and several of Cæsalpineæ, firmly cohering with the calyx, may account for the statement referred to. I am not, indeed, acquainted with any instance among Dicotyledonous plants of cohesion between a simple ovarium, which I consider that of Chrysobalaneæ to be, and the tube of the calyx.

The complete septum between the two ovula of Parinarium, existing before fecundation, is a peculiar structure in a simple ovarium; though in some degree analogous to the moveable dessepiment of Banksia and Dryandra, and to the complete, but less regular, division of the cavity that takes place after fecundation in some species of Persoonia.‖

MELASTOMACEÆ. Four plants only of this order occur in the collection.

The first is a species of *Tristemma,* very nearly related to *T. hirtum* of M. de Beauvois.¶

---

* *Juss. Gen.* 342. Parinari, *Aublet Guian.* 514. Petrocarya, *Schreb. Gen.* 629.

† *Reisc nach Guinea, p.* 54.   ‡ *Voyage au Senegal,* 175.   § *Gen. Plant.* 342.

‖ *Linn. Soc. Transact.* 10, *p.* 35.   ¶ *Flore d'Oware,* 1, *p.* 94, *t.* 57.

The second is perhaps not distinct from *Melastoma decumbens*, of the same author.*

The third and fourth are new species referable to *Rhexia*, as characterised by Ventenat,† though not to that genus as established by Linneus; and in some respects differing from the species that have been since added to it, all of which are natives of America.

In the original species of *Tristemma*‡ there are, in the upper part of the tube of the calyx, two circular ciliated membranous processes, from which the name of the genus is derived; the limb of the calyx itself being considered as constituting the third circle. The two circular membranes are also represented as complete in T. hirtum.

But in the species from Congo, which may be named *T. incompletum*, only one circular membrane exists, with the unilateral rudiment of the second.

The rudiment of the inferior membrane in this species, points out the relation between the apparently anomalous appendage of the calyx in Tristemma, and the ciliated scales irregularly scattered over its whole surface in Osbeckia; the analogy being established by the intermediate structure of an unpublished plant of this order from Sierra Leone, in Sir Joseph Banks's herbarium, in which the nearly similar squamæ, though distinct, are disposed in a single complete circle; and by *Melastoma octandra* of Linneus, in which they are only four in number, and alternate with the proper divisions of the calyx.

The two species here referred, though improperly, to Rhexia, agree with a considerable part of the species published in the monograph of that genus by M. Bonpland, and with some other genera of the order, in the peculiar manner in which the ovarium is connected with the tube of the calyx. This cohesion, instead of extending uniformly over the whole surface, is limited to ten longitudinal equidistant lines or membranous processes, apparently originating from the surface of the ovarium; the interstices, which are tubular, and gradually narrowing towards the base, being entirely free.

The function of these tubular interstices is as remarkable as their existence.

In Melastomaceæ, before the expansion of the corolla, the tops of the filaments are inflected, and the antheræ are pendulous and parallel to the lower or erect portion of the filament; their tips reaching, either to the line of complete cohesion between the calyx and ovarium, where that exists; or, where

---

* *Op. citat.* 1, *p.* 69, *t.* 49.   † *Mém. de l'Institut. sc. phys.* 1807, *prem. semest. p.* 11.

‡ Tristemma virusana, *Vent. Choix de Plantes* 35.

this cohesion is partial, and such as I have now described, being lodged in the tubular interstices; their points extending to the base of the ovarium.   From these sheaths, to which they are exactly adapted, the antheræ seem to be disengaged in consequence of the unequal growth of the different parts of the filament; the inflected portion ceasing to increase in length at an early period, while that below the curvature continues to elongate considerably until the extrication is complete, when expansion takes place.

It is singular that this mode of cohesion between the ovarium and calyx in certain genera of Melastomaceæ, and the equally remarkable æstivation of antheræ accompanying it, should have been universally overlooked, especially in the late monograph of M. Bonpland; as both the structure and economy certainly exist in some, and probably in the greater part, of the plants which that author has figured and described as belonging to Rhexia.

On the limits, structure, and generic division of Melastomaceæ, I may remark,

1st. That *Memecylon*, as M. du Petit Thouars has already suggested,* and *Petaloma* of Swartz† both belong to this order, and connect it with *Myrtaceæ*, from which they are to be distinguished only by the absence of the pellucid glands of the leaves and of other parts, existing in all the genera really belonging to that extensive family.

2dly. There are very few Melastomaceæ in which the ovarium does not in some degree cohere with the tube of the calyx; *Meriana*, properly so called, being, perhaps, the only exception.

And in the greater number of instances where, though the ovarium is coherent, the fruit is distinct, it becomes so from the laceration of the connecting processes already described.

3dly. That the generic divisions of the whole order remain to be established. On examination, I believe, it will be found that the original species of the Linnean genera, *Melastoma* and *Rhexia*, possess generic characters sufficiently distinguishing them from the greater part of the plants that have been since added to them by various authors.   In consequence of these additions, however, their botanical history has been so far neglected, that probably no genuine species of Melastoma, and certainly none of Rhexia, has yet been published in M. Bonpland's splendid and valuable monographs of these two genera.

* *Mélanges de Botanique; Observ. address. à M. Lamarck, p.* 57.
† *Flor. Ind. Occid.* 2, *p.* 831, *tab.* 14.

Of RHIZOPHOREÆ,[*] as I have formerly proposed to limit it, namely, to Rhizophora, Bruguiera, and Carallia, the collection contains only one plant, which is a species of Rhizophora, the Mangrove of the lower part of the river, and probably of the whole line of coast, but very different both from that of America, and from those either of India or of other equinoctical countries that have been described. There is, however, a plant in the collection which, though not strictly belonging to this order, suggests a few remarks on its affinities.

I referred *Carallia*[†] to Rhizophoreæ, from its agreement with them in habit, and in the structure of its flower. It is still uncertain whether its reniform seed is destitute of albumen; the absence of which, however, does not seem necessary to establish its affinity with the other genera of this order: for plants having the same remarkable economy in the germination of the embryo as that of Rhizophora, may belong to families which either have or are destitute of albumen.

The plant referred to from Congo, may be considered as a new species of *Legnotis*, having its petals less divided than those of the original species of that genus, and each cell of its ovarium containing only two pendulous ovula. The genus *Legnotis* agrees with Carallia in habit, especially in having opposite leaves with intermediate stipules; in the valvular æstivation of its calyx, and in several other points of structure of its flower. It differs in its divided petals; in its greater number of stamina, disposed, however, in a simple series; and in its ovarium not cohering with the calyx. It is therefore still more nearly related to *Richœia* of M. du Petit Thouars,[‡] from which perhaps it may not be generically distinct. The propriety of associating Carallia[§] with Rhizophoreæ is not perhaps likely to be disputed; and its affinity to Legnotis, especially to the species from Congo, appears very probable. It would seem therefore that we have already a series of structures connecting Rhizophora on the one hand with certain genera of *Salicariæ*, particularly with *Antherylium*, though that genus wants the intermediate stipules; and on the other with *Cunoniaceæ*,[||] especially with the simple leaved species of *Cerato-*

---

[*] *Flinders's Voy.* 2, *p.* 549.   [†] *Roxburgh. Coromand.* 3, p. 8, *t.* 211.

[‡] *Nov. Gen. Madagasc. n.* 84.

[§] Or Barraldeia, *Du Petit Thouars, Nov. Gen. Madagasc. n.* 82.

[||] *Flinders's Voy.* 2, *p.* 548.

petalum. While Loranthus and Viscum associated with Rhizophora, by M. de Jussieu, appear to form a very distinct family, and which, as it seems to me, should even occupy a distant place in the system.

HOMALINÆ.   In the collection from Congo a plant occurs evidently allied, and perhaps referable, to *Homalium,* from which it differs only in the greater number of glands alternating with the stamina, whose fasciculi are in consequence decomposed ; the inner stamen of each fasciculus being separated from the two outer by one of the additional glands.   This plant was first found on the banks of the Gambia, by Mr. Park, from whose specimens I have ascertained that the embryo is inclosed in a fleshy albumen.

The same structure of seed may be supposed, from very obvious affinity, to exist in *Astranthus* of Loureiro, to which *Blackwellia* of Commerson ought perhaps to be referred ; in *Napimoga* of Aublet, probably not different from Homalium ; and in *Nisa,** a genus admitting of subdivision, and which M. du Petit Thouars has referred to Rhamneæ.   All these genera appear to me sufficiently different from Rosaceæ, where M. de Jussieu has placed them, and from every other family of plants at present established.

Their distinguishing characters as a separate order, are, the segments of the perianthium disposed in a double series, or an equal number of segments nearly in the same series ; the want of petals ; the stamina being definite and opposite to the inner series of the perianthium, or to the alternate segments where they are disposed apparently in a simple series ; the unilocular ovarium (generally in some degree coherent with the calyx) having three parietal placentæ, with one, two, or even an indefinite number of ovula ; and the seeds having albumen, as inferred from its existence in the genus from Congo. The cohesion of the ovarium with the tube of the perianthium, though existing in various degrees in all the genera above enumerated, is probably a character of only secondary importance in Homalinæ.   For an unpublished genus found by Commerson in Madagascar, which in every other respect agrees with this family, has ovarium superum.   This genus at the same time seems to establish a considerably affinity between Homalinæ and certain genera, either absolutely belonging to *Passifloreæ,* especially *Paropsia* of M. du Petit Thouars,† or

---

* *Nov. Gen. Madagasc. n.* 81.

† *Hist. des Véget. des Isles de l'Afrique,* 59.

nearly related to them as *Erythrospermum*, well described and figured by the same excellent botanist.*

The increased number of stamina in Homalium, and particularly in the genus from Congo, instead of presenting an objection to this affinity, appears to me to confirm it. It may be observed also that there are two genera referable to Passifloreæ, though they will form a separate section of the order, which have a much greater, and even an indefinite, number of perfect stamina, namely, *Smeathmania*, an unpublished genus of equinoctial Africa, agreeing in habit, in perianthium, and in fruit, with Paropsia; and *Ryania* of Vahl,† which appears to me to belong to the same family.

In Passifloreæ the stamina, when their number is definite, which is the case in all the genera hitherto considered as belonging to them, are opposite to the outer series of the perianthium; a character, which, though of general importance, and here of practical utility in distinguishing them from Homalinæ, is not expressed in any of the numerous figures or descriptions that have been published of the plants of this order.

Passifloreæ and Cucurbitaceæ, though now admitted as distinct families, are still placed together by M. de Jussieu; and he considers the floral envelope in both orders as a perianthium or calyx, whose segments are disposed in a double series.‡

These views of affinity and structure are in some degree confirmed by Homalinæ, in which both ovarium inferum and superum occur; and in one genus of which, namely, *Blackwellia*, the segments of the perianthium, though the complete number, in relation to the other genera of the order, be present, are all of similar texture and form, and are disposed nearly in a simple series. If the approximation of these three families be admitted, they may be considered as forming a class intermediate between Polypetalæ and Apetalæ, whose principal characters would consist in the segments of the calyx being disposed in a double series, and in the absence of petals: the different orders nearly agreeing with each other in the structure of their seeds, and to a considerable degree in that of the ovarium.

The formation of this class, however, connected on the one hand with

---

* *Op. citat.* 65.   † *Eclog.* 1, *p.* 51, *t.* 9.

‡ *Annal. du Mus. d'Hist. Nat.* 6, *p.* 102.

Apetalæ by Samydeæ,* and on the other, though as it seems to me less intimately, with Polypetalæ by Violeæ, would not accord with any arrangement of natural orders that has yet been given. While the admission of the floral envelope being entirely calyx; and of the affinity of the class with Violeæ, would certainly be unfavourable to M. de Candolle's ingenious hypothesis of petals in all cases being modified stamina.

VIOLEÆ.† This order does not appear·to me so nearly related to Passifloreæ as M. du Petit Thouars is disposed to consider it: for it not only has a genuine polypetalous corolla, which is hypogynous, but its antheræ differ materially in structure, and its simple calyx is divided to the base. The irregularity both of petals and stamina in the original genera of the order, namely, Viola, Pombalia,‡ and Hybanthus, though characters of considerable importance, are not in all cases connected with such a difference in habit as to prevent their union with certain regular flowered genera, which it has lately been proposed to associate with them.

The collection from Congo contains two plants belonging to the section of Violeæ with regular flowers. One of these evidently belongs to *Passalia*, an unpublished genus in Sir Joseph Banks's herbarium, and described in the manuscripts of Solander from a plant found by Smeathman at Sierra Leone, which is perhaps not specifically distinct from that of Congo, or from *Ceranthera dentata* of the Flore d'Oware. But *Ceranthera*,§ which M. de Beauvois, being unacquainted with its fruit, has placed in the order Meliaceæ, is not different from *Alsodeia*, a genus published somewhat earlier, and from more perfect materials, by M. du Petit Thouars,‖ who refers it to Violeæ. The latter generic name ought of course to be adopted, and with a change in the termination (*Alsodineæ*) it may also denote the section of this order with regular flowers.

Physiphora of Sir Joseph Banks's herbarium, discovered by himself in Brazil, differs from Alsodeia only in its filaments being very slightly connected at base, and in the form and texture of its capsule, which is membranaceous, and, as the name imports, inflated.

* *Ventenat in Mém. de l'Instit. Sc. Phys.* 1807, 2 *sem. p.* 142.
† *Juss. Gen Pl.* 295. *Ventenat Malmais.* 27.
‡ *Vandelli Fasc Pl. p.* 7, *t.* 1. Ionidium, *Venten. Malmais.* 27.
§ *Flore d'Oware*, 2, *p.* 10.   ‖ *Hist. des Véget. des Isles de l'Afrique* 55.

Five species belonging to this section of Violeæ occur in Aublet's History of the Plants of Guiana, where each of them is considered as forming a separate genus. Of three of these genera, namely, *Conohoria*, *Rinorea*, and *Riana*, the flowers alone are described; the two others, *Passura* and *Piparea*, were seen in fruit only.

From the examination of flowers of Aublet's original specimens of the three former genera, in Sir Joseph Banks's herbarium, and of the fruit of *Conohoria*, which entirely agrees with that of *Passura*, and essentially with that of *Piparea*, I have hardly a doubt of these five plants, notwithstanding some differences in the disposition of their leaves, actually belonging to one and the same genus: and as they agree with *Physiphora* in every respect, except in the texture and form of the capsule, and with the *Passalia* of Sierra Leone and Congo, except in having their stamina nearly or entirely distinct, it appears that all these genera may be referred to Alsodeia.

I have also examined, in Sir Joseph Banks's herbarium, a specimen of *Pentaloba sessilis* of the Flora Cochinchinensis, which was sent so named, by Loureiro himself, and have found it to agree in every important point with Alsodeia, even as to the number of parietal placentæ. Loureiro, however, describes the fruit of Pentaloba as a five-lobed, five-seeded berry, and if this account be correct, the genus ought to be considered as distinct; but if, which is not very improbable, the fruit be really capsular, it is evidently referable to Alsodeia; with the species of which, from Madagascar and the west coast of equinoctial Africa, it agrees in the manifest union of its filaments.

It appears therefore that the ten genera now enumerated, and perhaps also *Lauradia* of Vandelli, may very properly be reduced to one; and they all at least manifestly belong to the same section of Violeæ, though at present they are to be found in various, and some rather distant, natural orders.

M. de Jussieu, in adopting Aublet's erroneous description of the stamina of Rinorea and Conohoria, has referred both these genera to Berberides,* to which

---

* The genera belonging to BERBERIDEÆ are *Berberis* (to which Ilex Japonica of Thunberg belongs); *Leontice* (including *Caulophyllum*, respecting which see *Linn. Soc. Transac.* 12, p. 145) *Epimedium*; and *Diphylleia* of Michaux. *Jeffersonia* may perhaps differ in the internal structure of its seeds, as its does in their arillus, from true Berberideæ, but it agrees with them in the three principal characters of their flower, namely, in their stamina being equal in number and opposite to the petals; in the remarkable dehiscence of antheræ; and in the structure of the ovarium. *Podophyllum* agrees with

he has also annexed Riana, adding a query whether Passura may not belong to the same genus.   With M. de Beauvois, he refers Ceranthera to Meliaceæ; and Pentaloba of Loureiro he reduces also to the same order.*   Piparea is, together with Viola, annexed to Cistinæ in his Genera Plantarum, and is therefore the most correctly placed, though its structure is the least known, of all these supposed genera.

An unpublished genus of New Holland, which I have named *Hymenanthera*, in Sir Joseph Banks's herbarium, agrees with Alsodeia in its calyx, in the insertion, expansion, and obliquely imbricate æstivation of its petals, and especially in the structure of its antheræ, which approach more nearly to those of Violeæ properly so called.   It differs, however, from this order in having five squamæ alternating with the petals; and especially in its fruit, which is a bilocular berry, having in each cell a single pendulous seed, whose internal structur resembles that both of Violeæ and Polygaleæ, between which I am inclined to think this genus should be placed.

CHAILLETEÆ.   The genus *Chailletia* was established by M. de Candolle† from a plant found by Martin in French Guiana, and which, as appears by specimens in Sir Joseph Banks's herbarium, had been many years before named *Patrisia* by Von Rohr, who discovered it in the same country.   At a still earlier period, Solander, in his manuscripts, preserved in the library of Sir Joseph Banks, described this genus under the name of *Mestotes*, from several species found by Smeathman at Sierra Leone.   Both *Dichapetalum* and *Leucosia* of M. du Petit Thouars ‡ appear to me, from the examination of authentic specimens, to belong to the same genus: and in Professor Smith's herbarium there is at least one additional species of Chailletia different from those of Sierra Leone.

Diphylleia in habit, and in the fasciculi of vessels of the stem being irregularly scattered; essentially in the floral envelope, and in the structure of the ovarium; its stamina, also, though numerous, are not altogether indefinite, but appear to have a certain relation both in number and insertion to the petals: in the dehiscence of antheræ, and perhaps also in the structure of seeds, it differs from this order, to which, however, it may be appended.   *Nandina* ought to be included in Berberideæ, differing only in its more numerous and densely imbricate bracteæ, from which to the calyx and even to the petals, the transition is nearly imperceptible; and in the dehiscence of its antheræ.

* *Mém. du Mus. d'Hist. Nat. 3, p.* 440.   † *Annal. du Mus. d'Hist. Nat.* 17, *p.* 153.
‡ *Nov. Gen. Madagasc. n.* 78 *et* 79.

Of the two generic names given by M. du Petit Thouars, and published somewhat earlier than M. de Candolle's Memoir, Leucosia will probably be considered inadmissable, having been previously applied by Fabricius to a genus of Crustacea; and Dichapetalum is perhaps objectionable, as derived from a character not existing in the whole genus, even allowing it to be really polypetalous. It seems expedient therefore to adopt the name proposed by M. de Candolle, who has well illustrated the genus in the memoir referred to. It appears to me that Chailletia, a genus nearly related to it from India with capsular fruit, and *Tapura* of Aublet (which is *Rohria* of Schreber,) form a natural order, very different from any yet established. The principal characters of this order may be gathered from M. de Candolle's figure and description of Chailletia, to which, however, must be added that the cells of the ovarium, either two or three in number, constantly contain two collateral pendulous ovula; and that in the regular flowered genera there exist within, and opposite to, the petal like bodies an equal number of glands, which are described by M. du Petit Thouars in Dichapetalum, but are unnoticed by him in Leucosia, where, however, they are equally present.

It may seem paradoxical to associate with these genera *Tapura*, whose flower is irregular, triandrous, and apparently monopetalous. But it will somewhat lessen their apparent differences of structure to consider the petal-like bodies, which, in all the genera of this order, are inserted nearly or absolutely in the same series with the filaments, as being barren stamina; a view which M. de Candolle has taken of those of Chailletia, and which M. Richard had long before published respecting Tapura.* It is probable also that M. de Candolle at least will admit the association here proposed, as his *Chailletia sessiliflora* seems to be merely an imperfect specimen of *Tapura guianensis*.

The genera to which Chailleteæ most nearly approach appear to me to be *Aquilaria* of Lamarck † and *Gyrinops* of Gærtner. But these two genera themselves, which are not referable to any order yet established, may either be regarded as a distinct family, or perhaps, to avoid the too great multiplication of families, as a section of that at present under consideration, and to which I

---

* *Dict. Elem. de Botanique par Bulliard, revu par L. C. Richard, ed.* 1802, *p.* 34.

† Or *Ophiospermum* of the Flora Cochinchinensis, as I have proved by comparison with a specimen from Loureiro himself.

should then propose to apply the name of AQUILARINÆ in preference to Chailleteæ.

The genus Aquilaria itself has been referred by Ventenat to *Samydeæ.* From this order, however, it is sufficiently distinct, not only in the structure of its ovarium and seeds, but in its leaves being altogether destitute of glands, which are not only numerous in Samydeæ, but consisting of a mixture of round and linear pellucid dots, distinguish them from all the other families* with which there is any probability of their being confounded.

Sir James Smith† has lately suggested the near affinity of Aquilaria to Euphorbiaceæ. But I confess it appears to me at least as distinct from that order as from Samydeæ: and I am inclined to think, paradoxical as it may seem, that it would be less difficult to prove its affinity to Thymeleæ than to either of them; a point however which, requiring considerable details, I do not mean to attempt in the present essay.

Of EUPHORBIACEÆ there are twenty species in the collection, or one twenty-eighth part of its Phænogamous plants. This is somewhat greater than the intratropical proportion of the order as stated by Baron Humboldt, but rather smaller than that of India or of the northern parts of New Holland.

The most remarkable plants of Euphorbiaceæ in the Congo herbarium are; a new species of the American genus *Alchornea;* a plant differing from *Ægopricon*, a genus also belonging to America, chiefly in its capsular fruit; two new species of *Bridelia*, which has hitherto been observed only in India; and an unpublished genus that I have formerly alluded to,‡ as in some degree explaining the real structure of Euphorbia, and from the consideration of which also it seems probable that what was formerly described as the hermaphrodite flower of that genus, is in reality a compound fasciculus of flowers.§ From the same species of this unpublished genus a substance resembling Caoutchouc is said to be obtained at Sierra Leone.

According to Mr. Lockhart a frutescent species of Euphorbia, about eight

---

* The only other genus in which I have observed an analogous variety of form in the glands of the leaves, is *Myroxylon*, (to which both *Myrospermum* and *Toluifera* belong,) in all of whose species this character is very remarkable, the pellucid lines being much longer than in Samydeæ.　　† *Linn. Soc. Transact.* 11, *p.* 230.

‡ *Flinders's Voy.* 2, *p.* 557.　　　　　§ *Linn. Soc. Transact.* 12, *p.* 99.

feet in height, with cylindrical stem and branches, was observed, planted on the graves of the natives near several of the villages; but of this, which may be what Captain Tuckey has called Cactus quadrangularis in his Narrative (p. 115) there is no specimen in the herbarium.

COMPOSITÆ. It is unnecessary here to enter into the question whether this family of plants, of which upwards of 3000 species are already known, ought to be considered as a class or as an order merely; the expediency of subdividing it, and affixing proper names to the divisions, being generally admitted. The divisions or tribes proposed by M. Cassini, in his valuable dissertations on this family appear, to be the most natural, though as yet they have not been very satisfactorily defined.

The number of Compositæ in the collection is only twenty-four, more than half of which are referable to *Heliantheæ* and *Vernoniaceæ* of M. Cassini. The greater part of these are unpublished species, and among them are five new genera. The published species belong to other divisions, and are chiefly Indian: but one of them, *Ageratum conyzoides*, is common to America and India; the *Struchium* (or Sparganophorus) of the collection does not appear to me different from that of the West Indies; and *Mikania chenopodifolia*, a plant very general on this line of coast, though perhaps confined to it, belongs to a genus of which all the other species are found only in America.

Baron Humboldt has stated * that Compositæ form one sixth of the Phænogamous plants within the tropics, and that their proportion gradually decreases in the higher latitudes until in the frigid zones it is reduced to one thirteenth. But in the herbarium from Congo Compositæ form only one twenty-third, and both in Smeathman's collection from Sierra Leone and in Dr. Roxburgh's Flora Indica, a still smaller part, of the Phænogamous plants. In the northern part of New Holland they form about one sixteenth; and in a manuscript catalogue of plants of equinoctial America, in the library of Sir Joseph Banks, they are nearly in the same proportion.

In estimating the comparative value of these different materials, I may, in the first place, observe that though the herbarium from Congo was collected in the dry season of the country, there is no reason to suppose on that account that the proportion of this family of plants, in particular, is materially or even in

* In *op. cilat.*

any degree diminished, nor can this objection be stated to the Sierra Leone collection, in which its relative number is still smaller.

To the Compositæ in Dr. Roxburgh's Flora Indica, however, a considerable addition ought, no doubt, to be made ; partly on the ground of his having apparently paid less attention to them himself, and still more because his correspondents, whose contributions form a considerable part of the Flora, have evidently in a great measure neglected them. This addition being made, the proportion of Compositæ in India would not differ very materially from that of the north coast of New Holland, according to my own collection, which I consider as having been formed in more favourable circumstances, and as probably giving an approximation of the true proportions in the country examined. Baron Humboldt's herbarium, though absolutely greater than any of the others referred to on this subject, is yet, with relation to the vast regions whose vegetation it represents, less extensive than either that of the north coast of New Holland, or even of the line of the Congo. And as it is in fact as much the Flora of the Andes as of the coasts of intratropical America, containing families nearly or wholly unknown on the shores of equinoctial countries, it may be supposed to have several of those families which are common to all such countries, and among them Compositæ, in very different proportion. At the same time it is not improbable that the relative number of this family in equinoctial America, may be greater than in the similar regions of other intratropical countries; while there seems some reason to suppose it considerably smaller on the west coast of Africa. This diminished proportion, however, in equinoctial Africa would be the more remarkable, as there is probably no part of the world in which Compositæ form so great a portion of the vegetation as at the Cape of Good Hope.

RUBIACEÆ   Of this family there are forty three species in the collection, or about one fourteenth of its Phænogamous plants.   I have no reason to suppose that this proportion is greater than that existing in other parts of equinoctial Africa ; on the contrary, it is exactly that of Smeathman's collection from Sierra Leone.

Baron Humboldt, however, states the equinoctial proportion of Rubiaceæ to Phænogamous plants to be one to twenty-nine, and that the order gradually diminishes in relative number towards the poles.

But it is to be observed that this family is composed of two divisions, having very different relations to climate; the *first*, with opposite, or more rarely verticillate leaves and intermediate stipules, to which, though constituting the great mass of the order, the name Rubiaceæ cannot be applied, being chiefly equinoctial; while the *second*, or *Stellatæ*, having verticillate or very rarely opposite leaves, but in no case intermediate stipules, has its maximum in the temperate zones, and is hardly found within the tropics, unless at great heights.

Hence perhaps we are to look for the minimum in number of species of the whole order, not in the frigid zone, but, at least in certain situations, a few degrees only beyond the tropics.

In conformity to this statement, M. Delile's valuable catalogue of the plants of Egypt* includes no indigenous species of the equinoctial division of the order, and only five of *Stellatæ*, or hardly the one hundred and sixtieth part of the Phænogamous plants. In M. Desfontaines' Flora Atlantica, Rubiaceæ, consisting of fifteen Stellatæ and only one species of the equinoctial division, form less than one ninetieth part of the Phænogamous plants, a proportion somewhat inferior to that existing in Lapland.

In Professor Thunberg's Flora of the Cape of Good Hope, where Rubiaceæ are to Phænogamous plants, as about one to one hundred and fifty, the order is differently constituted; the equinoctial division, by the addition of *Anthospermum*, a genus peculiar to southern Africa, somewhat exceeding Stellatæ in number. And in New Holland, in the same parallel of latitude, the relative number of Stellatæ is still smaller, from the existence of *Opercularia*, a genus found only in that part of the world, and by the addition of which the proportion of the whole order to the Phænogamous plants is there considerably increased.

More than half the Rubiaceæ from Congo belong to well known genera, chiefly to Gardenia, Psychotria, Morinda, Hedyotis, and Spermacoce.

Of the remaining part of the order, several form new genera.

The *first* of these is nearly related to Gardenia, which itself seems to require subdivision.

The *second* is intermediate between Rondeletia and Danais, and probably includes Rondeletia febrifuga of Afzelius.†

* *Flor. Egypt. Illustr. in Descript. de l'Egypte, Hist. Nat. v. 2. p.* 49.

† *In Herb. Banks.* This is the "New sort of Peruvian Bark" mentioned in his Report, p. 174: which is probably not different from the Bellenda or African Bark of Winterbottom's Account of Sierra Leone, vol. 2, p. 243.

The *third* has the inflorescence and flowers of *Nauclea,* but its ovaria and pericarpia are confluent, the whole head forming a compound spherical fleshy fruit, which is, I suppose, the country-fig of Sierra Leone, mentioned by Professor Afzelius.*

The *fourth* is a second species of *Neurocarpœa,* a genus which I have named, but not described, in the catalogue of Abyssinian plants appended to Mr. Salt's Travels.†

The *fifth* genus is intermediate between Rubiaceæ and Apocineæ. With the former it agrees in habit, especially in its interpetiolary stipules; and in the insertion and structure of its seeds, which are erect, and have the embryo lodged in a horny albumen forming the mass of the nucleus; while it resembles Apocineæ in having its ovarium entirely distinct from the calyx : its capsule in appearance and dehiscence is exactly like that of Bursaria.

The existence of this genus tends to confirm what I have formerly asserted respecting the want of satisfactory distinguishing characters between these two orders, and to prove that they belong to one natural class: the ovarium superum approximating it to Apocineæ; the interpetiolary stipules and structure of seeds connecting it, as it appears to me, still more intimately with Rubiaceæ.

The arguments adduced by M. de Jussieu‡ for excluding *Usteria* from Rubiaceæ and referring it to Apocineæ, are, its having ovarium superum, an irregular corolla, fleshy albumen, and only one stamen; there being no example of any reduction in the number of stamina in Rubiaceæ, (in which Opercularia and Pomax are not included by M. de Jussieu,) while one occurs in the male flowers of Ophioxylum, a genus belonging to Apocineæ. From analogous reasoning he at the same time decides in referring *Gœrtnera* of Lamarck§ to Rubiaceæ, though he admits it to have ovarium superum; its flowers being regular, its albumen more copious and horny, and its embryo erect. But all these characters exist in the new genus from Congo. These two genera therefore, together with *Pagamea* of Aublet, *Usteria, Geniostoma* of Forster (which is *Anasser* of Jussieu) and *Logania,*¶ might, from their mere agreement in the situation of ovarium, form a tribe intermediate between Rubiaceæ

* *Sierra Leone Report for* 1794, *p.* 171, *n.* 32.

† *Voyage to Abyssinia, append. p. lxvi.*   ‡ *Annal. du Mus. d'Hist. Nat.* 10, *p.* 323.

§ *Illustr, Gen. tab.* 167.   ¶ *Prodr. Flor. Nov. Holl.* 1, *p.* 455.

and Apocineæ. This tribe, however, would not be strictly natural, and from analogy with the primary divisions admitted in Rubiaceæ, as well as from habit, would require subdivision into at least four sections: but hence it may be concluded that the only combining character of these sections, namely, ovarium superum, is here of not more than generic value: and it must be admitted also that the existence or absence of stipules is in Logania * of still less importance.

APOCINEÆ. There are only six plants in the collection belonging to this order:

The *first* of these, together with some other species from Sierra Leone, constitutes an unpublished genus, the fruit of which externally resembles that of *Cerbera*, but essentially differs from it in its internal structure, being polyspermous. The Cream fruit of Sierra Leone mentioned by Professor Afzelius,† probably belongs to this genus, of which an idea may be formed by stating its flower to resemble that of Vahea, figured, but not described by M. Lamarck,‡ and its fruit, that of Voacanga§ of M. du Petit Thouars, from which bird lime is obtained in Madagascar, or of Urceola‖ of Dr. Roxburgh, the genus that produces the caoutchouc of Sumatra.

The *second* belongs to a genus discovered at Sierra Leone by Professor Afzelius, who has not yet described it, but has named it *Anthocleista*. This genus, however, differs from *Potalia* of Aublet (the Nicandra of Schreber) solely in having a four celled berry; that of Potalia being described both by Aublet and Schreber as trilocular, though according to my own observations it is bilocular. M. de Jussieu has appended *Potalia* to his Gentianeæ, partly determined, perhaps, from its being described as herbaceous. The species of *Anthocleista* from Congo, however, according to the account given me by Mr. Lockhart, the gardener of the expedition, is a tree of considerable size, and its place in the natural method is evidently near *Fagræa*.

Whether these genera should be united with Apocineæ, or only placed near them, forming a fifth section of the intermediate tribe already proposed, is somewhat doubtful.

In the perfect hermaphrodite flowers of Apocineæ, no exception occurs either

---

* *Prodr. Flor. Nov. Holl.* 1, p. 455.　　† *Sierra Leone Report*, 1794, p. 173, n. 47.
‡ *Illustr. Gen. tab.* 169.　　§ *Nov. Gen. Madagasc.* n. 32.
‖ *Asiat. Resear.* 5, p. 169.

to the quinary division of the floral envelopes and corresponding number of stamina, or to the bilocular or double ovarium: and in *Asclepiadeœ*, which are generally referred by authors to the same order, something like a necessary connection may be perceived between these relative numbers of stamina and pistilla, and the singular mode of fecundation in this tribe.   But in Potalia and Anthocleista there is a remarkable increase in the number of stamina and segments of the corolla, and at the same time a reduction in the divisions of the calyx.   The pistillum in Potalia, however, if my account of it be correct, agrees in division with that of Apocineæ : and the deviation from this division in Anthocleista is only apparent; the ovarium, according to the view I have elsewhere given of this organ,* being composed of two united ovaria, again indeed subdivided by processes of the placenta, but each of the sub-divisions or partial cells containing only one half of an ordinary placenta, and that not originating from its inner angle, as would be the case were the ovarium composed of four confluent organs.

Of ASCLEPIADEÆ there are very few species in the collection, and none of very remarkable structure. The *Periploca* of equinoctial Africa alluded to in my essay on this family,† was one of the first plants observed by Professor Smith at the mouth of the river; and a species of *Oxystelma*, hardly different from *O. esculentum* of India,‡ was found, apparently indigenous, on several parts of its banks.

The ACANTHACEÆ of the collection, consisting of sixteen species, the far greater part of which are new, have a much nearer relation to those of India than to the American portion of the order.   Among these there are several species of *Nelsonia*§ and *Hypoestes*;‖ a new species of *Ætheilema*,¶ a genus from which perhaps *Phaylopsis* of Willdenow is not different, though its fruit is described by Wendland** as a legumen, and by Willdenow, with almost equal impropriety, as a siliqua; a plant belonging to a genus I have formerly alluded to as consisting of *Ruellia uliginosa* and *R. balsamea* ;†† and

---

  * *Linn. Soc. Transact.* 12, *p.* 89.   † *Wernerian Nat. Hist. Soc. Trans.* 1, *p.* 40.

  ‡ Periploca esculenta, *Roxb. Coromand.* 1, *p.* 13, *t.* 11.

  § *Prodr. Flor. Nov. Holl.* 1, *p.* 480.   ‖ *Op. citat.* 1, *p.* 474.

  ¶ *Prodr. Flor. Nov. Holl.* 1, *p.* 478.   ** Micranthus, *Wend. Botan. Beobacht.* 38.

  †† *Loc. citat.*

a new species of *Blepharis*. All these genera exist in India, and none of them have yet been found in America.

CONVOLVULACEÆ. The herbarium of Professor Smith contains twenty-two species of this order, among which however there is no plant that presents any thing remarkable in its structure ; the far greater part belonging to Ipomæa, the rest to Convolvulus.

In the herbarium there is a single species of *Hydrolea*, nearly related to Sagonea palustris of Aublet, which would also be referred to this order by M. de Jussieu. But Hydrolea * appears, to me, to constitute, together with Nama, a distinct family (*Hydroleæ*) more nearly approaching to Polemoniaceæ than to Convolvulaceæ.

SCROPHULARINÆ. The collection contains only ten plants of this family, of which two form new genera, whose characters depend chiefly on the structure of antheræ and form of corolla.

The LABIATÆ of the herbarium consist of seven species, three of which belong to *Ocymum*, a genus common to equinoctial Asia and Africa, but not extending to America; an equal number to *Hyptis*, which is chiefly American, and has not been observed in India: the seventh is a species of *Hoslundia*, a genus hitherto found only on the west coast of Africa, and which in its inflorescence and in the verticillate leaves of one of its species, approaches to the following order.

VERBENACEÆ, together with Labiatæ form one natural class,† for the two orders of which it has already become difficult to find distinguishing characters.

In the Congo herbarium there are seven Verbenaceæ, consisting of three beautiful species of Clerodendron; two new species of Vitex ; Stachytarpheta indica of Vahl; and a new species of *Lippia*, which, from its habit and structure, confirms the union of Zapania with that genus, suggested by M. Richard.‡ This species from the Congo has its leaves in threes, and has nearly the same fragrance as Verbena triphylla, whose affinity to Lippia,

* *Vid. op. citat. p.* 482.   † *Flinders' Voy.* 2, *p.* 565.
‡ In *Mich. Flor. Bor.-Amer.* 2, *p.* 15.

notwithstanding the difference in calyx and inflorescence, is further confirmed by a peculiarity in the æstivation of its corolla, which extends only to Lippia and Lantana.

OLACINÆ. The herbarium contains a species of Olax differing from all the plants at present referred to that genus, in its calyx not being enlarged after fecundation, but in its original annular form surrounding the base only of the ripe fruit. The existence of this species, which agrees with those of New Holland, and with *Fissilia* of Commerson, in having only five petals, and in its barren stamina being undivided, while in habit it approaches rather more nearly to the original species *O. Zeylanica* and to *O. scandens* of Roxburgh, both of which I have examined, seems to confirm the union I have formerly proposed,* of all these plants into one genus. When I first referred *Fissilia* to this genus, I only presumed from the many other points of agreement that it had also the same structure of ovarium, on which, not only the generic character of Olax, but its affinities, seemed to me in a great measure to depend. M. Mirbel, however, has described the ovarium of Fissilia as trilocular.† I can only reconcile this statement with my own observations, by supposing him to have formed his opinion from a view of its transverse section; for on examining one of Commerson's specimens of *Fissilia disparilis,* communicated by M. de Jussieu, I have found its ovarium, like that of all the species of Olax, to be really unilocular; the central columnar placenta, at the top of which the three pendulous ovula are inserted, having no connection whatever with the sides of the cavity.

It was chiefly the agreement of Olax and Santalaceæ in this remarkable, and I believe, peculiar structure of ovarium, that induced me to propose, not their absolute union into one family, but their approximation in the natural series. I at the same time,‡ however, pointed out all the objections that M. de Jussieu has since stated to this affinity.§

Of these objections the two principal are the double floral envelope, and ovarium superum of Olax, opposed to the simple perianthium and ovarium inferum in Santalaceæ.

The first objection loses much of its importance, both on considering that

* *Prodr. Flor. Nov. Holl.* 1, *p.* 357.   † *Nouv. Bullet.* 3, *p.* 378.
‡ *Prodr. Flor. Nov. Holl.* 1, *p.* 351.   *Flinders' Voy.* 2, *p.* 570.
§ *Mém. du Mus. d'Hist. Nat.* 2. *p.* 439.

Quinchamalium, a genus in every other respect resembling Thesium, has an outer floral envelope surrounding its ovarium, and having more the usual appearance of calyx than that of Olax; and also in adverting to the generally admitted association of Loranthus and Viscum, of which the former is provided with both calyx and corolla, the latter, in its male flowers at least, with only a single envelope, and that analogous to the corolla of Loranthus.*

The second objection seems to be equally weakened by the obvious affinity of Santalaceæ to *Exocarpus,* which has not only ovarium superum, but the fleshy receptacle of whose fruit, similar to that of Taxus, perfectly resembles, and may be supposed in some degree analogous to, the enlarged calyx of certain species of Olax.

To these objections M. de Jussieu has added a third, which, were it well founded, would be more formidable than either of them, namely, that the ovarium of Santalaceæ is monospermous;† a statement, however, which I conclude must have proceeded from mere inadvertency.

URTICEÆ. In the collection the plants of this family, taking it in the most extensive sense, and considering it as a class rather than an order, belong chiefly to *Ficus,* of which there are seven species. One of these is very nearly related to Ficus religiosa; and like that species in India, is regarded as a sacred tree on the banks of the Congo.

A remarkable tree, called by the natives *Musanga,* under which name it is repeatedly mentioned in Professor Smith's Journal, forms a genus intermediate between Coussapoa of Aublet and Cecropia; agreeing with the latter in habit, and differing from it chiefly in the structure and disposition of its monandrous male flowers, and in the form of its female amenta.

In the inflorescence, and even in the structure of its male flowers, *Musanga* approaches very nearly to *Myrianthus* of M. de Beauvois,‡ which it also resembles in habit. But the fruit of Myrianthus, as given in the Flore d'Oware, is totally different, and, with relation to its male flowers, so remarkable, that a knowledge of the female flowers is wanting to fix our ideas both of the structure and affinities of the genus. This desideratum the expedition to Congo has not supplied, the male plant only of Myrianthus having been observed by Professor Smith.

---

* *Prodr. Flor. Nov. Holl.* 1, *p.* 352. † *Mem. du Mus. d'Hist. Nat.* 2, *p.* 439.

‡ *Flore d'Oware,* 1, *p.* 16, *tabb.* 11 *et* 12.

In *Artocarpeæ*, to which *Musanga* belongs, and in *Urticeæ* strictly so called, the ovulum, which is always solitary, is erect, while the embryo is inverted or pendulous. By these characters, as well as by the separation of sexes, they are readily distinguished from those genera of *Chenopodeæ* and of monosper-mous *Illecebreæ*,* in which the albumen is either entirely wanting or bears but a small proportion to the mass of the seed. And hence also *Celtis* and *Mertensia*,† in both of which the ovulum is pendulous, are to be excluded from Urticeæ, where they have been lately referred by M. Kunth. The same characters, of the erect ovulum and inverted embryo, characterise Polygoneæ,‡ as I have long since remarked, and exist in *Piperaceæ* and even in *Coniferæ*, if my notions of that remarkable family be correct. But from all those orders Urticeæ are easily distinguished by other obvious and important differences in structure.

PHYTOLACEÆ. In describing Chenopodeæ, in the Prodromus Floræ Novæ Hollandiæ, I had it particularly in view to exclude Phytolacca, Rivina, Microtea, and Petiveria, which I even then considered as forming the separate family now for the first time proposed.

In *Chenopodeæ* the stamina never exceed in number the divisions of the perianthium, to which they are opposite. In *Phytolaceæ* they are either in-definite, or when equal in number to the divisions of the perianthium, alternate with them. This disposition of stamina in Phytolaceæ, however, uniting genera with fruits so different as those of Phytolacca and Petiveria, it would be satisfactory to find in the same order a structure intermediate between the multilocular ovarium of the former and the monospermous ovarium, with lateral stigma, of the latter.

Two plants in the herbarium from Congo assist in establishing this con-nection.

The *first* is a species of *Phytolacca*, related to P. abyssinica, whose quin-quelocular fruit is so deeply divided, that its lobes cohere merely by their inner angles, and I believe ultimately separate.

The *second* is a species of *Giseckia*, a genus in which the five ovaria are

---

* *Prodr. Flor. Nov. Holl.* 1, *pp.* 405, 413, *et p.* 416. Paronychiearum sect. ii. *Jussieu in Mém. du Mus. d'Hist. Nat.* 2, *p.* 388.

† *Nov. Gen. et Sp. Pl. Orb. Nov.* 2, *p.* 30.   ‡ *Prodr. Flor. Nov. Holl.* 1, *p.* 419.

entirely distinct. This genus is placed by M. de Jussieu in Portulaceæ; but the alternation of its stamina with the segments of the perianthium, a part of its structure never before adverted to, as well as their insertion, seem to prove its nearer affinity to Phytolacca.*

Still, however, the lateral stigma, the spiral cotyledons, and want of albumen in Petiveria, remove it to some distance from the other genera of Phytolaceæ, and at the same time connect it with *Seguieria,* with which also it agrees in the alliaceous odour of the whole plant.

The affinity of *Seguieria* has hitherto remained undetermined, and is here proposed from the examination of three species lately discovered in Brazil, one of which has exactly the habit of Rivina octandra, and all of which agree with that plant, as well as with several others belonging to the order, in the very minute pellucid dots of their leaves.

Petiveria and Seguieria may therefore form a sub-division of Phytolaceæ: And another section of this order exists in New Holland, of which the two genera differ from each other in number of stamina as remarkably as Petiveria and Seguieria.

Of the Monocotyledonous orders, the first on which I have any remarks to offer, is that of

PALMÆ. The collection, however, contains no satisfactory specimens of any plant of this family except of *Elæis guineensis,* the *Maba* of the natives, or Oil Palm, which appears to be common along the whole of this line of coast. In Professor Smith's journal it is stated that a single plant of the Maba Palm† was cut down, from which Mr. Lockhart informs me that both the male and female spadices preserved in the collection were obtained. This fact seems to decide that Elæis is monœcious, which, indeed, Jacquin, by whom the genus

---

* *Ancistrocarpus* of M. Kunth (Nov. Gen. et Sp. Pl. Orb. Nov. 2, p. 186) belongs to Phytolaceæ, though its stamina are described to be opposite to the segments of the calyx: and it is not improbable that *Miltus* of Loureiro (Flor. Cochin. p. 302) whose habit, according to the description, is that of Giseckia, from which it differs nearly as Ancistrocarpus does from Microtea, or Rivina octandra from the other species of its genus, may also belong to this order.

† *Maba* is, perhaps, rather applied to the fruit than to the tree: *Emba* being, according to Merolla, the name of the single nut, and *Cachio* that of the entire cluster: for the Palm itself, he has no name. *Vide Piccardo Relaz. p.* 122.

was established, concluded it to be, though from less satisfactory evidence.* It was first described as diœcious by Gærtner, whose account has been adopted, probably without examination, by Schreber, Willdenow, and Persoon.

In Sir Joseph Banks's collection, however, from which Gærtner received the fruits he has described and figured, and where he may be supposed to have likewise obtained all the original information he had on the subject, there is no proof of the male and female spadices of Elæis guineenses belonging to different individuals.

Gærtner has fallen into a still more important mistake respecting the structure of the fruit of Elæis, the foramina of whose putamen, which are analogous to those of the Cocoa nut, being, according to his description, at the base, as in that genus, whereas they are actually at the apex. It is probable that *Alfonzia oleifera* of Humboldt, Bonpland, and Kunth, belongs to Elæis, and possibly it may not even differ from the African species.

It is a remarkable fact respecting the geographical distribution of Palmæ, that *Elæis guineensis,* which is universally, and I believe justly, considered as having been imported into the West India colonies from the west coast of Africa, and *Cocos indica,* which there is no reason to doubt is indigenous to the shores of equinoctial Asia and its islands, should be the only two species of an extensive and very natural section of the order, that are not confined to America.

To this section, whose principal character consists in the originally trilocular putamen having its cells when fertile perforated opposite to the seat of the embryo, and when abortive indicated by foramina cæca, as in the cocoa nut, the name Cocoinæ may be given; though it has been applied by M. Kunth† to a more extensive and less natural group, which includes all palms having trilocular ovaria, and the surface of whose fruit is not covered with imbricate scales. I may also remark that from the fruits of *Cocoinæ* only, as I have here proposed to limit the section, the oil afforded by plants of this family, is obtained.

Professor Smith in his journal frequently mentions a species of *Hyphæne,* by which he evidently intended the Palm first seen abundantly at the mouth of the river, and afterwards occasionally in the greater part of its course, especially near the Banzas, where it is probably planted for the sake of the wine obtained from it.

According to the gardener's information, this is a Palm of moderate height

* *Hist. Stirp. Amer. p.* 281.   † *Nova Gen. et Sp. Orb. Nov.* 1, *p.* 241.

with fan-shaped fronds and an undivided caudex. It therefore more probably belongs to Corypha than to Gærtner's Hyphænc, one species of which is the Cucifera of Delile, the Doom of Upper Egypt; the second, *Hyphæne coriacea*, is a native of Melinda, and probably of Madagascar, and both are remarkable in having the caudex dichotomous, or repeatedly divided.

As the Palm on the banks of the Congo was seen in fruit only, it is not difficult to account for Professor Smith's referring it rather to Hyphæne than to Corypha; Gærtner having described the embryo of the latter as at the base of the fruit, probably, however, from having inverted it, as he appears to have done in Elæis. It is at least certain that in *Corypha Taliera* * of the continent of India, which is very nearly allied to C. umbraculifera, the embryo is situated at the apex as in Hyphæne.

The journal also notices a species of Raphia, which is probably *Raphia vinifera* of M. de Beauvois,† the *Sagus Palma-pinus* of Gærtner.

The collection contains fronds similar to those of *Calamus secundiflorus* of M. de Beauvois,‡ which was also found at Sierra Leone by Professor Afzelius; and a male spadix very nearly resembling that of *Elate sylvestris* of India.

The Cocoa Nut was not observed in any part of the course of the river.

Only five species of Palms appear therefore to have been seen on the banks of the Congo. On the whole continent of Africa thirteen species, including those from Congo, have been found; which belong to genera either confined to this continent and its islands, or existing also in India, but none of which have yet been observed in America, unless perhaps Elæis, if Alfonsia oleifera of Humboldt should prove to be a distinct species of that genus.

CYPERACEÆ. In the collection there are thirty-two species belonging to this order, which forms therefore about one eighteenth of the Phænogamous plants. This is very different from what has been considered its equinoctial proportion, but is intermediate to that of the northern part of New Holland, where, from my own materials, it seems to be as 1:14; and of India, in which according to Dr. Roxburgh's Flora it is about 1:25.

In other intratropical countries the proportion may be still smaller; but I

* *Roxb. Coromand.* 3, *tabb.* 255 *et* 256.   † *Flore d'Oware* 1, *p.* 75, *tabb.* 44, 45, *et* 46.
‡ *Op. citat.* 1, *p.* 15, *tabb.* 9 *et* 10.

can neither adopt the general equinoctial ratio given by Baron Humboldt, namely, that of 1:60, nor suppose with him that the minimum of the order is within the tropics.   For Cyperaceæ, like Rubiaceæ, and indeed several other families, is composed of tribes or extensive genera, having very different relations to climate.   The mass of its equinoctial portion being formed of Cyperus and Fimbristylis, genera very sparingly found beyond the torrid zone; while that of the frigid and part of the temperate zones consists of the still more extensive genus Carex, which hardly exists within the tropics, unless at great heights.   Hence a few degrees beyond the northern tropic, on the old continent at least, the proportion of Cyperaceæ is evidently diminished, as in Egypt, according to M. Delile's valuable catalogue ;* and the minimum will, I believe, be found in the Flora Atlantica of M. Desfontaines and in Dr. Russel's catalogue of the plants of Aleppo.† It is not certain, however, that the smallest American proportion of the order exists in the same latitude.   And it appears that in the corresponding parallel of the southern hemisphere, at the Cape of Good Hope and Port Jackson, the proportion is considerably increased by the addition of genera either entirely different from, or there more extensive than those of other countries.

Among the Cyperaceæ of the Congo herbarium there are fifteen species of Cyperus, of which C. *Papyrus*, appears to be one.   The abundance of this remarkable species, especially near the mouth of the river, is repeatedly noticed in Professor Smith's journal, but from the single specimen with fructification in the collection, its identity with the plant of Egypt and Sicily, though very probable, cannot be absolutely determined.   I perceive a very slight difference in the sheaths of the radii of the common umbel, which in the plant from Congo are less angular and less exactly truncated, than in that of Egypt; in other respects the two plants seem to agree.   I have not seen C. laxiflorus, a species discovered in Madagascar by M. du Petit Thouars, and said to resemble C. Papyrus except in the vaginæ of the partial umbels.‡

Among the species of Cyperaceæ in the collection, having the most extensive range, are *Cyperus articulatus*, which is common to America, India, and

---

* *Flor. Ægypt. Illustr. in Descrip. de l'Egypte, Hist. Nat. 2, p. 49.*

† *Nat. Hist. of Aleppo, 2d. ed. vol. 2, p. 242.*

‡ *Encyc. Method. Botan. vol. 7, p. 270.*

Egypt; *Fuirena umbellata* and *Eleocharis capitata*,* both of which have been found in America, India, and New Holland; and *Cyperus ligularis* indigenous to other parts of Africa and to America.

Hypœlyptum argenteum, a species established by Vahl from specimens of India and Senegal, and since observed in equinoctial America by Baron Humboldt, is also in the collection.

The name *Hypœlyptum*, under which I have formerly described the genus that includes *H. argenteum*,† was adopted from Vahl, without enquiry into its origin. It is probably, however, a corruption of *Hypœlytrum*,‡ by which M. Richard, as he himself assures me, chiefly intended another genus, with apparently similar characters, though a very different habit, and one of whose species is described by Vahl in Hypælyptum; his character being so constructed as to include both genera. M. Kunth has lately published *H. argenteum* under the name of Hypælytrum;§ b t in adopting the generic character given in the Prodromus Floræ Novæ Hollandiæ, he has in fact excluded the plants that M. Richard more particularly meant to refer to that genus. It is therefore necessary, in order to avoid further confusion, to give a new name to Hypælyptum as I have proposed to limit it, which may be *Lipocarpha*, derived from the whole of its squamæ being deciduous.

In describing *Lipocarpha* (under the name of Hypælyptum) in the work referred to, I have endeavoured to establish the analogy of its structure to that of *Kyllinga*; the inner or upper squamæ being in both genera opposite to the inferior squama, or anterior and posterior, with relation to the axis of the spikelet: while the squamæ of Richard's Hypælytrum being lateral, or right and left with respect to the axis of the spikelet,‖ were compared to those of the female flowers of *Diplacrum*, to the utriculus or nectarium of *Carex*, and to the lateral bracteæ of *Lepeyrodia*, a genus belonging to the nearly related order Restiaceæ.¶ But as in *Hypœlytrum*, according to M. Richard's description, and I believe also in his *Diplasia*,** there are sometimes more than two inner squamæ, which are then imbricate, they may in both these genera be considered as a spikelet reduced to a single flower, as in several other genera

---

* *Prodr. Flor. Nov. Holl.* 1, *p.* 225. Scirpus capitatus *Willd. sp. pl.* 1, *p.* 294, exclus. syn. Gronovii.

† *Prodr. Flor, Nov. Holl.* 1, *p.* 219.

‡ *Persoon Syn. Plant.* 1, *p.* 70.

§ *Nov. Gen. et Sp. Plant.* 1, *p.* 218.

‖ *Prodr. Flor. Nov. Holl.* 1, *p.* 219.

¶ *Flinders's Voy.* 2, *p.* 579.

** *Persoon Syn. Pl.* 1, *p.* 70.

of Cyperaceæ, and in Lipocarpha itself, from which, however, they are still sufficiently different in their relation to the including squamæ and to the axis of the spike.

This view of the structure of Hypælytrum, of which there is one species in the Congo herbarium, appears to me in some degree confirmed by a comparison with that of *Chondrachne* and *Chorizandra*;* for in both of these genera the lower squamæ of the ultimate spikelet are not barren, but monandrous, the central or terminating flower only being hermaphrodite.

GRAMINEÆ. Of this extensive family there are forty-five species from the Congo, or one twelfth of the Phænogamous plants of the collection. This is very nearly the equinoctial proportion of the order as given by Baron Humboldt, namely, one to fifteen, with which that of India seems to agree On the north coast of New Holland, the proportion is still greater than that of Congo.

The two principal tribes which form the far greater part of Gramineæ, namely, *Poaceæ* and *Paniceæ* have, as I have formerly stated,† very different relations to climate, the maximum both in the absolute and relative number of species of Paniceæ being evidently within the tropics, that of Poaceæ beyond them.

I have hitherto found this superiority of Paniceæ to Poaceæ, at or near the level of the sea within the tropics, so constant, that I am inclined to consult the relative numbers of these two tribes, in determining whether the greater part of any intratropical Flora belongs to level tracts, or to regions of such elevation as would materially affect the proportions of the principal natural families: and in applying this test to Baron Humboldt's collection, it is found to partake somewhat of an extratropical character, Poaceæ being rather more numerous than Paniceæ. While in conformity to the usual equinoctial proportion, considerably more than half the Grasses in the Congo herbarium consist of Paniceæ.

Among the Paniceæ of the collection, there are two unpublished genera. The *first* is intermediate, in character, to Andropogon and Saccharum, but with a habit very different from both. The *second*, which is common to other

* *Prodr. Flor. Nov. Holl.* 1, p. 220.
† *Prodr. Fl. Nov. Holl.* 1, p. 169. *Obs. II. Flinders's Voy.* 2, p. 583.

parts of the coast and to India, appears to connect in some respects Saccharum with Panicum.

The remarks I have to make on the *Acotyledonous Plants* from Congo, relate entirely to

FILICES, of which there are twenty-two species in the collection. The far greater part of these are new, but all of them are referable to well established genera, particularly to Nephrodium, Asplenium, Pteris, and Polypodium. There are also among them two new species of *Adiantum*, a genus of which no species had been before observed on this line of coast. *Trichomanes* and *Hymenophyllum* are wanting in the collection, and these genera, which seem to require constant shade and humidity, are very rare in equinoctial Africa. Of *Osmundaceæ*, the herbarium contains only one plant, which is a new species of *Lygodium*, and the first of that genus that has been noticed from the continent of Africa.

Among the few species common to other countries, the most remarkable is Gleichenia Hermanni,[*] which I have compared and found to agree with specimens from the continent of India, from Ceylon, New Holland, and even from the Island of St. Vincent.

Acrostichum stemaria of M. de Beauvois,[†] which hardly differs from A. alcicorne of New Holland, and of several of the islands of the Malayan Archipelago, was also observed; and *Acrostichum aureum*, which agrees with specimens from equinoctial America, was found growing in plenty among the mangroves near the mouth of the river.

I have formerly observed that the number of Filices, unlike that of the other Cryptogamous orders, (Lycopodineæ excepted,) is greatest in the lower latitudes; and, as I then supposed, near or somewhat beyond the tropics. The latter part of this statement, however, is not altogether correct; the maximum of the order, both in absolute and relative number of species, being more probably within the tropics, though at considerable heights.

The degree of latitude alone being given, no judgment can be formed respecting the proportion of Filices: for besides a temperature somewhat

[*] *Prodr. Flor. Nov. Holl.* 1, *p.* 161. Mertensia dichotoma *Willd. Sp. Pl.* 5, *p.* 71.

[†] *Flore d'Oware* 1, *p.* 2, *t.* 2.

inferior, perhaps, to that of equinoctial countries of moderate elevation, a humid atmosphere and protection from the direct rays of the sun, seem to be requisite for their most abundant production.

When all these conditions co-exist, their equinoctial proportion to Phæno-gamous plants is probably about one to twenty, even on continents where the tracts most favourable to their production form only a small part, their number being increased according as such tracts constitute a more considerable portion of the surface.

Hence their maximum appears to exist in the high, and especially the well wooded, intratropical islands. Thus in Jamaica, where nearly two hundred species of Ferns have been found, their proportion to Phænogamous plants is probably about one to ten. In the Isles of France and Bourbon, from the facts stated by M. du Petit Thouars,* they appear to be about one to eight.

In Otaheite, according to Sir Joseph Banks's observations, they are as one to four. And in St. Helena, from Dr. Roxburgh's Catalogue,† they exceed one to two.

This high proportion extends to the islands considerably beyond the southern tropic. Thus in the collection formed by Sir Joseph Banks in New Zealand, they are about one to six : in Norfolk Island, from my friend Mr. Ferdinand Bauer's observations, they exceed one to three : and in Tristan Da Cunha, both from the Catalogue published by M. du Petit Thouars,‡ and the still more complete Flora of that Island, for which I am indebted to Captain Dugald Carmichael, they are to the Phænogamous plants as two to three.

The equinoctial proportion of Ferns in level and open tracts, is extremely dif-ferent from those already given ; and it is not improbable that as the maximum of this order is equinoctial, so its minimum will also be found either within or a few degrees beyond the tropics. Thus in several of the low Islands in the Gulf of Carpentaria, having a Flora of upwards of two hundred Phænogamous plants, not more than three species of Ferns were found, and those very sparingly. In Egypt it appears, both by Forskål's catalogue and the more extensive Flora of M. Delile, that only one Fern § has been observed.

---

* *Mélanges de Bot. Observ. add. à M. de Lamarck, p. 6, et 38.*

† *Beatson's Tracts relative to St. Helena, p. 295.*   ‡ *Mélanges de Botanique.*

§ Named *Adiantum capillus veneris* by both these authors; but possibly a nearly related species that has often been confounded with it. Of the species I allude to,

In Russel's catalogue of the plants of Aleppo two only are noticed: and even in M. Desfontaines' Flora Atlantica not more than eighteen species occur, or with relation to the Phænogamous plants, about one to one hundred.

The Ferns in the herbarium from Congo, are to the Phænogamous plants as about one to twenty-six, which agrees nearly with their proportion in Forskål's catalogue of the plants of Arabia, with that of the north coast of New Holland, according to my own observations, and which is probably not very different from their proportion in India.

In concluding here the subject of the proportional numbers of the Natural Orders of plants contained in the herbarium from Congo, I may observe, that the ratios I have stated, do not always agree with those given in Baron Humboldt's learned dissertation, so often referred to. I have ventured, however, to differ from that eminent naturalist with the less hesitation, as he has expressed himself dissatisfied with the materials from which his equinoctial proportions are deduced. Whatever may be the comparative value of the facts on which my own conclusions depend, I certainly do not look upon them as completely satisfactory in any case. And it appears to me evident, that with respect to several of the more extensive natural orders, other circumstances besides merely the degrees of latitude and even the mean temperature must be taken into account in determining their relative numbers. To arrive at satisfactory conclusions in such cases, it is necessary to begin by ascertaining the geographical distribution of genera, a subject, the careful investigation of which may likewise often lead to important improvements in the establishment or sub-divisions of these groups themselves, and assist in deciding from what regions certain species, now generally diffused, may have originally proceeded.

To the foregoing observations on the principal Natural Orders of Plants from the banks of the Congo, a few remarks may be added on such families as are general in equinoctial countries, but which are not contained in the collection.

which may be called *Adiantum africanum*, I have collected specimens in Madeira, and have seen others from Teneriffe, St. Jago, Mauritius or Isle de Bourbon, and Abyssinia. Adiantum africanum has also been confounded with *A. tenerum* of Jamaica, and other West India Islands, and the latter with *A. capillus veneris*, which has in consequence been supposed common to both hemispheres, to the old and new continent, and to the torrid and temperate zones.

These are Cycadeæ, Piperaceæ, Begoniaceæ, Laurinæ (Cassytha excepted,) Passifloreæ, Myrsineæ, Magnoliaceæ, Guttiferæ, Hesperideæ, Cedreleæ, and Meliaceæ.

Cycadeæ, although not found in equinoctial Africa, exist at the Cape of Good Hope and in Madagascar.

Piperaceæ, as has been already remarked by Baron Humboldt,[*] are very rare in equinoctial Africa; and indeed only two species have hitherto been published as belonging to the west coast: the first, supposed to be *Piper Cubeba*, and certainly very nearly related to it, is noticed by Clusius;[†] the second is imperfectly described by Adanson in his account of Senegal. A third species, of Piper, however, occurs in Sir Joseph Banks's herbarium, from Sierra Leone: and we know that at least one species of this genus and several of Peperomia, exist at the Cape of Good Hope.

The extensive genus *Begonia*, which it is perhaps expedient to divide, may be considered as forming a natural order, whose place, however, among the Dicotyledonous families, is not satisfactorily determined. Of *Begoniaceæ*,[‡] no species has yet been observed on the continent of Africa, though several have been found in Madagascar and the Isles of France and Bourbon, and one in the Island of Johanna.

No genus of *Laurinæ*, is known to exist in any part of the continent of Africa, except the paradoxical Cassytha, of which the only species in the Congo collection can hardly be distinguished from that of the West Indies, or from *C. pubescens* of New Holland. The absence of Laurinæ on the continent of Africa is more remarkable, as several species of Laurus have been found both in Teneriffe and Madeira, and certain other genera belonging to this family exist in Madagascar and in the Isles of France and Bourbon.

Passifloreæ. A few remarkable plants of this order have been observed on the different parts of the west coast of Africa, especially Modecca of the Hortus Malabaricus and Smeathmania, an unpublished genus already mentioned in treating of Homalinæ.

Myrsineæ. No species of any division of this order, has been met with in equinoctial Africa, though several of the first section, or Myrsineæ,

---

[*] *Nov. Gen. et Sp. Pl Orb. Nov.* 1, *p.* 60.

[†] Piper ex Guinea, *Clus. exot. p.* 184, who considers it as not different from the Piper caudatum, figured on the same page, and which is no doubt Piper Cubeba of the Malayan Archipelago.   [‡] *Bonpland Malmais.* 151.

properly so called, exist both at the Cape of Good Hope and in the Canary Islands.*

Magnoliaceæ and *Cedreleæ*, which are common to America and India, have not been found on the continent of Africa, nor on any of the adjoining Islands.

Guttiferæ and *Hesperideæ* exist, though sparingly, on other parts of the coast.

A few plants really belonging to *Meliaceæ* have been found on other parts of western equinoctial Africa, and a species of *Leea* (or *Aquilicia*, for these are only different names for the same genus) which was formerly referred to this order, occurs in the herbarium from Congo.

M. de Jussieu, who has lately had occasion to treat of the affinity of Aquilicia,† does not venture to fix its place in the system. Its resemblance to Viniferæ in the singular structure of seeds, in the valvular æstivation of the corolla, in the division of its leaves, the presence of stipules, and even in inflorescence, appears to me to determine, if not its absolute union, at least its near affinity to that order. Of *Viniferæ*, Vitis is at present the only certain genus; for *Cissus* and *Ampelopsis* having, as Richard has already observed, exactly the same structure of ovarium, namely, two cells with two erect collateral ovula in each, should surely be referred to it; nor is there any part of the character or description of *Botria* of Loureiro, which prevents its being also included in the same genus.

Lusianthera of M. de Beauvois,‡ referred by its author to Apocineæ, but

---

* To the first section belong *Myrsine*, *Ardisia*, and *Bladhai*. The second, including *Embelia*, and perhaps also *Othera* of Thunberg, differs from the first merely in its corolla being polypetalous. *Ægiceras* may be considered as forming a third section, from the remarkable evolution of its embryo and consequent want of albumen. In the æstivation of calyx and corolla it agrees with *Jacquinia*, which together with *Theophrasta*, (or *Clavija* of the Flora Peruviana,) forms the fourth section; characterised by the squamæ, more or less distinct, of the faux of the corolla, and by generally ripening more than one seed. The fifth, includes only *Bæobotrys* of Forster (the *Mæsa* of Forskål) which, having ovarium inferum and five barren filaments alternating with the segments of the corolla, bears the same relation to the other genera of this order, that Samolus does to Primulaceæ. On the near affinity, and slight differences in fructification, between this family and Myrsineæ, I have formerly made a few remarks in the Prodr. Flor. Nov. Holl. 1, p. 533.

† *Mém. du Mus. d'Hist. Nat. 3, p. 437 et 441.*   ‡ *Flore d'Oware, 1, p. 85.*

which M. de Jussieu has lately suggested may belong to Viniferæ,* is too imperfectly known to admit of its place being determined.

III. In the third part of my subject I am to compare the vegetation of the line of the river Congo with that of other equinoctial countries, and with the various parts of the continent of Africa and its adjoining Islands.

The first comparison to be made, is obviously with the other parts of the *West coast of equinoctial Africa.*

The most important materials from this coast to which I have had access are contained in the herbarium of Sir Joseph Banks, and consist chiefly of the collections of Smeathman from Sierra Leone, of Brass from Cape Coast (Cabo Corso), and the greater part of the much more numerous discoveries of Professor Afzelius already referred to. Besides these, there are a few less extensive collections in the same herbarium, especially one from the banks of the Gambia, made by Mr. Park in returning from his first journey into the interior; and a few remarkable species brought from Suconda and other points in the vicinity of Cape Coast, by Mr. Hove. The published plants from the west coast of Africa are to be found in the splendid and interesting *Flore d'Oware et Benin* of the Baron de Beauvois; in the earlier volumes of the Botanical Dictionary of the Encyclopedie Methodique by M. Lamarck, chiefly from Sierra Leone and Senegal; in the different volumes of Willdenow's Species Plantarum from Isert; in Vahl's Enumeratio Plantarum from Thonning; a few from Senegal in the Genera Plantarum of M. de Jussieu; and from Sierra Leone in a memoir on certain genera of Rubiaceæ by M. de Candolle, in the Annales du Museum d'Histoire Naturelle. Many remarkable plants are also mentioned in Adanson's Account of Senegal, and in Isert's Travels in Guinea.

On comparing Professor Smith's herbarium with these materials, it appears that from the river Senegal in about 16° N. lat. to the Congo which is in upwards of 6° S. lat. there is a remarkable uniformity in the vegetation, not only as to the principal natural orders and genera, but even to a considerable extent in the species of which it consists. Upwards of one third part of the plants in the collection from Congo had been previously observed on other parts of the coast, though of these the greater part are yet unpublished.

* *Loc. cit.*

Many of the Trees, the Palms, and several other remarkable plants, which characterise the landscape, as *Adansonia, Bombax pentandrum, Anthocleista, Musanga* of the natives (the genus related to Cecropia,) *Elœis guineensis, Raphia vinifera,* and *Pandanus Candelabrum,* appear to be very general along the whole extent of coast.

Sterculia acuminata, * the seed of which is the *Cola,* mentioned in the earliest accounts of Congo, exists, and is equally valued, in Guinea and Sierra Leone, and what is remarkable, has the same name in every part of the west coast.

The *Ordeal Tree* noticed in Professor Smith's journal under the name of Cassa; and in Captain Tuckey's narrative erroneously called a species of Cassia, if not absolutely the same plant as the *Red Water Tree* of Sierra Leone,† and as it is said also of the Gold Coast, belongs at least to the same genus.

A species of the *Cream Fruit,* mentioned by Professor Afzelius,‡ remarkable in affording a wholesome and pleasant saccharine fluid, used by the natives of Sierra Leone even to quench their thirst, though the plant belongs to Apocineæ, a family so generally deleterious, was also met with.

The *Sarcocephalus* of the same author,§ which is probably what he has noticed under the name of the country-fig of Sierra Leone,‖ was found, and seems to be not uncommon, on the banks of the Congo.

Anona Senegalensis, whose fruit, though smaller than that of the cultivated species of the genus, has, according to Mr. Lockhart, a flavour superior to any of them, was every where observed, especially above Embomma, and appears to be a very general plant along the whole extent of coast:

And *Chrysobalanus Icaco,* or a species very nearly related to it, which is equally common from Senegal to Congo, was found abundantly near the mouth of the river.

The remarks I have to make on *Esculent Plants,* my knowledge of which is chiefly derived from the journals of Captain Tuckey and Professor Smith,

* *De Beauvois. Flore d'Oware* 1, p. 41, t. 24.
† *Winterbottom's Sierra Leone* 1, p. 129.
‡ *Sierra Leone Report for* 1794, p. 173, n. 47.
§ *In Herb. Banks.* ‖ *Op. cit. p.* 171, n. 32.

and the communications of Mr. Lockhart, may be here introduced; the cultivated as well as the indigenous species being very similar along the whole of the west coast.

On the banks of the Congo, as far as the expedition proceeded, the principal articles of vegetable food were found to be Indian Corn or Maize (*Zea Mays*) Cassava, both sweet and bitter, (*Iatropha Manihot L.*); two kinds of Pulse, extensively cultivated, one of which is *Cytisus Cajan* of Linneus, the other not determined, but believed to be a species of *Phaseolus;* and Ground Nuts (*Arachis hypogæa L.*)

The most valuable fruits seen were Plantains (*Musa sapientum;*) the Papaw (*Carica papaya*) Pumpkins (*Cucurbita Pepo;*) Limes and Oranges (*Citrus medica et aurantium;*) Pine Apples (*Bromelia Ananas;*) the common Tamarind (*Tamarindus indica;*) and *Safu,* a fruit the size of a small plum, which was not seen ripe.

One of the most important plants not only of Congo, but of the whole extent of coast, is *Elæis guineensis* or the *Oil Palm,* from which also the best kind of Palm Wine is procured. Wine is likewise obtained from two other species of Palms, which are probably *Raphia vinifera,* and the supposed *Corypha,* considered as an Hyphæne by Professor Smith.

Among the other Alimentary Plants which are either of less importance or imperfectly known, may be mentioned the " *Shrubby Holcus,*" noticed by Captain Tuckey (p. 138); the common *Yam,* which Mr. Lockhart informs me he saw only near Cooloo; and another species of *Dioscorea* found wild only, and very inferior to the Yam, requiring, according to the narrative, "four days boiling to free it from its pernicious qualities." On the same authority, "Sugar Canes of two kinds " were seen at Embomma, and Cahbages at Banza Noki: a kind of Capsicum or Bird Pepper, and Tobacco, were both observed to be generally cultivated: and I find in the herbarium, a specimen of the *Malaguetta Pepper,* or one of the species of Amomum, confounded under the name of *A. Granum Paradisi.*

Mr. Lockhart believes there was also a second kind of Ground Nut or Pea, which may be that mentioned by Merolla, under the name of *Incumba,** and the second sort perhaps noticed in Proyart's account of Loango,† which is

---

* *Piccardo Relaz. del Viag. nel Reg. di Congo, p.* 119.   † *P.* 18.

probably *Glycine subterranea* of Linneus, the Voandzeia of M. du Petit Thouars,* or Voandzou of Madagascar, where it is generally cultivated.†

Of the indigenous fruits, Anona senegalensis, Sarcocephalus, a species of Cream fruit, and Chrysobalanus Icaco, have been already mentioned, as trees common to the whole line of coast.

A species of *Ximenia* was also found by Professor Smith, who was inclined to consider it as not different from *X. americana:* its fruit, which, according to his account, is yellow, the size of a plum and of an acid, but not disagreeable taste, is in the higher parts of the river called Gangi, it may therefore probably be the *Ogheghe* of Lopez,‡ by whom it is compared to a yellow plum, and the tree producing it said to be very generally planted.

An *Antidesma*, probably like that mentioned by Afzelius, as having a fruit in size and taste resembling the currant, is also in the herbarium.

It is particularly deserving of attention, that the greater part of the plants now enumerated, as cultivated on the banks of the Congo, and among them nearly the whole of the most important species, have probably been introduced from other parts of the world, and do not originally belong even to the continent of Africa. Thus it may be stated with confidence that the Maize, the Manioc or Cassava, and the Pine Apple, have been brought from America, and probably the Papaw, the Capsicum, and Tobacco; while the Banana or Plantain, the Lime, the Orange, the Tamarind, and the Sugar Cane, may be considered as of Asiatic origin.

In a former part of this essay, I have suggested that a careful investigation of the geographical distribution of genera might in some cases lead to the determination of the native country of plants at present generally dispersed. The value of the assistance to be derived from the source referred to, would amount to this; that in doubtful cases, where other arguments were equal, it would appear more probable that the plant in question should belong to that country in which all the other species of the same genus were found decidedly indigenous, than to that where it was the only species of the genus known to exist. It seems to me that this reasoning may be applied with advantage

---

* *Nov. Gen. Madagasc. n.* 77.   † *Flacourt Madagasc. pp.* 114 *et* 118.

‡ *Pigafetta, Hartwell's Translat. p.* 115.

towards determining the original country of several of the plants here enumerated, especially of the Banana, the Papaw, the Capsicum, and Tobacco.

The *Banana* is generally considered to be of Indian origin : Baron Humboldt, however, has lately suggested * that several species of *Musa* may possibly be confounded under the names of Plantain and Banana ; and that part
of these species may be supposed to be indigenous to America.   How far the
general tradition said to obtain both in Mexico, and Terra Firma, as well as the
assertion of Garcilasso de la Vega respecting Peru, may establish the fact of the
Musa having been cultivated in the new continent before the arrival of the
Spaniards, † I do not mean at present to enquire.   But in opposition to the
conjecture referred to, it may be advanced that there is no circumstance in the
structure of any of the states of the Banana or Plantain cultivated in India,
or the islands of equinoctial Asia, to prevent their being all considered as
merely varieties of one and the same species, namely *Musa sapientum ;* that
their reduction to a single species is even confirmed by the multitude of varieties
that exist ;‡ by nearly the whole of these varieties being destitute of seeds ;
and by the existence of a plant indigenous to the continent of India,§ producing perfect seeds ; from which, therefore, all of them may be supposed to
have sprung.

To these objections to the hypothesis of the plurality of species of the Banana,
may be added the argument referred to as contributing to establish its Asiatic
origin ; for we are already acquainted with at least five distinct species of
Musa in equinoctial Asia, while no other species has been found in America ;
nor does it appear that the varieties of Banana, cultivated in that continent,
may not equally be reduced to Musa sapientum as those of India : and lastly,
it is not even asserted that the types of any of those supposed species of
American Banana, growing without cultivation, and producing perfect seeds,
have any where been found.‖

* *Nouv. Espag. vol.* 2, *p.* 360.

† *Op. cit. p.* 361. It may be observed, however, that this is not the opinion in every
part of the continent of South America, for with respect to Brazil, Marcgraf and Piso
assert that both the Banana and Plantain are considered as introduced plants, and the
latter apparently from Congo. (*Marcg. p.* 137, *et Piso Hist. Nat. Bras. p.* 154.)

‡ Musa sapientum, *Rox. Corom. tab.* 275.

§ M. Desvaux, in a dissertation on the genus Musa (*in Journ. de Botanique appl. vol.* 4,

That the Bananas now cultivated in equinoctial Africa, come originally from India, appears to me equally probable, though it may be allowed that the *Ensete* of Bruce * is perhaps a distinct species of this genus, and indigenous only to Africa.

The *Papaw* (Carica papaya), from analogous reasoning, may be regarded as of American origin ; there being several other decidedly distinct species natives of that continent, while no species except the cultivated Papaw, nor any plant nearly related to this singular genus, is known to exist either in Asia or Africa. But in the present case, the assistance derived from the argument adduced, may perhaps be considered as unnecessary ; for the circumstance of there being no Sanscrit name for so remarkable a plant as the Papaw,† is nearly decisive of its not being indigenous to India. And in the Malay Islands, the opinion of the inhabitants, according to Rumphius,‡ is that it was there introduced by the Portuguese.

The same argument may be extended to *Capsicum,* of which all the known species probably belong to the new continent ; for the only important exception stated to this genus being wholly of American origin, namely *C. frutescens,*

p. 1), has come to the same conclusion respecting the original country of the cultivated Banana, and also that its numerous varieties are reducible to one species. In this dissertation he takes a view of the floral envelope of Musa peculiar to himself. The perianthium in this genus is generally described as consisting of two unequal divisions or lips. Of these, one is divided at top into five, or more rarely into three segments, and envelopes the other, which is entire, of a different form and more petal-like texture. The enveloping division M. Desvaux regards as the calyx, the inner as the corolla. It seems very evident to me, however, that the deviation in Musa from the regular form of a Monocotyledonous flower, consists in the confluence of the three divisions of the outer series of the perianthium, and in the cohesion, more or less intimate, with these of the two lateral divisions of the inner series ; the third division of this series, analogous to the labellum in the Orchideæ, being the inner lip of the flower. This view seems to be established by the several modifications observable in the different species of Musa itself, especially in *M. superba* of Roxburgh, (*Plants of Coromand.* 3, *tab.* 223) and in the flower of Musa figured by Plumier, (*Nov. gen. t.* 34.), but still more by the irregularity confined to the inner series in Strelitzia, and by the near approach to regularity, even in this series, in Ravenala (or Urania), both of which belong to the same natural order.

* *Travels, vol.* 5, *p.* 36.

† *Fleming in Asiat. Resear.* ii. *p.* 161.   ‡ *Herb. Amboin.* i. *p.* 147.

seems to be set aside merely by the appellations of *Tchilli* and *Lada Tchilli*, as given to it in the Malay Islands ; *Chilli*, either simply, or in composition, being the Mexican name for all the species and varieties of this genus.*

All the species of *Nicotiana* appear to be American, except *N. Australasiæ* (the *N. undulata* of Ventenat and Prod. Flor. Nov. Holl. but not of Flora Peruviana,) which is certainly a native of New Holland. The exception here, however, does not materially invalidate the reasoning, *N. Australasiæ* differing so much from the other species as to form a separate section of the genus.

The same argument might perhaps be applied to other plants of doubtful origin, as to *Canna indica,* which it would derive from America.

It is certainly not meant, however, to employ this reasoning in every case, and in opposition to all other evidence ; and instances may be found, even among the alimentary plants, where it is very far from being satisfactory. Thus the Cocoa Nut, though it will probably be considered as indigenous to the shores and islands of equinoctial Asia, is yet the only species of its genus that does not belong exclusively to America.

Cytisus Cajan, may be supposed to have been introduced from India. This plant, which is very generally cultivated in the vicinity of the Congo, I conclude is the *Voando,* mentioned by Captain Tuckey as being ripe in October; and as Mr. Lockhart understood from the natives, that Cytisus Cajan continues to bear for three years, it is probably Merolla's *Ovvando,* of which he gives a similar account.†

Whether *Arachis hypogœa* be indigenous or introduced, cannot now perhaps be satisfactorily determined. This remarkable plant, whose singular structure and economy were first correctly described by M. Poiteau,‡ and which was every where seen in abundance, as far as the river was examined, appears to form an important article of cultivation along the whole of the west coast of Africa, and probably also on the east coast, on several parts of which it was found by Loureiro.§

According to the same author, it is also universally cultivated in China and Cochinchina.

---

* *Hernandez, Rer. Medic. Nov. Hispan. Thesaur. p.* 134, *et Nieremb. Hist. Nat. p.*363.

† *Piccardo Relaz. p.* 120.   ‡ *Mém. de l'Instit. Sc. Phys. Sav. Etrang.* 1, *p.* 455.

§ *Flor. Cochin.* 430.

From China it has probably been introduced into the continent of India, Ceylon, and the Malayan Archipelago, where, though now generally cultivated, there is reason to believe, particularly from the names given to it, that it is not indigenous. I think it not very improbable that it may have been carried from Africa to various parts of equinoctial America, though it is noticed in some of the early accounts of that continent, especially of Peru and Brazil.

According to Professor Sprengel,* it is mentioned by Theophrastus as cultivated in Egypt: but it is by no means evident that Arachis is the plant intended in the passage of Theophrastus referred to; and it is probable that had it been formerly cultivated in Egypt, it would still be found in that country; it is not, however, included either in Forskål's Catalogue, or in the more extensive Flora Egyptiaca of M. Delile.

There is nothing very improbable in the supposition of Arachis hypogæa being indigenous to Asia, Africa, and even America; but if it be considered as originally belonging to one of those continents only, it is more likely to have been brought from China through India to Africa, than to have been carried in the opposite direction.

Glycine subterranea, however, which is extensively cultivated in Africa, Madagascar, and several parts of equinoctial America, is probably of African origin; it is stated, at least both by Marcgraf and Piso, to have been introduced into Brazil from Angola or Congo.†

The *Holcus* noticed by Captain Tuckey, of which the specimens in the herbarium do not enable me to determine whether it be a distinct species, or a variety only of *H. sorghum* or *saccharatus*, may be considered as indigenous, or at least as belonging to Africa. According to Mr. Lockhart, it is very generally found wild, and it is only once mentioned as cultivated: it may, however, have been formerly cultivated, along with other species of Millet, to a much greater extent; its place being now supplied by the Maize, which gives probably both a more productive and a more certain crop.

The *Dioscorea* or bitter Yam, which was observed only in a wild state, may be presumed to be a native species; and if ever it has been cultivated, it may

---

* *Hist. Rei Herb.* 1, p. 98.

† Mandubi d'Angola. *Marcg. Hist. Nat. Brasil.* 43. Mandobi, *Piso, Hist. Nat. Brasil.* p. 256.

in like manner be supposed to have been superseded by the Manioc or Cassava.

The *Safu*,* which Mr. Lockhart understood from the natives was one of their most esteemed fruits, he observed to be very generally planted round the villages, especially from Embomma upwards, and to be carefully preserved from birds: its importance is perhaps increased from its ripening in October, a season when the general supply of vegetable food may be supposed to be scanty.

There seems no reason to doubt that this tree, whose probable place in the system I have stated in my remarks on Amyrideæ, belongs originally to the west coast of Africa,

Elæis guineensis, of which the oil is distinctly described in the beginning of the sixteenth century by *Da Ca da Mosto*, in his account of Senegal,† is without doubt indigenous to the whole extent of this coast; as is *Raphia vinifera*, of which the remarkable fruit also very early attracted attention;‡ and the supposed species of *Corypha*.

Of Alimentary Plants, whether cultivated or indigenous, that are known or supposed to belong to the west coast of equinoctial Africa, but which were not seen on the banks of the Congo, a few of the more important may be mentioned.

Among these are the Cocoa Nut and Rice, the former, according to the natives, not being found in the country. The absence of these two valuable plants is the more remarkable, as the Cocoa Nut is said to exist in the neighbouring kingdom of Loango; and according to Captain Tuckey, a certain portion of land was seen on the banks of the river well adapted to the production of Rice, which is mentioned as cultivated in some of the earlier accounts of Congo.

The Sweet Potatoe (*Convolvulus Batatas*), also noticed by the Portuguese Missionaries, was not met with.

The Butter and Tallow Tree of Afzelius, which forms a new genus belonging to Guttiferæ; the Velvet-Tamarind of Sierra Leone (Codarium acutifolium;§) and the Monkey Pepper, or Piper Æthiopicum of the shops (*Unona æthiopica* of Dunal), which is common on many parts of the coast, were not observed.

* Probably the *Zaffo* of some of the earlier accounts of Congo, vide *Malte-Brun Precis de la Geogr.* 5, *p.* 9.

† *Ramusio* 1, *p.* 104. *Gryn. Nov. Orb.* 28.   ‡ Palma-Pinus, *Lobel. advers. p.* 450.

§ *Afzel. Gen Plant. Guineen. par. prim.p.* 23. Codarium nitidum *Vahl. enum.* 1, *p.* 302.

Two remarkable plants, the *Akee** and the *Jamaica* or *American Nutmeg*,† now cultivated in the West India colonies; and the former undoubtedly, the latter probably, introduced from Africa by the Negroes, were neither met with on the banks of the Congo, nor have they been yet traced to any part of the west coast.

The relation which the vegetation of the *Eastern shores of equinoctial Africa* has to that of the west coast, we have at present no means of determining; for the few plants, chiefly from the neighbourhood of Mozambique, included in Loureiro's Flora Cochinchinensis, and a very small number collected by Mr. Salt on the same part of the coast, do not afford materials for comparison

The character of the collections of *Abyssinian Plants* made by Mr. Salt in his two journeys, forming part of Sir Joseph Banks's herbarium, and amounting to about 260 species, is somewhat extratropical, and has but little affinity to that of the vegetation of the west coast of Africa.

To the Flora of *Egypt*, that of Congo has still less relation, either in the number or proportions of its natural families: the herbarium, however,

---

* Blighia sapida, *König in Annals of Bot.* 2, *p.* 571. *Hort. Kew. ed.* 2*da. vol.* 2, *p.* 350.

At the moment that this sheet was about to have been sent to the press, Sir Joseph Banks received a small collection of specimens and figures of plants, observed in the late Mission to Cummazee, the capital of Ashantee; and among them a drawing of the fruit and leaf of a plant, there called *Attueah* or *Attuah*, which is no doubt the *Akee*, whose native country is therefore now ascertained.

† Monodora myristica, *Dunal Annonac. p.* 80. *Decand. Syst. Nat. Reg. Véget.* 1, *p.* 477. Anona myristica, *Gært. Sem.* 2, *p.* 194, *t.* 125, *p.* 1. *Lunan Hort. Jamaic.* 2, *p.* 10. This remarkable plant is very properly separated from Anona, and considered as a distinct genus by M. Dunal in his monograph of Anonaceæ. The character given of this new genus, however, is not altogether satisfactory, M. de Candolle's description, from which it is derived, having probably been taken from specimens which he had it not in his power to examine completely. Both these authors have added to this genus Annona microcarpa of Jacquin (*Fragm. Bot p.* 40, *t.* 44, *f.* 7), established by that author from the fruit of my *Cargillia australis* (*Prodr. Flor. Nov. Holl.* 1, p. 527) which belongs to the very different family of Ebenaceæ.

Long, in his History of Jamaica (*vol.* 3, *p.* 735.) has given the earliest account of *Monodora Myristica*, under the name of the *American Nutmeg*, and considers it to have been probably introduced from South America: according to other accounts, it comes from the Mosquito shore; but there is more reason to suppose that it has been brought by the Negroes from some part of the west coast of Africa.

includes several species which also belong to Egypt, as Nymphæa Lotus, Cyperus Papyrus and articulatus, Sphenoclea zeylanica, Glinus lotoides, Ethulia conyzoides, and Grangea maderaspatana.

Of the many remarkable genera and orders characterising the vegetation of *South Africa,* no traces are to be found in the herbarium from Congo. This fact is the more worthy of notice, because even in Abyssinia a few remains, if I may so speak, of these characteristic tribes, have been met with; as the *Protea abyssinica,** observed by Bruce, and *Pelargonium abyssinicum* and *Geisorrhiza abyssinica* † found by Mr. Salt.

Between the plants collected by Professor Smith in the island of *St. Jago* and those of the Congo herbarium, there is very little affinity; great part of the orders and genera being different, and not more than three species, of which Cassia occidentalis is one, being common to both. To judge from this collection of St. Jago, it would seem that the vegetation of the Cape Verd Islands is of a character intermediate between that of the adjoining continent and of the Canary Islands, of which the Flora has, of course, still less connection with that of Congo.

It might perhaps have been expected that the examination of the vicinity of the Congo would have thrown some light on the origin, if I may so express myself, of the Flora of *St. Helena.* This, however, has not proved to be the case; for neither has a single indigenous species, nor have any of the principal genera, characterising the vegetation of that Island, been found either on the banks of the Congo, or on any other part of this coast of Africa.

There appears to be some affinity between the vegetation of the banks of the Congo and that of *Madagascar* and the *Isles of France* and *Bourbon.* This affinity, however, consists more in a certain degree of resemblance in several natural families and extensive or remarkable genera, than in identity of species, of which there seems to be very few in common.

The Flora of Congo may be compared with those of equinoctial countries still more remote.

With that of *India,* it agrees not only in the proportions of many of its principal families, or in what may be termed the equinoctial relation, but also, to a certain degree, in the more extensive genera of which several of these

* Gaguedi *Bruce's Travels* 5, *p.* 52.
† *Salt's Travels in Abyssinia, append. p. lxiii. and lxv.*

families consist: and there are even about forty species common to these distant regions.

To the vegetation of *Equinoctial America* it has certainly much less affinity. Several genera, however, which have not yet been observed in India or New Holland, are common to this part of Africa and America:* and there are upwards of thirty species in the Congo herbarium, which are also natives of the opposite coasts of Brazil and Guiana.

As the identity of species, especially of the Dicotyledonous division, common to equinoctial America and other intratropical countries, has often been questioned, I have subjoined two lists of plants included in the Congo herbarium, of which the first consists of such species as are common to America and India: and the second, of such as are found in America only.

I have given also a third list, of species common to Congo and India, or its Islands, but which have not been observed in America:

And a fourth is added, consisting of doubtful plants, to which I have, in the mean time, applied the names of those species they most nearly resemble, and to which they may really belong, without, however, considering their identity as determined

I. *List of Plants common to Equinoctial Africa, America, and Asia.*

Gleichenia Hermanni *Prodr. Flor. Nov. Holl.*
  Mertensia dichotoma *Willd.*
Agrostis Virginica L.
Cyperus articulatus L. } *ead. sp.*
 —— niloticus Vahl.
Lipocarpha argentea *Nob.*
 Hypælyptum argenteum *Vahl.* }
Eleocharis capitata *Prod. Fl. Nov. Holl.*
Fuirena umbellata *L. fil.*
Pistia Stratiotes L.

Boerhaavia mutabilis *Prodr. Flor. Nov. Holl.*
Ipomœa pes capræ *Nob.*
 Convolvoulus pes capræ L. } *ead. sp.*
 ————— brasiliensis L.
Ipomœa pentaphylla *Jacqu.*
Scoparia dulcis L.
Heliotropium indicum L.
Sphenoclea zeylanica *Gært.*
Ageratum conyzoides L.

---

* Namely, Elæis *Jacqu.* Rivina *L.* Telanthera *Nob* (Alternantheræ pentandræ.) Alchornea *Sw.* Blechum *Prodr. Flor. Nov. Holl.* (Blechi *sp. Juss.*) Schwenckia *L.* Hyptis *Jacqu.* Vandellia *L.* Annona *L.* Banisteria *Nob.* (Banisteriæ *sp. L.*) Paullinia *Juss.* (Paulliniæ *sp. L.*) Vismia *Ruiz. et Pav.* Conocarpus *L.* Legnotis *Sw.* (Cassipourea *Aubl.*) Chailletia *Decand.*

Waltheria indica L.
——— americana L. } ead. sp.

Hibiscus tiliaceus L.
Sida periplocifolia L.
Cassia occidentalis L.

Guilandina Bonduc L.
——— Bonducella L. } ead. sp.

Abrus precatorius L.
Hedysarum triflorum L.

II. *Plants common to Equinoctial Africa and America: but not found in India.*

Octoblepharum albidum *Hedw.*
Acrostichum aureum L.

Eragrostis ciliaris.
Poa ciliaris L. }

Cyperus ligularis L.
Schwenckia americana L.
Hyptis obtusifolia *Nob.*
Struchium (americanum) *Br. jam.* 312.

Sida juncea *Banks. et Soland. Mss. Brasil.*

Urena americana L.
——— reticulata *Cavan.* } ead. sp.

Malachra radiata L.
Jussiæa erecta L.
Crotalaria axillaris *Hort. Kew. & Willd*
Pterocarpus lunatus L.

III. *Plants common to Equinoctial Africa and India: but not found in America.*

Roccella fuciformis *Achar. Lichenog.* 440.
Perotis latifolia *Soland. in Hort. Kew.*
Centotheca lappacea *Beauv.*
Eleusine indica *Gært.*
Flagellaria indica L.
Gloriosa superba L.
Celosia argentea L.
— margaritacea L. } ead. sp.
— albida ? *Willd.*
Desmochæta lappacea *Decand.*

Grangea (maderaspatana) *Adans.*
Lavenia erecta *Sw.*
Oxystelma esculentum *Nob.*
Periploca esculenta *Roxb.* }
Nymphæa Lotus L.
——— pubescens *Willd.* } ead. sp.
Hibiscus surattensis L.
Leea sambucina L.
Hedysarum pictum L.
Indigofera lateritia *Willd.*
Glinus lotoides L.

IV. *List of Species which have not been satisfactorily ascertained.*

Acrostichum alcicorne *Sw.*
——— stemaria *Beauv.* }
Imperata cylindrica *Prodr. Flor. Nov. Holl.*

Panicum crus-galli L.
Typha angustifolia L.
Giseckia pharnaceoides L.

Cassytha pubescens *Prodr. Flor. Nov.*   Hydrocotyle asiatica L.

  Holl.   Hedysarum adscendens *Sw.*

Celtis orientalis L.   Hedysarum vaginale L.

Cardiospermum grandiflorum *Sw.*   Pterocarpus Ecastophyllum L.

Paullinia pinnata L.

On these lists it is necessary to make some observations.

1st. The number of species in the three first lists taken together is equal to at least one-twelfth of the whole collection. The proportion, indeed, which these species bear to the entire mass of vegetation on the banks of the Congo is probably considerably smaller, for there is no reason to believe that any of them are very abundant except Cyperus Papyrus and Bombax pentandrum, and most of them appear to have been seen only on the lower part of the river.

2nd. The relative numbers of the species belonging to the primary divisions in the lists, is analogous to, and not very materially different from, those of the whole herbarium; Dicotyledones being to Monocotyledones nearly as 3 to 1; and Acotyledones being to both these divisions united as hardly 1 to 16: hence the Phænogamous plants of the lists alone form about one-thirteenth of the entire collection.

The proportions now stated are very different from those existing in the catalogue I have given of plants common to New Holland and Europe;* in which the Acotyledones form one-twentieth, and the Phænogamous plants only one-sixtieth part of the extra-tropical portion of the Flora; while the Monocotyledones are to the Dicotyledones as 2 to 1.

The great proportion of Dicotyledonous plants in the lists now given, and especially in the two first, which are altogether composed of American species, is singularly at variance with an opinion very generally received, that no well established instance can be produced of a Dicotyledonous plant, common to the equinoctial regions of the old and new continent.

3d. The far greater part of the species in the lists are strictly equinoctial; a few, however, have also been observed in the temperate zones, namely Agrostis virginica, belonging, as its name implies, to Virginia, and found also on the shores of Van Diemen's Island, in a still higher latitude; Cyperus Papyrus, and articulatus, Nymphæa Lotus, and Pistia Stratiotes, which are

* *Flinders' Voy.* 2. *p.* 592.

natives of Egypt; Glinus lotoides of Egypt and Barbary; and Flagellaria indica, existing on the east coast of New Holland, in as high a latitude as 32°. S.

4th. It may perhaps be suggested with respect to these lists, that they contain
or even chiefly consist of plants that during the constant intercourse which has
now subsisted for upwards of three centuries between Africa, America, and
India, may have, either from design or accidentally, been carried from one of
these regions to another, and therefore are to be regarded as truly natives of
that continent only from which they originally proceeded.

It appears to me, however, that there is no plant included in any of the lists
which can well be supposed to have been *purposely* carried from one continent to
another, unless perhaps *Chrysobalanus Icaco*, and *Cassia occidentalis*; both of
which may possibly have been introduced into America by the Negroes, from
the west coast of Africa; the former as an eatable fruit, the latter as an article
of medicine.   It seems at least more likely that they should have travelled
in this than in the opposite direction.   But I confess the mode of introduction
now stated, does not appear to me very probable, even with respect to these
two plants; both of them being very general in Africa, as well as in America;
though Chrysobalanus Icaco is considered of but little value as a fruit in either
continent; and for Cassia occidentalis, which exists also in India, another
mode of conveyance must likewise be sought.

Several species in the lists, however, may be supposed to have been *accidentally* carried, from adhering to, or being mixed with, articles of food or
commerce; either from the nature of the surface of their pericarpial covering,
ás Desmochæta lappacea, Lavenia erecta, Ageratum conyzoides, Grangea maderaspatana, Boerhaavia mutabilis, and Hyptis obtusifolia; or from the minute-
ness of their seeds, as Schwenckia americana, Scoparia dulcis, Jussiæa erecta,
and Sphenoclea zeylanica. That the plants here enumerated have actually been
carried in the manner now stated is, however, entirely conjectural, and the
supposition is by no means necessary: several of them, as Lavenia erecta,
Scoparia dulcis, and Boerhaavia mutabilis, being also natives of the intratropical part of New Holland; their transportation to or from which cannot be
supposed to have been effected in any of the ways suggested.

The probability, however, of these modes of transportation, with respect to
the plants referred to, and others of similar structure, being even admitted,
the greater part of the lists would still remain; and to account for the disper-

sion of these, recourse must be had to natural causes, or such as are unconnected with human agency. But the necessity of calling in the operation of these causes implies the adoption of that theory according to which each species of plants is originally produced in one spot only, from which it is gradually propagated. Whether this be the only, or the most probable opinion that can be held, it is not my intention to enquire : it may however be stated as not unfavourable to it, that, of the Dicotyledonous plants of the lists, a considerable number have the embryo of the seed highly developed, and at the same time well protected by the texture of its integuments.

This is the case in Malvaceæ, Convolvulaceæ, and particularly in Leguminosæ, which is also the most numerous family in the lists, and in several of whose species, as *Guilandina Bonduc*, and *Abrus precatorius*, the two conditions of developement and protection of the embryo coexist in so remarkable a degree, that I have no doubt the seeds of these plants would retain their vitality for a great length of time, either in the currents of the ocean,* or in the digestive organs of birds and other animals; the only means apparently by which their transportation from one continent to another can be effected : and it is deserving of notice that these seem to be the two most general plants on the shores of all equinoctial countries.

The Dicotyledonous plants in the lists which belong to other families have the embryo of the seed apparently less advanced, but yet in a state of considerable developement, indicated either by the entire want or scanty remains of albumen: the only exception to this being *Leea*, in which the embryo is many times exceeded in size by the albumen.

In the Monocotyledonous plants, on the other hand, consisting of Gramineæ, Cyperaceæ, Gloriosa, Flagellaria and Pistia, the embryo bears a very small proportion to the mass of the seed, which is formed of albumen, generally farinaceous. But it may here be observed that the existence of a copious albumen in Monocotyledones does not equally imply an inferior degree of

---

* Sir Joseph Banks informs me, that he received some years ago the drawing of a plant, which his correspondent assured him was raised from a seed found on the west coast of Ireland, and that the plant was indisputably *Guilandina Bonduc*. Linnæus also seems to have been acquainted with other instances of germination having taken place in seeds thrown on shore on the coast of Norway. *Vide Coloniæ Plantarum, p. 3, in Amœn. Acad. vol.* 8.

vitality in the embryo, but may be considered as the natural structure of that primary division ; seeds without albumen occurring only in certain genera of the paradoxical Aroideæ, and in some other Monocotyledonous orders which are chiefly aquatic.

5. Doubts may be entertained of the identity of particular species. On this subject I may observe, that for whatever errors may be detected in these lists, I must be considered as solely responsible ; the insertion of every plant contained in them being founded on a comparison of specimens from the various regions of which their existence in the particular lists implies them to be natives. The only exception to this being Lipocarpha argentea, of which I have not seen American specimens ; as a native of that continent therefore it rests on the very sufficient authority of Baron Humboldt and M. Kunth.

In my remarks on the natural orders, I have already suggested doubts with respect to certain species included in the lists, and shall here add a few observations on such of the others as seem to require it.

Acrostichum Aureum L. was compared, and judged to agree, with American specimens ; and I have therefore placed it in the 2d list, without, however, meaning to decide whether those plants originally combined with A. aureum, and now separated from it, should be regarded as species or varieties.

Fuirena umbellata L. fil. from Congo has its umbels somewhat less divided than either the American plant or that from the continent of India ; but from specimens collected in the Nicobar Islands, this would appear to be a variable circumstance.

Gloriosa superba L. which seems to be very general along the whole of the west coast of Africa, is considered as a variety of the Indian plant by M. Lamarck. This African variety has no doubt given rise to the establishment of the second species of the genus, namely *G. simplex*, which Linnæus adopted from Miller ;* and which Miller founded on the account sent to him by M. Richard, of the Trianon Garden, along with the seeds of what he called a new Gloriosa, brought from Senegal by Adanson, and having blue flowers. Miller had no opportunity of determining the correctness of this account ; for though the seeds vegetated, the plant died without flowering ; but he added a character not unlikely to belong to the seedling plants of G. superba, namely

---

* Gloriosa 2, *Mill. Dict. cd.* 7.

the want of tendrils. Adanson himself, indeed, notices what he considers a new species of Gloriosa in Senegal,* but he says nothing of the colour of its flowers, which he would hardly have omitted, had they been blue : that his plant, however, was not without tendrils, may be inferred from their entering into the character he afterwards gave of the genus,† as well as from M. Lamarck's account of his variety β of G. superba,‡ which he seems to have described from Adanson's specimens. And as no one has since pretended to have seen a species of this genus, either with blue flowers, or leaves without tendrils, *G. simplex*, which has long been considered as doubtful, may be safely left out of all future editions of the Species Plantarum. As the supposed G. superba of this coast, however, seems to differ from the Indian plant in the greater length and more equal diameter of its capsule, it may possibly be a distinct species, though at present I am inclined to consider it as only a variety.

Sphenoclea zeylanica Gært. I have compared this plant from Congo with specimens from India, Java, China, Cochinchina,§ Gambia, Demerary, and the island of Trinidad.

I was at one time inclined to believe, that Sphenoclea might be considered as an attendant on Rice, which it very generally accompanies, and with which I supposed it to have been originally imported from India into the various countries where it is found. This hypothesis may still account for its existence in the rice fields of Egypt;|| but as it now appears have to been observed in countries where there is no reason to believe that rice has ever been cultivated, the conjecture must be abandoned.

Hibiscus tiliaceus L. agrees with the plant of India, except in a very slight difference in the acumen of the leaf; but the specimens from America have their outer calyx proportionally longer.

Sida periplocifolia L. corresponds with American specimens; those in Hermann's herbarium, from which the species was established, have a longer acumen to the leaf: in other respects I perceive no difference.

---

* Nouvelle espèce de Methonica, *Hist. Nat. du Senegal, p.* 137.

† Mendoni, *Fam. des Plant.* 2. *p.* 48.

‡ *Encyc. Method. Botan.* 4. *p.* 134.

§ *Rapinia herbacea* of the Flora Cochinchinensis (p. 127) is certainly *Sphenoclea zeylanica*, as appears by a specimen sent to Sir Joseph Banks by Loureiro himself.

|| *Delile Flor. Egypt. illust. in op. cit.*

Waltheria indica L. I consider *W. americana* to be a variety of this sportive species, which seems to be common to all equinoctial countries.

Urena americana L. and *U. reticulata Cavan.* appear to me not to differ specifically; and the plant from Congo agrees with West India specimens.

Jussiæa erecta L. from Congo, agrees with West India specimens in having linear leaves; a specimen, however, from Miller's herbarium, which has been compared, and is said to correspond, with that in the Linnean collection, has elliptical leaves.

Chrysobalanus Icaco L. has its leaves more deeply retuse than any American specimens I have seen, but in this respect it agrees with Catesby's figure.

Guilandina Bonduc L. from which *G. Bonducella* does not appear to differ in any respect, is one of the most general plants on the shores of equinoctial countries.

Pterocarpus lunatus L. I have compared the plant from Congo with an authentic specimen from the Linnean herbarium, the examination of which proves that the appearance of ferruginous pubescence in the panicle, noticed in Linné's description, is the consequence of his specimen having been immersed in spirits.

Several of the plants included in the fourth list, I am inclined to consider varieties only of the species to which they are referred; but I have placed them among the more doubtful plants of this list, as their differences seem to be permanent, and are such as admit of being expressed. One of these is

Cardiospermum grandiflorum Sw. of which the specimens from Congo differ somewhat in inflorescence from the West India plant.

Paullinia pinnata L. is distinguished rather remarkably from the American plant by the figure of the leaflets, which approach to cuneiform, or widen upwards, but I can perceive no other difference.

Pterocarpus Ecastophyllum L. differs merely in the want of the very short acumen or narrow apex of the leaf, which I have constantly found in all the West India specimens I have examined.

Giseckia pharnaceoides L. from Congo, has nearly linear leaves; but I have seen specimens from Kœnig with leaves of an intermediate form.

I shall conclude this essay, already extended considerably beyond my original plan, with a general statement of the proportion of new genera and species contained in Professor's Smith's herbarium.

The whole number of species in the collection is about 620; but as specimens of about thirty of these are so imperfect as not to be referable to their proper genera, and some of them not even to natural orders, its amount may be stated at 590 species.

Of these about 250 are absolutely new : nearly an equal number exist also in different parts of the west coast of equinoctial Africa, and not in other countries; of which, however, the greater part are yet unpublished : and about 70 are common to other intratropical regions.

Of unpublished genera there are 32 in the collection; twelve of which are absolutely new, and three, though observed in other parts of this coast of equinoctial Africa, had not been found before in a state sufficiently perfect, to ascertain their structure; ten belong to different parts of the same line of coast; and seven are common to other countries.

No natural order, absolutely new, exists in the herbarium; nor has any family been found peculiar to equinoctial Africa.

The extent of Professor Smith's herbarium proves not only the zeal and activity of my lamented friend, but also his great acquirements in that branch of science, which was his more particular province, and to his excessive exertions in the investigation of which he fell a victim, in the ill-fated expedition to Congo.

Had he returned to Europe, he would assuredly have given a far more complete and generally interesting account of his discoveries than what is here attempted: and the numerous facts which he could no doubt have communicated respecting the habit, the structure, and the uses of the more important and remarkable plants, would probably have determined him to have followed a very different plan from that adopted in the present essay.

It remains only that I should notice the exemplary diligence of the Botanic Gardener, Mr. David Lockhart, the only survivor, I believe, of the party by whom the river above the falls was examined, in that disastrous journey which proved fatal to the expedition.

From Mr. Lockhart I have received valuable information concerning many of the specimens contained in the herbarium, and also respecting the esculent plants observed on the banks of the Congo.

APPENDIX. No. VI.

My Dear Sir,   *British Museum, Nov. 5th,* 1817.

In compliance with your desire, I have examined the specimens of rocks that were collected in the late expedition to Congo, and presented by the Lords Commissioners of the Admiralty to the British Museum. These specimens are all from the banks of the narrows of the Zaire, and are few in number; but they suffice to prove that the rock formation which prevails on the banks and islands of the lower parts of the river is primitive, and greatly resembles that beyond the ocean to the west: a circumstance which adds to the probability that the mountains of Pernambuco, Rio, and other adjacent parts of South America, were primevally connected with the opposite chains that traverse the plains of Congo and Loango. As you have already given, in your " General Observations," a satisfactory view of the broad geological features of the country along those banks, as far as they could be collected from the narratives of Captain Tuckey, and Dr. Smith, I shall confine myself to a few desultory remarks which offer themselves upon comparing the observations of these gentlemen with such of the specimens as have their respective localities affixed to them.

The specimens from the Fetish rocks exhibit a series of granitic compounds, in which the feldspar predominates: and most of them, especially the fine-grained varieties, contain disseminated a great quantity of minute noble garnets, some of them pellucid, others opaque, and of a reddish brown colour; and all belonging to the trapezoidal or leucite modification. Similar garnets also abound in the mica slate of Gombae. A few specimens of a siliceous rock, nearly compact, being composed of confluent particles of quartz, intermixed with minute scales of mica, are likewise ticketed as obtained from the Fetishes. This insulated group of rocks seems to represent a miniature of the stupendous granitic bulwark, which arises from the plain on the north side of the river Coanzo, near Cabazzo (the capital of Matamba), and of which an account, together with a good representation, has been given by Father Cavazzo. We are told by this author, that the name given to it by the Portuguese, is *il Presidio* (the fortress), and that the natives call it " *Maopongu*,"— a word which, making but little allowance for national difference of expressing the same sound in writing, is pretty like that of " *M'wangoo*," synonymously added to the sketch given of the Fetish rock by Captain Tuckey.

Boka M'Bomma, according to the same gentleman's account, consists entirely of shistus; but his own specimens from the S., S.E., N.E., and S. W parts of that island, are stratified granite or gneiss, in which the feldspar exists in very small proportion, and which, on the S.W. side, passes into a beautifully resplendent silvery variety, stained, towards the surface of the blocks or separated pieces, by brown iron ochre. It is in the variety of this gneiss from the last-mentioned part of the island, that laminar particles of a dark brown colour are seen, some of them exhibiting traces of the regular octohedral form, and which appear to be an iron oxydule. There is also, among the specimens from Boka Embomma, a fragment of primitive green-stone with embedded garnets.

The specimens from the creek of Banza M'bomma exhibit a mixture of fine granular hornblende and quartz; some of them are real hornblende-rock, and contain disseminated garnets. These specimens, among which there are also some varieties of reddish massive quartz, not unlike milk-quartz, were collected by Mr. Tudor.

Besides these primitive rocks, and those from Chesalla, near the Banza, which latter affords two varieties of gneiss with black and with yellow mica, we have, from the same neighbourhood, and particularly from the Chimoenga cliffs, a few specimens of sandstone: it is coarse-grained and ferruginous; its colour is grayish, and yellowish, with here and there some purplish specks ; and it appears to belong to the oldest formation of this rock. The plain on which the banza is situated, is covered by a bed of clay, which, according to a label accompanying the specimens, is two feet thick. It is of an ash-gray colour, and perfectly plastic.

The quartz mentioned by Captain Tuckey and Professor Smith, as being found in large masses, on the summit of Fidler's Elbow, belongs to the variety called fat-quartz : the fragments have mica adhering to them, and are here and there stained of a blood-red colour. Some specimens of brown-ironstone, massive and friable, have likewise been found on this hill. A ticket written by

Capt. Tuckey, and affixed to one of them from the highest summit, informs us that a globule of some metal, either gold or copper, has been seen adhering to one of the cavities, by Dr. Smith; but no such observation has been communicated in this Gentleman's journal: the lump, however, bears evident marks of having been exposed to the action of fire. There is scarcely any appearance of metals in the rocks near the lower parts of the Zaire; if, therefore, the accounts which the missionaries have given of the great abundance of every description of ores in Congo, extend to the banks of that river, it must be higher up; where, according to Dr. Smith's account, the rock-formation appears to adopt a different character. The specimens from Condo-Sono, Banza Nokki, and Benda consist of sienite, with green hornblende, and a rock composed of feldspar and quartz, with thickly-disseminated particles of magnetic iron stone, instead of mica or hornblende. It is probable that the primitive trapp occurs here in beds subordinate to the gneiss and mica-slate, of which a few specimens are sent, together with some others from the same parts, which appear to be flint-slate.

A rolled piece of sienite from the falls of Yellala, covered by a thin, shining, black crust, proves, that the action of the water of the Zaire is similar in its effects to that of the Oronoko. There are boulders of sienite in our collection, found by Baron Humboldt, at the cataracts of Alures, which are covered with exactly the same crust, and bear, externally, the most striking resemblance to meteoric stones. This black crust, both in the stones from the Zaire, and from the Oronoko, Mr. Children, to whom I communicated some particles, found to be a mixture of oxides of iron and manganese.

There are no specimens sent from above the Falls, except two varieties of compact limestone, one of them magnesian; but the places from which they came are not distinctly marked.

This is all I have to say on the scanty materials before me; and I leave it to you to make any use of it you please. Believe me, my dear Sir,

Your most obedient humble Servant,

CHARLES KONIG.

To John Barrow, Esq. Sec. to
the Admiralty, &c. &c. &c.

APPENDIX. No. VII.

Hydrographical Remarks from the Island of St. Thomas, to the Mouth of the Zaire.

1816. 16th May. At noon the land was reported to be seen from the mast-head, and immediately after from the deck : it was Prince's Island ; and bore S E b. S, 12 or 14 leagues distant. By our observations the centre of it lies in lat. 1° 35' N, long. 7° 17' 45 E ; variation about 21° W. We continued endeavouring to get to the southward, with light airs from S to S W, and a strong current setting to the northward ; and on the 19th, at daylight, saw the Island of St. Thomas, bearing S b. W 12 or 14 leagues. We remained two days in sight of it, making scarcely any way to the southward, which induced us to stand to the westward, hoping to get the wind more westerly, and less current ; in both of which expectations we were disappointed : nor did we succeed in getting to the southward of this island until the 27th, having been obliged to stand so far to the westward as 4° E, and then passed only about 4 leagues to the southward of it. The little island of Rolle, at the south end of St Thomas, lies on the equinoctial line, and in long. about 6° 44' 37" E. We continued working to the southward, taking every advantage of the wind, and frequently trying for soundings without obtaining them, until the 3d of June, when, at noon, we had 17½ fathoms greenish ouse, and immediately saw the coast of Africa. Lat. observed 2° 10' S, long. 9° 29' E. The land was about 4 leagues to the eastward, which would make the longitude of the coast in the above latitude in 9° 41' E. The land appeared low, but owing to the weather being hazy, and our standing off from it, we were not able to make any remarks. The bank of oundings here does not probably run farther off shore than ten leagues, deepening as follows : at 10 or 12 miles off shore 18 fathoms, greenish ouze. We then ran S W½W 5 miles, and had 30 fathoms, coarse sand ; then S W 3 miles further, 47 fathoms, sand and broken shells : then S W b. W 4 miles, 67 fathoms, same bottom, and S W b. W½ W 4 miles, no bottom with 130 fathoms, (these courses are from the true meridian). From this day (the 5th), we conti-

3 R

nued working out of soundings, making very little way to the southward, until the 17th, when we were by observation, 3° 12′ S, long. 9° 59′ 30″ E, from which time to 3 P. M. we ran E½ S (true) 12 miles, and sounded in 66 fathoms, coarse brown sand with red specks.   On the same course to 6 P. M. ran 10 miles, and had 56 fathoms, fine oazy gray sand.   By a lunar observation taken this day, the longitude reduced to noon was 8 miles to the eastward of that given by the chronometer from the midnight above, to 2 A. M. on the 18th.   We now ran S E b. E. 8 miles, and had 37 fathoms, fine gray sand. From 2 to 4 on the same course, 8 miles; had 33 fathoms, fine white sand.   From 4 to 7 the same course, 12 miles; had 25 fathoms, same bottom as last.   We then observed in lat 3° 24′ S. long. 10° 44′ 30″ E. about 5 leagues off shore; found by the chronometer we had experienced since the preceding noon a current running N. 56 W. 34 miles.   From noon to 4 P. M. we ran E S E¼ S. 18 miles, and sounded in 17 fathoms; from 4 to 6, E S E ½ E, 6 miles, 16 fathoms; then 3 miles farther on the same course, and tacked at 7 : from which time to 10, we ran W b. S ½ S 9 miles, and had 25 fathoms; then tacked again, and stood in east about 2 miles, when it falling nearly calm, came to with the stream anchor in 25 fathoms, red clay.   Found a current running due north by compass 1¼ mile an hour.

May 19.   At 4 A. M., a light breeze came off the land with drizzling rain, when we weighed, standing to the southward till 11, when it fell calm, and we came to with the kedge in 38 fathoms, slate-coloured sandy clay.   Found a current running N W b N. (compass)¼ mile an hour : weighed again and made sail on the western tack with the wind at south.   We were this day disappointed in our hopes of the sea breeze ; and about 10..30 P. M. it fell calm, when we came to with the stream in 37 fathoms, current setting N W b. N (compass) 1½ knot an hour.

May 21st. During this forenoon it was either calm or the wind so very light from the southward that had we weighed we should have lost ground considerably ; but at 3..30 it became pretty brisk from south, which we immediately took advantage of, standing to the eastward ; just before weighing, we saw the land plainly, from N E ½ E to E b. N ; at 5..30 the wind came round to the S S W, and at 6 sounded in 25 fathoms, ousy ground ; we had run since weighing S E b. E (compass) 8 miles ; at 7 had 23 fathoms, having run S E 2 miles, again at 8 had 23 fathoms, all the same bottom ; having run 2 miles S E b. E ;

at 9..30 it fell calm. We came to with the stream in 22 fathoms, ouse; very little current to the N W; we were now about 8 or 9 miles off shore

May 22d. At daylight calm and hazy; at 7 saw the land from N E ½ N to E b. N. might have seen much farther to the southward but for the haze; a very light drain of current to the N W. At 8 weighed with a light breeze from the southward, hauled on the western tack. At noon still very hazy; had 25 fathoms, ouse, land about Cape Mayumba N E to E N E ½ N. At 4 we were about the same place as at noon, with same soundings; at 6 had 23 fathoms, fine gray sand with red and black specks. Cape Mayumba N. 50 E. (compass) about 7 or 8 miles: at 8 we heard the roaring of the surf on the beach very loud; at 9 it fell calm, when we came to with the stream in 22 fathoms, greenish sand and shells.

May 23d. At daylight calm and hazy. This morning while at anchor caught several fish of the bream species of a reddish colour. At 9 a light breeze springing up from the S E, weighed and hauled on the western tack, and soon after saw Cape Mayumba, and the land on each side of it to some extent; at noon observed in lat. 3° 42′ S. Cape Mayumba N E ½ E. 8 or 9 miles had 25 fathoms fine white sand with red and black specks, this day and yesterday had no sights for chronometers, at 4 got a fresh breeze at S S W, which lasted about half an hour, and then became very light; about 9 it fell calm; came to with the stream in 18 fathoms, greenish sand; a very light drain of current to the N W.

Remark. During the nights we find a most perfect calm prevail throughout, and a very light drain of current going to the N W-ward, both by night and day; from sun set to its falling calm a moderate dew falls, which ceases directly the clouds collect over the land, which take place immediately the wind dies away.

May 24th. At 2 A. M. a light breeze sprung up at E S E. We weighed and made sail to the southward; at about 4 it fell calm, and we anchored again in 20 fathoms. At daylight, very hazy; saw the land from N N E to E b. S. This day many of the fish spoken of yesterday were caught on board both vessels. During the whole of the night we heard the surf roaring extremely loud on the beach, and we were at least 6 or 7 miles off shore. At 8 got a light breeze from the S E, weighed, and stood to the S W. At noon lat. observed 3..43. S, long. 11° 13′ E, Cape Mayumba N. 29 E, (compass) 8 or 9 miles; had 21 fathoms coarse brown sand ; we now stood in shore, the wind having drawn round to the S S W. a pretty brisk sea breeze ; at 4 we had 17 fathoms, fine gray sand. Cape Mayumba N b. E 9 or ten miles.   At 6, having run S E b. S (compass) 5 miles since 4 o'clock, we had 13 fathoms, small gravel and shells ; the land from N N E to S S E ; at S calm, came to in 10 fathoms, dark soft sandy ouse, and no current.

Remark.   We have hitherto remarked that as soon as it becomes calm in the evening, a very thick haze arises, and the dew falls much less than between this and sun set ; the surf, though not heard by day, then becomes very noisy.   The calm generally takes place between 8 and 10 P. M.

The latitude and longitude of Cape Mayumba from this day's observation is 3° 34′ S, 11° 13′ 36″ E, by chronometer.   With respect to the longitude, there appears so great a difference between that given by our chronometers and those assigned in the following charts, that it must remain for future navigators to decide.   Laurie and Whittle's chart places it in 10..16. E, and Arrowsmith in 10° 23′ E.

May 25th.   At daylight calm and hazy, a heavy surf rolling on the shore, from which we were about 3 miles distant.   At 9 a light air from the southward ; weighed and made sail on the western tack : at noon observed in lat. 3° 49′ S. long. 11° 5′ E. had 16 fathoms, fine brown sand with black specks, extremes of the land from S 34° E to N 8 W; at 2 P. M. had 17 fathoms; at 4 the same depth about 7 or 8 miles off shore ; at 6 had 11 fathoms, then tacked, standing off W b. N 4 miles till half past 7, when we had 13 fathoms, stiff black mud, and anchored.   Variation (azimuth) 25° 30′ W.

May 26th.   At 2 a light breeze sprung up at E N E : weighed and made sail to the southward ; soon after the wind chopped round N W, and continued gradually drawing round to S W, where it continued till 9, when some small rain fell ; it was very cloudy, and the wind suddenly veered round to S E. After weighing we ran on a S W b. S course 4 miles, and had 20 fathoms ; 8 miles further 29 fathoms ; then S ½ E. 4 miles, 35 fathoms ; all black mud and broken shells.   At noon observed in lat. 4° 8′ S, long. 11° 15′ 22″ E, and had 48 fathoms, same bottom ; we also found that a current had set us since weighing 10 miles north.   About 2 the wind drew round to S S W, wore to the S E, at 4 running on a S E course 3 knots an hour, we had 49 fathoms, at 5, 48., at 6, 47 fathoms, all muddy bottom, at 8, 40 fathoms, coarse sand, at 9, 35, and at 10, 33 fathoms, same bottom ; at 10..30 it fell calm. Came to in 32 fathoms, black mud. Found per log a current running N N W, $\frac{1}{4}$ knot an hour. Variation of the compass by amp. 25° 33′ W, by azimuth 25° 30′·

May 27th. At 9, a light breeze sprung up at S S W; weighed, and stood S E 3 miles, and had 28 fathoms, S E b. E, 2 miles 24 fathoms, and S E 1 mile 21 fathoms, all black mud and shells ; here we observed in lat. 4° 9′ S. lon. 11° 38′ 37″ E. Banda point E b. N 8 or 9 miles, extremes of the land from E b. S $\frac{1}{2}$ S. to N b. W. About 6 P. M. the wind drew round to W b. S, a fresh breeze; from noon we ran on a S E b. S course, shoaling gradually, with scarcely any variation in the bottom (chiefly ouse.) In running along shore the land is beautiful, appearing as if laid out in parks and pleasure grounds ; it is noted in the charts as being high ; which is certainly an error, as there has not been any part of it yet seen by us higher than the Lizard on our own coast. The latitude of Banda point by this day's observation is about 4° 4′ S, lon. 11° 46′ 2″ E. Arrowsmith in his chart of this part of the coast, places it in lat. 4..3 S. lon. 10..52..0, and Laurie and Whittle in 3° 53′ S. and 10° 30′ E.

May 28th. The breeze of last night continued at W S W till about 6 this morning, when it fell calm, and we came to in 15 fathoms, ouse ; found the current per log N b. W $1\frac{1}{4}$ mile an hour. At 8 a light breeze from the S E, with which we weighed, but falling calm almost immediately, came to again in 14 fathoms, ouse ; just before noon weighed again with a light breeze, but finding we lost ground came to directly in 12 fathoms, ouse, where we observed, in lat. 4° 24′ S, lon. 12° 11′ E, extremes of the land from S S E to N b. E current N b. W $1\frac{1}{2}$ mile an hour. About 2 P. M. the sea breeze came in moderate at W S W; weighed and made sail, but the ship in a most unaccountable manner, with all sail set and a good breeze on the quarter, refused to come higher than E S E, and lay like a log on the water, while the Congo, whose tow-rope we had cast off, was lying up south about 3 knots; at 2..30 finding we were drifting bodily in shore with the current, and had decreased our depth of water from 12 fathoms ouse, to $9\frac{1}{2}$ rocky bottom, let go the stream anchor ; and before we could bring up got into $7\frac{1}{2}$ rocky bottom. We were about 6 miles from the nearest shore, no point or known headland to set, but right abreast of us were some reddish cliffs ; and a little to the southward of them two holes in the land, apparently of the same quality as the cliffs ; these holes, both in size and shape, are much like the large chalk pit on Portsdown hill.

To the southward of these pits, there is a deep bay, the southern point of which is low at the extremity but rises gradually to moderately high land.

The above reef is certainly very alarming, the water shoaling very rapidly and the current setting right over it due north $2\frac{1}{2}$ miles an hour ; the rocks of which it is composed are soft, the lead always bringing up pieces sticking to it : it is probable this may be what is called in the charts Kilongo reef; if so, it is laid down much to the northward of its real situation.

At about half past 3 the breeze freshening we made all sail, hove in as much of the slack cable at we could, but having run over the anchor, and the cable becoming taught, cut it, leaving the anchor and about $\frac{3}{4}$ of the stream cable behind.

The bay before mentioned agrees precisely with the description of Loango Bay, as given by Grandpré, a French navigator ; and also in one of Laurie and Whittle's charts ; but the latitude differs so widely, as to make us doubt the reality of its being that bay. At 6 P. M. we had the point bearing E S E, consequently nearly on its parallel, at which time our latitude could not be more than 4° 33′ S, and Indian point, the south point of Loango Bay in Arrowsmith's chart, lies in latitude 4° 53′ and in Laurie and Whittle's 4° 45′ S, so that the latitudes assigned to it in the above charts must be extremely erroneous. At about 8 P. M. the sea breeze backed round to west, and at half past 11 it fell calm, when we came to with the kedge anchor in 22 fathoms, soft muddy bottom.   Found the current per log. to run N N W $1\frac{1}{2}$ knot an hour.

May 29th.   At daylight saw the land to a great extent.   Observed the point spoken of yesterday to bear a great resemblance to the Bill of Portland lengthened, which by Laurie and Whittle's chart appears to be the case with the south point of Loango Bay, or Indian Point.   At 10 a light breeze sprang up at S E, which, on our weighing died away, we therefore anchored again, and at noon observed in lat. 4° 44′ S. lon. 12° 14′ E, Indian point N 68 E, 10 miles, which would make its lat. 4° 37′ S, extremes of the land from N b E. to S E $\frac{1}{2}$ E, current running N b. W $1\frac{1}{4}$ mile an hour.

By the mean of yesterday and this day's observation Indian point lies in 4..35 S.

At half past one P. M. the sea breeze set in at W b. S. Made all sail ; found the ship slacked the cable. Up anchor, and steering a south course $4\frac{1}{2}$ knots, we gradually deepened our water to 28 fathoms, and then shoaled again to 23 fathoms, ouse, when we anchored at half past 9, being quite calm.   During these last 24 hours the water has had a deep tinge, like blood and water mixed,

Remark. For some days past we have invariably had a light breeze from S S W to S S E, which springs up about 5 o'clock in the morning, and generally ends about 10, though sometimes so late as 11; it is then calm till one or two in the afternoon, when the sea breeze sets in at S W, or S W b. W, light, and about an hour after it has commenced it generally carries us 3 or 3½ knots, gradually gathering strength and drawing round to W, and sometimes to W N W; again hauling round to W S W an hour or more before it falls calm. The time of its falling calm appears to depend on the time of the sea breeze setting this breeze generally lasts about 9 or 10 hours, but in one or two instances we had it 12 or 14.

May 30th. At daylight calm, dark, and cloudy; saw the land, but could not distinguish any particular or known point: current running N N E 1¼ knot an hour. At noon still calm; observed in lat. 5º 2' S. lon. 12º 15' E extremes of the land from S E½E to N E ½ N, about 11 or 12 miles, current the same as at daylight.

There can be no doubt of the point before described being Indian point, as we are this day to the southward of its situation in every chart; and have not seen the least appearance of such a bay as that of Loango, between the land we are now abreast of and the above point. Horsburgh, in his chart, places Indian point in lat. 5º 0' S, which makes it 25 miles, Arrowsmith 18, and Laurie and Whittle 10 miles all too far to the southward, supposing 4º 35' to be the latitude, which I am certain is very near the truth.

At 3 P. M. the sea breeze came in at W b. S, with which we weighed and made sail, at half past 10 falling little wind, came to in 11 fathoms, muddy bottom, scarcely any current.

June 1st. At daylight observed we were abreast of a river with a very fine entrance; a light breeze springing up, we shifted a little off shore, but falling calm, anchored again at 9; about 10, two canoes came along side with 8 natives in them, one of whom spoke tolerably good English, and said he came to inform us the Mafouk or governor of Malemba was coming on board; from the information of these people it appears this river is called Louango Louiza, and not Louiza Louango as in the charts; and on enquiring after the river Kacongo, they knew nothing of it, declaring there is no other river between Loango Bay and Malemba, than Louango Louiza: when we enquired what river the town of Kinghele was situated on, they called it Chimbélé, and said it stood on the bank of a little river called Bele, which is situated to the southward of Cabenda. If their account is correct, which there is no reason to doubt, a river is laid down in the charts that does not exist, a corroborative proof of which is, that in our run between Loango bay and where we now are, not the least appearance of one was seen : and the natives assure us there is not another till you come to the southward of Cabenda.

At noon observed in 5° 17′ S. lon. 12° 10′ 15″ E, south point of the river Louango Louiza E b. N $\frac{1}{2}$ N, 7 or 8 miles, which makes this point in 5° 12′ S, 12° 15′ 33″ E, current running N b. W$\frac{1}{4}$ mile an hour. About 2 the sea breeze set in at W b. S, with which we weighed, and at 10. 30 falling little wind, came to in 11 fathoms, current running N N W$\frac{1}{4}$ mile an hour : in running along shore the natives pointed out to us the point of Malemba ; it is a bluff cliff not easily distinguished, being considerably lower than the land at the back of it, which is but moderately high, and may be easily known, by its being full of red cliffs, like chalk pits.

June 2nd. At daylight, light airs from the southward with light rain and haze ; at 8, went in-shore with two boats to find out a bank, which Grandpré states to lie between Malemba and Cabenda, close on the north side of Cabenda bay ; two of the natives, who said they knew the bank, went with me ; one of whom said he was on board Maxwell's ship when she grounded on it ; I went into 5$\frac{1}{2}$ fathoms, which was about 2$\frac{1}{2}$ or 3 miles from Cabenda point, which depth was sufficient to disprove Grandpré's assertion that " After you have 7 fathoms, you will be on shore before you get another cast." I found the water to shoal gradually, and 7 fathoms was at least one mile outside of me. After weighing we stood W b. N 2 miles, then S b. E 3 miles, when we had 14 fathoms, from this we ran S b. W 3 miles, and had 13 fathoms, 3 miles further on the same course 12 fathoms, 2$\frac{1}{4}$ miles further 10 fathoms, and 2 miles farther 7 fathoms, where, at 5..40, we anchored, with Red point south 9 or 10 miles, Cabenda hook N E b. E $\frac{1}{4}$ E, about 3 miles off shore : current running N b. W 1$\frac{1}{4}$ mile an hour. I now went in the gig to sound, and found the water to shoal gradually all the way to the shore, at about $\frac{1}{4}$ mile distant from which had 3 fathoms : by the time I got in shore, had no other light than that afforded by the moon (which had just completed the first quarter), which enabled me to see some heavy breakers to the southward, I accordingly rowed towards them, and found a dangerous reef running off shore to the westward, about $\frac{1}{4}$ of a mile with 5 fathoms, close to the western edge; it lies about half way between Cabenda hook and Red point; about 8, I returned to the ship: the soundings regular as before.

June 3d. At noon observed in lat. 5° 37′ S, the high land of Cabenda, E b. N, and Red point S S E½E. For the last two days we have not been able to get sights for longitude. At 2 wore ship, and stood to the southward with a moderate sea breeze at W b. S; we continued sounding every hour, and had constantly 23 fathoms, until 11 o'clock, when it fell little wind, which obliged us to anchor in the above depth, sandy clay of a greenish hue.

June 4th. At daylight saw the land about Cabenda 5 or 6 leagues distant: during the whole of this day, the breezes were extremely light, and the weather dark and hazy, had no observation, current running N b.W, 1¼ mile an hour.

June 5th. The whole of this day was calm, dark, and cloudy; and we remained at anchor, current N N W, 1¾ mile an hour.

June 6th. The whole of this forenoon was calm, dark, and cloudy; at noon observed in 5° 40′ S, at 2..30, the sea breeze came in at W b. S, when we weighed and made all possible sail to the southward: at 4 the haze cleared off a a little, saw the land, the southern extreme of which bore S S E, sounded in 15 fathoms; from this time to 6 o'clock, we ran on a S b.W course, gradually shoaling to 13 fathoms; from 6 we kept on a S and S b.W. course going 3½ knots an hour till 9, keeping all the time the same depth of water: and then gradually deepening to 17 fathoms; shortly after getting this latter depth, we had no bottom with 160, fathoms; in about an hour after the wind headed us, and we observed an extremely great ripple all round the ship, making a noise like a mill sluice (apparently a very rapid current); we soon got out of this, and on trying for soundings without the hope of getting any, had 24 fathoms, muddy bottom; we immediately anchored at about half an hour after midnight, and found scarcely any current, what little there was ran to the southward; had it not been for this circumstance, should have concluded we had been drifted back again to the northward, which was Captain Tuckey's opinion; but considering we had always found a strong northerly current, when to the northward of the deep water channel, that now we must have crossed it, on finding it run to the southward, which proved to be the case.

From the above it will appear that the deep water is much narrower than is generally supposed, it is probably not more than 3 miles across ; we sounded every hour, and at 10, we had 16¼ fathoms, at 11, no bottom, at 11..30 the wind headed, and we fell off to S E, going only 2 knots, and at half past midnight we had 24 fathoms.

June 7th.   At daylight saw the land from S S E to N E, the Zaire apparently open abreast of us ; I went in-shore to examine the coast and ascertain whether the opening we saw was really it ; about 11 o'clock I observed the ships were under way, by which time I was sufficiently near to be satisfied of our being at the entrance of he river, and accordingly after taking a view of the coast returned on board to inform Captain Tuckey, which together with the observation they had taken on board, being 6° 5′ S, proved it beyond a doubt, we accordingly bore up and made sail ; and at half past 3, came to under Shark's point in 4¼ fathoms, the point bearing E S E about ¼ mile from the nearest shore.

London: Printed by W. Bulmer and Co.
Cleveland Row, St. James's.